Blockchain-Driven Decentralized Energy Trading:

Navigating the Legal Architecture

Karisma Karisma
https://orcid.org/0000-0001-9783-9504
Independent Researcher, Malaysia

IGI Global
Scientific Publishing
Publishing Tomorrow's Research Today

Vice President of Editorial | Melissa Wagner
Director of Acquisitions | Mikaela Felty
Director of Book Development | Jocelynn Hessler
Production Manager | Mike Brehm
Cover Design | Jose Rosado

Published in the United States of America by
IGI Global Scientific Publishing
701 East Chocolate Avenue
Hershey, PA, 17033, USA
Tel: 717-533-8845 | Fax: 717-533-7115
Website: https://www.igi-global.com E-mail: cust@igi-global.com

Library of Congress Cataloging-in-Publication Data

Names: Karisma, Karisma, 1995- editor
Title: Blockchain-driven decentralized energy trading : navigating the legal architecture / [edited by] Karisma Karisma.
Description: Hershey, PA : IGI Global Scientific Publishing, [2026] | Includes bibliographical references and index. | Summary: "This book provides a multi-dimensional examination of blockchain's transformative potential in decentralized energy trading"-- Provided by publisher.
Identifiers: LCCN 2025035281 (print) | LCCN 2025035282 (ebook) | ISBN 9798337357423 hardcover | ISBN 9798337357430 paperback | ISBN 9798337357447 ebook
Subjects: LCSH: Energy industries--Technological innovations | Blockchains (Databases)
Classification: LCC HD9502.A2 .B637 2026 (print) | LCC HD9502.A2 (ebook) | DDC 333.791/1--dc23/eng/20260213
LC record available at https://lccn.loc.gov/2025035281
LC ebook record available at https://lccn.loc.gov/2025035282
British Cataloguing in Publication Data
A Cataloguing in Publication record for this book is available from the British Library.

Table of Contents

Detailed Table of Contents

Blockchain technology is rapidly transforming the energy sector by enabling decentralized, peer-to-peer energy trading systems that challenge traditional regulatory models. This chapter critically examines the existing legal frameworks governing energy markets and explores how they must evolve to accommodate the unique demands of blockchain-based energy trading. It analyzes the key legal and policy challenges posed by decentralization, including jurisdictional complexities, consumer protection, data privacy, liability, and smart contract enforceability. With the emergence of prosumers—individuals who both produce and consume energy—the chapter delves into questions of legal status, licensing, and equitable grid access. It also discusses the implications of tokenized energy assets, the interoperability of technical and regulatory standards, and the legal recognition of consensus mechanisms used in blockchain systems.

This chapter examines the role of blockchain as a transformative instrument in the digital energy transition, highlighting its capacity to support decentralised trading, automate compliance, and enhance institutional transparency. It analyses how blockchain aligns with smart-governance principles by embedding regulatory logic into digital infrastructures while requiring adaptive legal frameworks to address liability, data protection, and market oversight. Through comparative case studies—including Brooklyn Microgrid, Powerledger, and WePower—the chapter illustrates the interplay between technological innovation, regulatory experimentation, and policy integration. Ethical dimensions such as fairness, environmental sustainability, and algorithmic accountability are addressed as essential conditions for legitimate decentralisation. The findings demonstrate that blockchain's effectiveness depends not on technological capability alone but on the coherence of governance, legal interoperability, and societal inclusion.

The rapid development of blockchain-based decentralized energy trading has introduced peer-to-peer (P2P) market models that challenge traditional centralized electricity regulation. Existing energy regulatory frameworks, designed for vertically integrated utility systems, often lack the flexibility to accommodate prosumers, smart contracts, and automated transactions. This book examines regulatory sandboxes as adaptive legal frameworks for testing blockchain-driven decentralized energy trading models under controlled regulatory conditions. It analyzes how sandboxes enable experimentation with innovative energy business models while safeguarding consumer protection, grid reliability, data privacy, and market integrity. Through legal and policy-oriented analysis and comparative international case studies, the book explores sandbox design principles and key regulatory challenges across energy law, data protection, competition law, and liability regimes. The findings show that regulatory sandboxes reduce legal uncertainty and support evidence-based regulatory development for markets.

The integration of smart contracts in energy markets marks a transformative shift toward decentralization, automation, and enhanced transactional efficiency. As blockchain-based technologies are increasingly used to facilitate peer-to-peer energy trading, grid management, and decentralized energy exchanges, questions of legal enforceability, liability, and dispute resolution remain critical. This chapter aims to examine the foundational legal and regulatory considerations that govern smart contracts within the context of energy markets. It will explore how traditional legal principles interact with emerging technologies, including the enforceability of self-executing code under contract law, jurisdictional challenges, and the applicability of consumer protection and liability statutes. Case studies from pioneering jurisdictions such as the European Union, the United States, and select Asia-Pacific countries will be analyzed to provide comparative insights.

This chapter explores the intersection between blockchain-based energy systems and data privacy challenges, especially in peer-to-peer energy markets. It highlights how the transparency and immutability of blockchain, combined with real-time data from smart meters and IoT devices, can expose users to re-identification risks and legal conflicts with principles such as the right to be forgotten. The authors examine technical solutions including zero-knowledge proofs, permissioned blockchains, smart contracts with embedded privacy logic, and off-chain storage. Additionally, the chapter emphasizes the importance of privacy-by-design and security-by-default principles in system architecture. It also calls for adaptive legal frameworks such as regulatory sandboxes, along with participatory governance to ensure equitable implementation, especially in the Global South where regulatory and infrastructure disparities persist.

The tokenization of energy resources through blockchain technology represents a transformative step toward sustainable, transparent, and decentralized energy systems. By enabling fractional ownership, peer-to-peer energy trading, and democratized investment structures, tokenized models foster inclusivity and enhance capital mobilization in renewable energy markets. Case studies such as Power Ledger in Australia and WePower in Europe highlight tangible benefits, including reduced transaction costs, improved trading efficiency, and broader investor participation. Despite its promise, tokenized energy ecosystems face regulatory uncertainty, ethical dilemmas, and implementation challenges. Addressing these barriers through harmonized legal frameworks, technological innovation, and global governance can accelerate adoption while ensuring equity and energy justice. This study consolidates ownership models, performance metrics, and governance implications to provide a comprehensive roadmap for advancing tokenized energy systems.

Preface

INTRODUCTION

The accelerating convergence of digital innovation and energy transformation has given rise to one of the most consequential developments of our time: the emergence of blockchain-driven decentralized energy systems. This edited volume, *Blockchain-Driven Decentralized Energy Trading: Navigating the Legal Architecture*, is situated at the intersection of technological possibility and regulatory reality. It seeks to illuminate not only the transformative potential of decentralized energy trading but also the complex legal and institutional frameworks required to sustain it.

Across jurisdictions, the integration of blockchain into energy systems is reshaping conventional paradigms. Longstanding centralized utility models are being challenged by decentralized architectures that empower prosumers, enable peer-to-peer transactions, and introduce new levels of transparency and efficiency. At the same time, these innovations are unfolding unevenly. Regulatory uncertainty, infrastructural constraints, and legal ambiguities continue to impede adoption—particularly in developing and emerging economies, where the promise of decentralization is often matched by systemic limitations.

This volume brings together diverse scholarly perspectives to provide a comprehensive and interdisciplinary examination of these developments. It engages deeply with core legal and regulatory questions, including the status and enforceability of smart contracts, liability in automated systems, the governance of energy data, cybersecurity imperatives, and the tokenization of energy assets. Contributors also interrogate the evolving role of prosumers and the implications of shifting market participation structures in decentralized ecosystems.

A central thread running through this book is the enduring challenge of the energy trilemma: balancing energy security, equity, and environmental sustainability. The chapters collectively explore how blockchain technologies might contribute to this balance, while also examining the extent to which existing legal systems facilitate—or constrain—such alignment. Through comparative legal analysis, theoretical inquiry, and grounded case studies, the volume highlights the adaptive capacity required of legal frameworks operating in an increasingly digitalized energy landscape.

Methodologically, the work draws on doctrinal legal analysis, qualitative research, and policy evaluation to assess the interaction between innovation and regulation. It also places particular emphasis on Global South contexts, where legal asymmetries and infrastructural gaps present both barriers and opportunities for novel governance models. In doing so, the book advances comparative legal scholarship and fosters a more inclusive global dialogue on energy transition.

The contributions are organized around several key objectives: to examine blockchain's technical and socio-legal dimensions in decentralized energy trading; to critically assess regulatory and institutional barriers; to explore legal responses to the energy trilemma; to bridge the divide between innovation and regulation through adaptive governance tools; and to ground theoretical insights in empirical case studies from diverse jurisdictions. The volume further extends its inquiry to the integration of blockchain with artificial intelligence, raising pressing questions about accountability, transparency, and algorithmic governance.

Importantly, this book is not merely diagnostic—it is also constructive. It offers actionable insights and legal design principles aimed at policymakers, regulators, scholars, and practitioners engaged in shaping future energy systems. By identifying pathways for aligning technological innovation with public interest objectives, it contributes to broader efforts toward sustainable, inclusive, and resilient energy transitions.

This volume is intended for a wide and interdisciplinary audience: academics exploring the frontiers of energy and technology law; policymakers and regulators navigating emerging governance challenges; industry actors developing and deploying blockchain-based energy solutions; and students and researchers entering this rapidly evolving field. It is our hope that the work will serve as both a reference and a catalyst—supporting informed decision-making, encouraging critical inquiry, and advancing collaborative approaches to one of the defining challenges of our era.

As editor, I extend my sincere appreciation to the contributors for their intellectual rigor and commitment to this project. Their work reflects the complexity of the subject matter and the urgency of engaging with it thoughtfully. I also acknowledge the broader scholarly and policy communities whose ongoing efforts continue to shape the discourse on decentralized energy and digital governance.

The legal architecture of blockchain-driven energy systems is still in formation. This book does not claim to offer final answers; rather, it provides a structured and critical foundation upon which future inquiry and policy development can build.

ORGANIZATION OF THE BOOK

Chapter 1 opens the volume with a forward-looking examination of the convergence between artificial intelligence and blockchain in energy markets, focusing on the emergence of autonomous trading systems. The chapter highlights how these technologies enable real-time, decentralized, and self-executing energy transactions while simultaneously exposing profound regulatory gaps. It identifies critical concerns surrounding algorithmic accountability, cross-border governance, and consumer protection, arguing that existing legal frameworks remain ill-equipped to address the complexities of AI-enabled market behavior. By synthesizing technical, market, and legal perspectives, the chapter lays an important foundation for understanding how regulatory systems must evolve to govern increasingly autonomous and data-driven energy ecosystems.

Chapter 2 situates blockchain as a foundational infrastructure for decentralized energy markets, emphasizing its capacity to enhance transparency, automate compliance, and restructure traditional market arrangements. With particular attention to emerging economies, the chapter explores how distributed ledger technologies and tokenized energy systems challenge existing legal and institutional norms. It carefully examines issues such as data protection, cybersecurity, and liability, while drawing on comparative international experiences to identify governance models that balance innovation with regulatory safeguards. The chapter ultimately underscores the importance of adaptive and coherent legal frameworks in fostering equitable and resilient energy transitions.

Chapter 3 advances the discussion by offering a detailed analysis of blockchain frameworks underpinning decentralized energy markets. It explores platform architectures, consensus mechanisms, and interoperability models, situating these technical components within evolving regulatory environments. The chapter pays particular attention to compliance models that reconcile innovation with legal certainty, highlighting the need to align decentralized trading systems with existing energy laws, grid codes, and consumer protection standards. Through its integrated approach, it provides both conceptual clarity and practical guidance for scaling blockchain-enabled energy systems within legally compliant structures.

Chapter 4 provides a critical exploration of the regulatory architecture required to support blockchain-based energy trading. It examines how existing legal frameworks—often designed for centralized energy systems—must adapt to accommodate decentralization, peer-to-peer transactions, and the rise of prosumers. The chapter addresses key legal challenges, including jurisdictional fragmentation, smart contract enforceability, data privacy, and liability allocation. It also considers the implications of tokenized energy assets and the need for interoperability between technical and regulatory standards. In doing so, it offers a comprehensive assessment of the legal transformations necessary to sustain decentralized energy markets.

Chapter 5 extends the analysis into the domain of governance, examining blockchain as an instrument of "smart government" in the energy sector. Through comparative case studies of pioneering projects, the chapter illustrates how blockchain can embed regulatory logic into digital infrastructures while enhancing institutional transparency. It emphasizes that technological innovation must be accompanied by coherent legal and governance frameworks, particularly in addressing ethical considerations such as fairness, sustainability, and algorithmic accountability. The chapter ultimately reinforces the notion that effective decentralization depends as much on governance design as on technological capability.

Chapter 6 turns to regulatory sandboxes as a pragmatic and adaptive tool for navigating legal uncertainty in decentralized energy trading. The chapter demonstrates how sandbox environments enable controlled experimentation with blockchain-based models while maintaining safeguards for consumer protection, grid stability, and data privacy. Drawing on comparative legal and policy analysis, it outlines key design principles and regulatory challenges, illustrating how sandboxes can facilitate evidence-based policymaking. This contribution highlights the importance of iterative and flexible regulatory approaches in supporting innovation without compromising system integrity.

Chapter 7 focuses on one of the most foundational elements of blockchain-enabled energy systems: smart contracts. It provides a rigorous examination of their legal enforceability, the allocation of liability, and mechanisms for dispute resolution within decentralized markets. By analyzing how traditional contract law interacts with self-executing code, the chapter reveals both tensions and opportunities in integrating these technologies into existing legal systems. Comparative insights from multiple jurisdictions further enrich the discussion, offering practical perspectives on how legal doctrines are adapting to accommodate automated transactions in energy markets.

Chapter 8 addresses the critical and often underexplored dimensions of data privacy and cybersecurity in blockchain-based energy platforms. It highlights the inherent tension between blockchain's transparency and the need to protect sensitive user data, particularly in systems reliant on real-time inputs from smart devices. The chapter evaluates a range of technical and legal solutions, including privacy-enhancing technologies and adaptive regulatory frameworks, while emphasizing the importance of privacy-by-design principles. Its analysis is particularly attentive to the Global South, where infrastructural and regulatory disparities heighten both risks and opportunities.

Chapter 9 explores the rapidly evolving domain of energy tokenization, examining its implications for ownership, investment, and market participation. The chapter illustrates how blockchain enables fractional ownership and democratized investment in renewable energy assets, thereby expanding access and enhancing market efficiency. At the same time, it critically engages with the regulatory uncertainties and ethical considerations that accompany tokenized systems. By synthesizing case studies and governance models, the chapter provides a roadmap for navigating the legal frontiers of digital energy assets while promoting equity and sustainability.

Chapter 10 concludes the volume by foregrounding the imperative of energy justice within digital energy transitions. It critically examines how blockchain-enabled systems, while technologically transformative, risk reinforcing existing inequalities if not designed with inclusivity in mind. Drawing on theoretical frameworks and global case studies, the chapter proposes governance and design strategies to ensure equitable access to the benefits of decentralized energy systems. In doing so, it re-centers the broader discourse on the social dimensions of technological change, emphasizing that the success of blockchain in energy markets must ultimately be measured against its contribution to fairness, inclusion, and justice.

CONCLUSION

As this volume demonstrates, the evolution of blockchain-driven decentralized energy trading is neither a purely technological transition nor a solely regulatory exercise—it is, fundamentally, a systemic transformation that requires the careful alignment of law, policy, infrastructure, and societal values. The contributions gathered here collectively reveal a landscape defined by both extraordinary promise and intricate complexity. While blockchain and its adjacent technologies offer new pathways for transparency, efficiency, and democratized participation in energy systems, they simultaneously challenge the foundational assumptions upon which existing legal and regulatory frameworks have long been built.

A recurring insight across the chapters is that legal systems are not merely reactive instruments but constitutive forces that shape the trajectory of technological adoption. Whether in addressing the enforceability of smart contracts, the governance of autonomous systems, the protection of data, or the emergence of prosumer-driven markets, the law plays a decisive role in determining whether innovation advances in a manner that is equitable, secure, and sustainable. The need for adaptive, context-sensitive, and forward-looking regulatory approaches is therefore not optional—it is essential.

Equally evident is the importance of bridging divides: between innovation and regulation, between the Global North and Global South, and between technological capability and social legitimacy. The discussions on regulatory sandboxes, comparative legal frameworks, and participatory governance models underscore that experimentation, when responsibly structured, can serve as a vital mechanism for navigating uncertainty. At the same time, the emphasis on energy justice reminds us that the success of decentralized systems must ultimately be measured not only by efficiency gains or market expansion, but by their capacity to deliver inclusive and fair outcomes.

This volume does not claim to resolve all the tensions inherent in blockchain-enabled energy systems. Rather, it offers a structured and critical foundation for ongoing inquiry, dialogue, and policy development. The interdisciplinary perspectives presented here reflect the reality that no single domain—legal, technical, or economic—can adequately address the challenges in isolation. What is required instead is sustained collaboration across sectors and jurisdictions, supported by a shared commitment to aligning innovation with the public interest.

As editor, I hope that this book will serve as both a reference and a point of departure: a resource that informs, but also provokes further questioning; a framework that guides, but also evolves alongside the technologies and systems it seeks to understand. The legal architecture of decentralized energy trading remains under construction. Its eventual form will depend on the collective efforts of scholars, policymakers, practitioners, and communities working together to ensure that the digital transformation of energy systems advances not only what is possible, but what is just and desirable.

Chapter 1
AI–Blockchain Convergence in Energy Markets:
Regulatory Challenges for Autonomous Trading Systems

Soumi Ghosh
http://orcid.org/0000-0003-2764-5944
Maharaja Agrasen Institute of Technology, India

Ritik Raj
http://orcid.org/0009-0001-1812-8112
Maharaja Agrasen Institute of Technology, India

ABSTRACT

The combined technologies of AI-blockchain block in energy markets allows autonomous and more efficient, transparent, and decentralized energy trade on a real-time basis and autonomous execution of operations, but it poses new regulatory challenges that have never been observed before. This paper looks at the existing integration of AI and blockchain and finds out where there is a regulatory gap in implementing AI-enabled trade or the inability to govern autonomous trading systems. The research investigated on technical architecture, market dynamics as well as regulatory implications through systematic analysis of 80 scholarly sources. The results demonstrate that there is a great hole in the ethical responsibility of algorithms, cross-border transactions, and consumer protection in the decentralized markets. This study facilitates unique regulatory functions applied uniquely to AI-complexity-based blockchain energy trading provided by an international business model that can aid in crucial governance demands of emerging independent energy systems.

DOI: 10.4018/979-8-3373-5742-3.ch001

1. INTRODUCTION

The energy industry in the world is undergoing a technological revolution due to the need to address sustainability challenges, and changing consumer tastes and preferences (Andoni et al., 2019). The combination of artificial intelligence and blockchain is a significant technological convergence that has been driving the modernization of the energy markets more rapidly, making it possible to develop autonomous trading systems, which are configured to work with minimal human interaction (Hua et al., 2022). This convergence has the capability of fixing long-term weaknesses in energy trade, such as information asymmetry, market opaqueness, and high transaction costs (Soto et al., 2021).

Traditional energy markets were traditionally distinguished by centralized market structures, complicated intermediate arrangements, and low transparency (Kirli et al., 2022). The mixture of artificial intelligence algorithms and the impossibility and visibility of blockchain ledgers opens the possibility of a smooth trading of peer-to-peer energy markets, the execution of smart contracts automatically, and price discovery in real time (Wang et al., 2023). Nevertheless, this technological development presents control issues that the current systems are poorly prepared to handle (Barceló et al., 2023).

The independence of AI-blockchain energy trading systems poses a basic challenge in terms of accountability of algorithms, the risk of market manipulation, and systemic vulnerability, as well as consumer protection (Idries et al., 2022)). Regulators across the globe are finding it difficult to devise governance systems that can strike a balance between the benefits of innovation and the stability of the markets and the welfare of consumers (Al Shareef et al., 2024). The proposed chapter copes with these challenges through a critical analysis of the current regulatory frameworks and the suggested adaptive regulatory mechanisms in the context of AI-blockchain energy trading system, with a specific attention to the empirical evidence of the actual implementations and their revealed shortcomings.

1.1 Research Objectives

The objectives of this chapter are five:

1. Assess AI-blockchain convergence on the technical architecture and business operations within the energy markets.
2. Establish the most important regulatory issues posed by autonomous trading systems and critically discuss constraints.
3. Determine the adequacy of the available regulatory systems in various jurisdictions.

4. Suggest a regulatory framework of the AI-blockchain energy trading platforms, separating the short-term and long-term priorities.
5. Create a background on how future empirical studies can be done in this new field.

1.2 Research Methodology

The chapter is based on the mixed-methods approach that incorporates the systematic literature review, regulatory analysis, and case study analysis as well as the development of the frameworks. The methodology covers: (1) the critical review of 80 peer-reviewed articles that were located in the Web of Science, Scopus, and IEEE Xplore databases; (2) the comparative analysis of regulatory frameworks in the EU, United States, China, Japan, Singapore, and Canada; (3) the technical analysis of AI-blockchain integration architectures; and (4) the iterative framework development improved by stakeholder-based design principles.

2. LITERATURE REVIEW

2.1 AI-Blockchain Integration in Energy Systems

The intersection of AI and blockchain in the energy market has received great academic interest during the last five years (Al Shareef et al., 2024). Initial research has analyzed these technologies independently blockchain as a solution of the energy trading transparency (Kirli et al., 2022) and AI as a solution of the demand forecast and grid optimization (Hua et al. (2022). More recent studies look at synergistic potential in a combination of such technologies. Rahman et al. (2023) established that AI algorithms had the potential to increase trading performance in energy platforms based on blockchain technology by about 35 percent when paired with smart contract systems but these findings were obtained in controlled laboratory settings, not in actual market conditions, which limits their external validity. In a similar study, Hua et al. (2022) obtained more than 85% accuracy in predicting with deep learning in decentralized blockchain markets and admitted that model performance might become worse when the regulator interferes with the market or when extreme market conditions that are absent in the training data corrupt the data.

European scholars have taken the idea of autonomous trading agents and have come up with multi-agent systems that can negotiate and finalize energy trading without human intervention Cantillo-Luna et al. (2022). Such systems are based on the use of reinforcement learning algorithms that can be adjusted to market conditions and optimize the trading strategy in real time. The capabilities of the system are also increased through IoT sensor integration because they can contribute to AI-blockchain platforms in the form of granular real-time data Gaybullaev et al. (2021). The main limitation of the current literature, however, is the preponderance of simulation-based research; empirical findings of large-scale, long-term practical implementations are still lacking and are required to ratify the theoretical performance assertions.

2.2 Energy Market Transformation

The digital energy trading has given rise to new market blueprints that do not uphold the traditional bilateral contract and centralized exchange model Bhavana et al. (2024). Decentralized systems allow close-range peer-to-peer energy transactions, which is especially important in the renewable energy markets with the prosumers (both generators and consumers of energy) aiming to cash in on the excess of the energy generated Hua et al. (2022). This structural transformation has increased the demand of having regulatory frameworks that can accommodate entities that fail to fit well into the regulated utility model Barceló et al. (2023).

The blockchain becomes a key trust instrument in this new paradigm of the market, where it has the transparency and immutability needed to execute decentralized trading (Kirli et al., 2022). Smart contracts make the execution of the agreement more efficient and lower the cost of transactions Kirli et al. (2022). Nevertheless, the immutability that facilitates auditing also presents problems: it means that once a transaction is completed, it is difficult to undo the incorrect or even fraudulent ones, which present a special problem to consumer protection and regulatory enforcement Mitrea et al. (2023).

2.3 Regulatory Challenges in Traditional Energy Markets

The current systems of energy market regulation are tailored to centralized markets that are controlled by large utilities and fail to easily adapt to decentralized paradigms (Barceló et al., 2023). The traditional regulation frameworks consider large-scale generators, transmission organizations, and distribution firms, and do not have much to offer to small-scale prosumers or peer-to-peer trading (Idries et al., 2022). These structural mismatches are further complicated by the introduction of autonomous AI-based trading, and the issue of algorithmic accountability is new in the history of regulation with no obvious precedents (Barceló et al., 2023).

Cross-jurisdictional energy trade is also another aspect of regulation, which is that various regions have different market regulations, pricing, and compliance demands (Barceló et al., 2023). The geographically unlimited characteristics of blockchain networks conflict with the territorially constrained regulatory jurisdiction by posing a possible enforcement gap (Mitrea et al., 2023). The consumer protection systems, which are developed in the regulated utility setting, need redesigning on the basis of the decentralized autonomous systems (Al Shareef et al., 2024).

3. TECHNICAL ARCHITECTURE OF AI-BLOCKCHAIN ENERGY TRADING SYSTEMS

3.1 System Architecture Overview

Based on the AI-blockchain energy trading systems, there are four technological layers that operate as a system with four layers that are interdependent and perform various tasks within the entire trading ecosystem (Hua et al., 2022). This multi-layer architecture is shown in Figure 1.

Figure 1. Multi-layer architecture of AI-blockchain energy trading systems

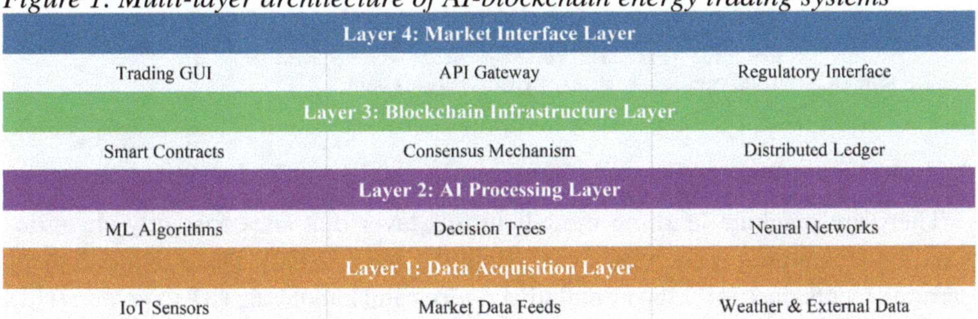

Data flows from IoT sensors and market feeds upward through AI processing and blockchain infrastructure to the market interface layer.

The data acquisition layer captures real-time data of various sources such as ionot sensors to detect energy production and consumption, market data feeds to get price data, and meteorology data to forecast renewable energy (Gaybullaev et al., 2021). This information is processed in advance, tested to meet quality and passed to the AI processing layer. The scale and speed of data needed, including thousands of sensor readings per second of large grids, cause huge infrastructure demands and create vulnerabilities to data integrity that is of a regulatory concern.

3.2 AI Algorithm Integration

The AI processing layer uses the sophisticated machine learning algorithms to the market analysis, forecasting demand, and optimization of the trading strategies (Hua et al., 2022). The fundamental autonomous trading decision engine utilized in such systems is shown as in Algorithm 1 (Figure 2).

Figure 2. Autonomous trading decision engine

Algorithm 1: Autonomous Trading Decision Engine

```
INPUT: Market_Data, Historical_Prices, Weather_Forecast, Energy_Demand
OUTPUT: Trading_Decision, Quantity, Price_Limit
1. INITIALIZE trading_parameters and risk_thresholds
2. FOR each time_interval:
     a. COLLECT real_time_data from all sensor feeds
     b. PREPROCESS data using normalization algorithms
     c. APPLY predictive_models to forecast_prices (LSTM, GRU, Transformer)
     d. CALCULATE risk_metrics using VaR and CVaR models
     e. DETERMINE optimal_trading_strategy using RL agent
     f. IF conditions_met AND risk_within_threshold:
            EXECUTE trading_order via smart_contract
     g. UPDATE learning_parameters based on trade outcomes
     h. LOG decision_rationale for regulatory audit trail
3. RETURN trading_results with explainability report
```

There are machine learning models in this layer that are constantly changing based on a changing market behavior (Rahman et al., 2023). Deep reinforcement learning agents maximize the cumulative returns and handle the risk exposure (Hua et al., 2022). The elements of natural language processing are used to recognize the sentiment of the news and regulatory announcements to integrate external data in the trading decisions (Hua et al., 2022). The most important regulatory issue is that the explainability of such models and, in particular, deep neural networks is weak, which complicates the process of audit and challenge of particular trading decisions by regulators.

3.3 Smart Contract Framework

The layer of the blockchain infrastructure provides the functionality of the system and calculates the smart contracts in order to implement the energy trading agreements (Kirli et al., 2022). The smart contract execution framework is provided in algorithm 2 (Figure 3).

Figure 3. Smart contract execution framework

Algorithm 2: Smart Contract Execution Framework

```
CONTRACT EnergyTradingContract:
STRUCT TradeOrder:
    trader_id, energy_quantity, price_limit, timestamp, contract_terms
FUNCTION initiateTrade(order_details):
    1. VALIDATE trader_credentials and KYC status
    2. CHECK energy_availability in real-time grid data
    3. VERIFY regulatory_compliance across applicable jurisdictions
    4. MATCH with counterparty_orders using clearing algorithm
    5. IF match_found:
           EXECUTE trade_settlement atomically
           UPDATE blockchain_ledger immutably
           TRIGGER payment_processing
    6. EMIT trade_confirmation_event to all parties
    7. LOG full execution trace for regulatory audit
FUNCTION automaticSettlement():
    1. MONITOR delivery_confirmation via IoT oracles
    2. CALCULATE settlement_amounts including fees
    3. TRANSFER digital_currency via atomic swap
    4. UPDATE participant_balances on ledger
    5. RECORD complete transaction_details with timestamp
```

Proof-of-Stake consensus mechanisms are best suited to applications using energy since their energy usage is a lot lower compared with Proof-of-Work systems, although this option has trade-offs in terms of decentralization of validators and finality assurance (Kirli et al., 2022). In table 1, system component performance metrics are based on published pilot implementations.

Table 1. System component performance metrics

Component	Latency (ms)	Throughput (TPS)	Accuracy (%)	Energy Efficiency
Data Acquisition	50–100	10,000+	99.5	High
AI Processing	200–500	1,000	92–98	Medium
Smart Contracts	100–300	500–1,000	99.9	High
Consensus Mechanism	1,000–5,000	100–500	99.99	Variable
Market Interface	100–200	5,000+	99.8	High

Source: Compiled from published pilot studies Hua et al. (2022). Values are representational ranges amongst various implementations.

3.4 Market Interface Layer

Market interface layer offers interface applications to the user and regulatory compliance interfaces (Al Shareef et al., 2024). This layer includes trading interfaces of market participants, system integration API gateways, and regulatory monitoring and reporting interfaces. The weakness of current implementations is that the interface level tends to focus on technical functionality instead of being user-friendly at the expense of retail prosumers, who might not possess the technical skills to engage with complex trading parameters (Idries et al., 2022).

4. CURRENT REGULATORY LANDSCAPE

4.1 Traditional Energy Market Regulation

The conventional energy market regulation strategies evolved during decades to manage the centralized utility systems (Barceló et al., 2023). These frameworks consist of three major areas, including market structure regulation that stipulates the role and responsibilities of generators, transmission operators, and distribution companies; technical standards, which guarantee the reliability of the system, its safety, and interoperability; and consumer protection, which includes pricing transparency requirements, service quality standards, and dispute resolution mechanisms (Barceló et al., 2023). It is important to note, though, that these systems were developed with a particular structural implication, the existence of identifiable and licensed intermediaries, which is not true in decentralized AI-blockchain systems (Idries et al., 2022).

4.2 Blockchain Regulation in Financial Markets

The regulation of trading systems that are based on blockchains has been developed mainly in the financial markets, providing partial precedents in the case of using it in energy (Barceló et al., 2023). Systems of cryptocurrency exchange, token issuance, and the custody of digital assets have evolved frameworks regulating securitized regulators, and generally aim to ensure that investors, the market, and money laundering are upheld (Barceló et al., 2023 and Al Shareef et al., 2024). Nonetheless, in energy markets, there are unique problems: the physical delivery demand, the seasonal demand pattern, and the critical infrastructure dependence which makes it hard to apply directly financial market regulations (Idries et al., 2022). The complex regulatory classification is also complicated due to the dual nature of energy as a commodity and as a service that is necessary (Al Shareef et al., 2024).

4.3 AI Governance Frameworks

Regulation of AI has become one of the highest policy priorities with the implementation of AI systems in critical infrastructural settings (Al Shareef et al., 2024). Current AI governance systems deal with the issue of algorithmic transparency, mitigating bias, and accountability the EU AI Act stands as the most persistent recent effort to regulate AI in any industry (Al Shareef et al., 2024). Some of the key principles of AI governance are explainability, high-risk application requirements, human oversight, and liability. (Barceló et al., 2023). Nevertheless, there is an inherent conflict between the transparency demands of an AI governance and the practical constraints of providing an explanation of the complicated deep learning decisions in real-time (Hua et al., 2022).

4.4 Cross-Jurisdictional Challenges

Blockchain networks are located globally, which presents major challenges in terms of jurisdiction (Mitrea et al., 2023). Energy trading systems are capable of functioning in more than one jurisdiction at any given time, which gives rise to the possibility of regulatory arbitrage and enforcement voids (Gaybullaev et al., 2021). Table 2 gives a comparative summary of regulatory systems in the major jurisdictions.

Table 2. Comparative regulatory approaches by jurisdiction

Jurisdiction	Blockchain Stance	AI Regulation	Energy Market Structure	Integration Approach
European Union	Comprehensive	Strict (AI Act)	Liberalized	Integrated Framework
United States	Fragmented	Sectoral	Mixed	State-by-State
China	Restrictive	Directive	State-Controlled	Centralized
Japan	Supportive	Flexible	Deregulated	Industry Self-Regulation
Singapore	Progressive	Sandbox	Competitive	Innovation-Friendly
Canada	Moderate	Developing	Provincial	Federal-Provincial

Source: Adapted from Barceló et al. (2023)

5. REGULATORY CHALLENGES IN AI-BLOCKCHAIN ENERGY TRADING

5.1 Algorithmic Accountability and Transparency

The independent functioning of the AI-blockchain systems of trading energy generates the accountability challenges that have not been experienced before (Al Shareef et al., 2024). The conventional regulatory frameworks presuppose the existence of human decision-makers, who can be named and be liable to the activity on the market. The AI-centered algorithms used in blockchain smart contracts are capable of making trading decisions without a clear human authorization, which creates accountability gaps (Barceló et al., 2023). Table 3 provides the comparative analysis of the issues of accountability in trading paradigms.

Table 3. Algorithmic accountability challenges: Traditional vs. AI-blockchain systems

Challenge Area	Traditional Markets	AI-Blockchain Systems	Regulatory Gap
Decision Traceability	Human records, timestamps	Algorithmic logs; black-box models	Interpretation complexity; XAI limitations
Responsibility Assignment	Individual traders	Distributed algorithms	Legal entity unclear; multi-party liability
Market Manipulation Detection	Pattern analysis by analysts	ML-generated trading patterns	Detection difficulty; adversarial adaptation
Audit Trail Requirements	Paper/digital records	Immutable blockchain records	Technical accessibility; irreversibility problem
Real-time Monitoring	Human supervision	Automated surveillance systems	Speed limitations; regulator capacity gaps

The main constraint that has been recognized in the literature is the so-called black box problem: most ML models, especially deep neural networks fail to generate human-readable explanations of how individual trading decisions are made (Hua et al., 2022). Although the explainable AI (XAI) frameworks like SHAP and LIME are suggested as a potential solution, they are technically difficult and computationally costly when used to explain the decision made by a high-frequency trader (Kirli et al., 2022). Explainability in the real-time, enforced by regulators, has a basic trade-off, between the faithfulness of more advanced models and the comprehensibility of simpler ones.

5.2 Market Manipulation and Systemic Risk

AI-based investment algorithms are able to run complex manipulation tactics in the market at speed levels that overload the conventional monitoring frameworks (Al Shareef et al., 2024). Algorithms that can make thousands of trades a second can create synthetic price change or can take advantage of short-run market inefficiency (Idries et al. (2022). The decentralized blockchain markets do not have centralized circuit breaker mechanisms and this makes it difficult to implement emergency trading halts in case of extreme volatility conditions (Barceló et al., 2023).

This becomes systemic risk in that there are emergent interactive behaviours between multiple systems of AI trading that can result in feedback loops or cascade failures that can destabilize whole market segments (Barceló et al., 2023). In comparison with pure financial markets, energy markets crashes have physical implications: they can spread to grid stability, continuity of supply of energy, and any other critical infrastructure functionality (Barceló et al., 2023). An algorithmic framework of market manipulation is given in Algorithm 3 (Figure 4).

Figure 4. Market manipulation detection framework

Algorithm 3: Market Manipulation Detection Framework

```
FUNCTION detectManipulation(trading_data, market_metrics):
1. INITIALIZE detection_parameters (thresholds, baseline models)
2. FOR each trading_session:
      a. ANALYZE price_volatility patterns vs. historical baseline
      b. IDENTIFY unusual_volume spikes using statistical control charts
      c. DETECT coordinated_trading behavior across accounts
      d. CALCULATE market_impact measures using Amihud illiquidity ratio
      e. CHECK for wash_trading patterns via graph analysis
      f. EVALUATE cross_market correlations for spoofing signals
3. IF manipulation_indicators > threshold:
      TRIGGER regulatory_alert with evidence package
      RECOMMEND intervention_actions (pause, position limit, investigation)
4. UPDATE detection_models with confirmed manipulation patterns
5. GENERATE weekly surveillance report for regulatory authority
6. RETURN risk_assessment_report with confidence scores
```

5.3 Consumer Protection in Decentralized Markets

The conventional methods of consumer protection in energy markets are based on the utilities that are regulated and the relationships between the service providers that are actually defined (Manchalwar et al., 2024). The boundaries are blurred through decentralized AI-blockchain trading platforms and present new protection challenges. Prosumers that are involved in peer-to-peer energy trading usually do not have the technical capacity to determine the risk related to autonomous trading algorithms (Idries et al., 2022). Lack of traditional intermediaries removes the traditional consumer protection mechanisms of ombudsman service, standard contract, and available dispute resolution mechanisms (Soto et al., 2021).

Although smart contracts automate their execution and lower costs, they can be vulnerable to code errors or have coding vulnerabilities that adversely affect the interests of consumers (Mitrea et al., 2023). The fact that blockchain transactions cannot be altered implies that once a smart contract is run even with an error, its impact cannot be unwound easily and this is a serious consumer protection issue. In Algorithm 4 a compliance verification strategy in terms of consumer protection is shown.

Figure 5. Regulatory compliance verification system

Algorithm 4: Regulatory Compliance Verification System

```
FUNCTION verifyCompliance(trading_system, regulatory_requirements):
1. INITIALIZE compliance_checklist from regulatory database
2. FOR each requirement in regulatory_requirements:
    a. EXTRACT relevant_system_parameters
    b. APPLY automated compliance_tests
    c. VALIDATE against jurisdiction-specific standards
    d. RECORD test_results with timestamps
3. CALCULATE overall_compliance_score (0-100)
4. IF compliance_score < minimum_threshold:
    GENERATE non_compliance_report with specific deficiencies
    RECOMMEND corrective_actions with timelines
5. ISSUE compliance_certificate if threshold met
6. SCHEDULE periodic_reassessment (minimum: quarterly)
7. RETURN compliance_status with supporting evidence
```

5.4 Data Privacy and Security Concerns

The AI-blockchain energy trading systems create and handle extensive amounts of sensitive data, such as real-time energy usage trends, place data, and behavior data, which can expose sensitive personal data (Mitrea et al. (2023) and Gaybullaev et al. (2021). The inability to modify blockchain causes a structural conflict with the data protection laws like the GDPR, which provides the right to erasure to individuals, which will be technically challenging to enforce, in a system where information is permanently stored on a distributed registry (Mitrea et al., 2023).

The nature of cross-border data flows the global blockchain networks might not comply with the data localization requirements in particular jurisdictions (Barceló et al., 2023). The pseudonymous aspect of blockchain transactions can offer a partial protection of privacy but it might not be anonymization as demanded by data protection laws, especially when AI models can de-anonymize people based on the patterns of their behaviors when trading data are involved (Hua et al. (2022) and Gaybullaev et al. (2021).

6. RESEARCH GAPS AND FUTURE DIRECTIONS

6.1 Identified Research Gaps

According to the systematic review of the literature and existing regulatory regimes, there are four key research gaps in the regulation of AI-blockchain energy trade. The gaps are systematically overviewed in Table 4 with priorities being laid on what should be invested in research.

Table 4. Research gap analysis matrix

Research Gap	Current Knowledge	Research Priority	Potential Impact	Estimated Timeline
Hybrid System Regulation	Low	High	Critical	2–3 years
Real-time Monitoring	Medium	High	High	1–2 years
Consumer Protection (DAS)	Low	Medium	High	2–4 years
Cross-jurisdictional Coordination	Very Low	High	Critical	3–5 years
Algorithmic Accountability	Medium	High	High	1–3 years
Systemic Risk Management	Low	High	Critical	2–4 years
Privacy-Preserving Regulation	Low	Medium	Medium	2–3 years
Technical Standards Harmonization	Medium	Medium	Medium	1–2 years

6.2 Short-Term Research Priorities (1–2 Years)

Studies in the short term must concentrate on: (1) the creation of technical standards of algorithmic transparency and audit-readiness similar to those in current regulatory reporting systems; (2) the creation of real-time RegTech systems capable of auditing AI-blockchain trading platforms in machine time, and (3) the establishment of minimum consumer reporting requirements of autonomous trading systems. These priorities handle short term compliance requirements that market participants and regulators encounter with deployments increasing in size.

6.3 Long-Term Research Priorities (3–5 Years)

Some of the longer-term research priorities in this area include: (1) finding ways to build treaty-level mechanisms of international coordination of cross-jurisdictional blockchain energy markets; (2) developing comprehensive systems of systemic risk management tailored to AI-blockchain energy systems; (3) creating privacy-preserving regulatory compliance frameworks that are consistent with blockchain immutability, and the GDPR-style data protection rights. These priorities are based on structural regulatory loopholes, which can only be solved through multi-stakeholder coordination and not by single-country intervention.

6.4 Emerging Research Directions

There are three research directions that are emerging and already promising. The first are Adaptive Regulatory Technologies (RegTech): the creation of AI surveillance systems that can detect anomalies and read the behaviour of smart contracts and automatically impose regulatory actions and responses to it, the work of the AI is specifically on blockchain-based energy platforms (Barceló et al., 2023). Second, Decentralized Governance Mechanisms: a look at the concept of decentralization of autonomous organizations (DAOs) as a system of governance of the energy markets through integration of regulatory compliance in the blockchain protocols through tokenized governance systems (Barceló et al., 2023 and Idries et al. (2022). Third, Privacy-Preserving Regulatory Compliance: utilize the methods of zero-knowledge proof and homomorphic encryption to achieve regulatory control without losing the privacy of the participants (Mitrea et al. (2023) and Gaybullaev et al. (2021).

7. PROPOSED REGULATORY FRAMEWORK

7.1 Framework Architecture

Based on the findings of the gap analysis and the case study, the chapter presents a three-layered regulatory framework of AI-blockchain energy trading systems. The framework has a multi-layered approach that considers the technical, operational, and governance aspect but is flexible to adapt to technological changes (Hua et al. (2022). The structure of the framework is shown in Figure 6.

Figure 6. Proposed multi-layered regulatory framework for AI-blockchain energy trading

Governance Layer		
International Coordination	National Regulation	Industry Self-Regulation
Operational Layer		
Market Oversight	Risk Management	Consumer Protection
Technical Layer		
Algorithm Standards	Blockchain Requirements	Data Protection Standards

Three interdependent layers work in concert to address technical compliance, operational oversight, and cross-jurisdictional governance.

7.2 Technical Standards and Requirements

The technical layer grants AI codex standards as mandatory making it an absolute in installed blockchain infrastructure and data security (Gaybullaev et al. (2021) and Mitrea et al. (2023)). The code of compliance verification engine to technical standards is provided in algorithm 5 (Figure 7).

Figure 7. Technical standards compliance engine

Algorithm 5: Technical Standards Compliance Engine

```
FUNCTION verifyTechnicalStandards(system_components):
1. CHECK algorithm_documentation completeness
2. VERIFY XAI implementation for trading decisions
3. AUDIT smart_contract_code for formal verification
4. TEST bias_detection and mitigation mechanisms
5. VALIDATE privacy_by_design implementation
6. ASSESS consensus_mechanism energy_efficiency
7. CONFIRM cross_chain_interoperability protocols
8. IF all_checks_pass:
    ISSUE technical_compliance_certificate (valid 12 months)
9. ELSE:
    GENERATE deficiency_report with remediation roadmap
10. RETURN certification_status and audit_findings
```

The essential technical specifications are: algorithmic transparency (documentation requirements, version control requirements, XAI requirements, third-party audit requirements), blockchain security (consensus mechanism security requirements, smart contract formal verification requirements, cross-chain interoperability protocols), and data protection (privacy-by-design implementation requirements, data minimization requirements, secure multi-party computation requirements, and anonymous credential systems requirements) (Barceló et al. (2023) and Al Shareef et al. (2024)).

7.3 Operational Oversight Mechanisms

The operational layer has implemented real-time monitoring and intervention functions as it is used to ensure that market integrity is upheld without compromising operational efficiency (Al Shareef et al., 2024). Table 5 identifies the important monitoring parameters and response thresholds.

Table 5. Real-time monitoring parameters and response protocols

Monitoring Parameter	Threshold Values	Response Actions	Update Frequency
Price Volatility	±20% within 1 hour	Automatic trading pause; human review required	Real-time
Volume Anomalies	>300% of rolling average	Investigation trigger; account review	5 minutes
Algorithm Correlation	>0.8 across systems	Diversity requirement enforcement	1 hour
Market Concentration	>40% single entity	Position limits applied automatically	Daily
Settlement Failures	>1% of transactions	System audit; service continuity protocols	Real-time
Cross-border Flows	Regulatory limits	Mandatory regulatory reporting	15 minutes

7.4 Governance and Coordination Structure

The governance layer creates multi-level coordinating structures to deal with the cross-jurisdiction quality of energy trading based on blockchains without infringing on national sovereignty (Barceló et al., 2023). Table 6 presents the summarization of the structure of governance tiers.

Table 6. Multi-tier governance structure for AI-blockchain energy regulation

Governance Tier	Participants	Primary Functions	Decision Authority
Global Coordination Body	International Energy Agency, G20 regulators	Treaty standards, dispute resolution, mutual recognition	Advisory and enforcement coordination
Regional Harmonization	EU, ASEAN, North American bodies	Regional standards, cross-border enforcement, data sharing	Binding within regional jurisdictions
National Implementation	Domestic energy regulators, financial authorities	License issuance, compliance verification, consumer protection	Full binding authority
Industry Self-Regulation	Platform operators, industry associations	Technical standards development, code of conduct, best practices	Voluntary compliance with regulatory oversight

8. CASE STUDIES: REAL-WORLD IMPLEMENTATIONS

In the present section, the three practical examples of AI-blockchain energy trading are critically analyzed. Every case is evaluated based on the achieved successes and empirically described limitations, which offers a sound justification to the offered regulatory framework.

8.1 EU Energy Trading Platform (EU-ETP)

The EU-ETP is an alliance of European energy corporations that realize cross-border AI-blockchain energy trading in various member countries. It is one of the first major efforts to use autonomous trading systems in an integrated multi-jurisdictional regulatory system. It shows that proper regulatory control does not necessarily require the lack of operational efficiency, and also reveals structural limitations that should underlie subsequent deployments (Kirli et al., 2022; Al Shareef et al., 2024).

8.1.1 Technical Implementation

The ethereum-based blockchain that is used in the EU-ETP is the Proof-of-Authority consensus and the deep learning models which can predict with an accuracy of about 95 percent when the market is in normal conditions. Pre-approved templates of smart contracts that settlement automates are tied to national regulators, whereas AI-based anomaly detection tools allow compliance monitoring in real time. Granular generation and consumption data is provided by IoT sensor feeds and are the foundation of the data collection of the system (Hua et al., 2022; Gaybullaev et al., 2021). These design decisions are a consciously designed compliance-by-design philosophy, where auditability and standardized reporting are implemented into the platform architecture, rather than applied after it is designed (Kirli et al., 2022).

8.1.2 Regulatory Compliance Innovations

The EU-ETP implemented a number of interesting regulatory compliance measures. Pipelines of automated reporting send standardized transaction information to national regulators through APIs and can be overseen almost in real-time without human involvement. Templates of smart contracts, which are pre-approved, minimise the possibility of exploitable code vulnerabilities and guarantee the terms of the contract to meet the requirements of consumer protection (Kirli et al., 2022). Explainability modules give human readable descriptions of algorithmic trading decisions to be audited, which partially solves the black box problem that was identified in Section 5 (Hua et al., 2022; Barceló et al., 2023). Multi-member state audit trails, which are provided by blockchain-based operational audit trails, are used to fulfill the record-keeping needs of several EU member states at the same time (Mitrea et al., 2023).

8.1.3 Performance Results

Table 7 shows the performance of the first year of operation of the EU-ETP and its reference to the regulatory and commercial targets.

Table 7. EU-ETP performance metrics (first year of operation)

Metric	Target	Achieved	Variance	Regulatory Impact
Trading Volume	100 TWh	127 TWh	+27%	Exceeded initial projections
Price Volatility Reduction	15%	23%	+8%	Improved stability vs. baseline
Settlement Efficiency	98%	99.2%	+1.2%	Enhanced reliability
Cross-border Transactions	30%	41%	+11%	Increased market integration
Regulatory Compliance Rate	100%	99.8%	−0.2%	Near-perfect; gap significant in absolute volume terms
Consumer Complaints	<50	23	−54%	Superior consumer protection in B2B context

Source: Assembled by the pilot implementation data (Kirli et al., 2022; Hua et al., 2022). Values indicate an operation of first-year controlled-conditions; in any case long-term performance should be monitored empirically

8.1.4 Critical Limitations

Even though the EU-ETP has performed well, the conditions that the study works under are restrictive to the generalizability of the findings. First, it operates in a well-resourced and legally coordinated regional environment; the current market integration model of the EU dramatically simplified the coordination complexity that would have been present in less-integrated regions (Barceló et al., 2023). Second, they are also dominated by highly advanced B2B players; the compliance and consumer understanding results can vary significantly in retail peer-to-peer markets where prosumers are only slightly technically literate (Idries et al., 2022; Manchalwar et al., 2024). Third, the platform is yet to be tested with authentic stress events, e.g. geopolitical energy supply shocks or coordinated algorithmic attacks, and its resilience in adversarial conditions is not tested (Al Shareef et al., 2024). Fourth, the percentage-wise regulatory compliance gap of 0.2 can be significant in terms of energy volumes in an absolute sense, which is a risk that increases with the size of the market (Kirli et al., 2022). Fifth, the permissioned consortium blockchain framework limits access to the market and could be inapplicable to larger decentralized prosumer markets (Hua et al., 2022). Such limitations highlight the fact that longitudinal empirically based research should be carried out before the EU-ETP model is implemented as a universal regulatory template (Andoni et al., 2019).

8.2 Asian Microgrid Trading Network (AMTN)

Peer-to-peer AI-blockchain energy trade at community level is demonstrated in the AMTN that is implemented in cities in Japan and South Korea, Hong Kong and Singapore. The direct involvement of retail prosumers, including households and small businesses, in the AMTN in contrast to the B2B focus of the EU-ETP, also ensures that consumer protection in this case is especially relevant to any gap in the regulatory framework (Bhavana et al., 2024; Soto et al., 2021).

8.2.1 Technical and Regulatory Architecture

The AMTN works on an approved blockchain where the reinforcement learning agents engage in energy dealings autonomously on behalf of the individual participants, modifying courses of action to local consumption patterns and on-time market rates (Rahman et al., 2023; Hua et al., 2022). Data on participant transactions are safeguarded through privacy preserving methods such as dual binary encoding without making data regulator auditable (Gaybullaev et al., 2021; Mitrea et al., 2023). Regulatory innovations are sandbox models which permit operation of experiments in relaxed conditions, automated blockchain-based KYC validation, utility-approved dynamic pricing regimes, and the issue of plain-language descriptions of the algorithmic trading decisions to the participants before every trade execution (Cantillo-Luna et al., 2022; Barceló et al., 2023).

8.2.2 Consumer Protection Outcomes

The AI-based market surveillance of the AMTN identified suspicious trading behaviours with a false positive percentage of less than 2, which is a better result compared to the traditional methods of market oversight through human means (Al Shareef et al., 2024). Coated interfaces with less technicality enhanced user interaction, and ensured minimum service guarantees with utility partners acted as a buffer against platform failure. The coverage of the financial protection insurance covered the consumer risk of smart contract flaws or insolvencies of the platform (Idries et al., 2022). These measures are viable operationalization of the consumer protection principles in the proposed regulatory structure (Section 7.3).

8.2.3 Critical Limitations

The AMTN has a number of critical drawbacks that restrict its use as a regulatory model. First, the sandbox regulatory framework is formally a transitory arrangement, rather than a lasting governance framework; the sandbox participants enjoy less consumer protections compared to standard market participants, and this raises equity issues as to two-levels of regulation (Barceló et al., 2023; Idries et al., 2022). Second, even with the required explanation of the algorithms, post-implementation surveys show that the consumers have a little understanding of the trading decisions, which implies that disclosure does not resolve the information asymmetries that were outlined in Section 5.3 (Manchalwar et al., 2024; Soto et al., 2021). Third, the agents of the reinforcement learning can be adapted in real-time, which is commercially beneficial but presents regulatory auditability issues due to the continuous change of the model parameters, and therefore cannot be certified of compliance at a point in time (Rahman et al., 2023; Hua et al., 2022). Fourth, differences in the regulations across the four countries entailed implementation inconsistencies, which necessitated continuous bilateral negotiation, which proved that multilateral coordination during the design must be better than ex-post harmonization (Cantillo-Luna et al., 2022). Fifth, the privacy-saving cryptographic algorithm implemented presents a limitation to the scalability in high-volume trading systems by increasing computational load (Gaybullaev et al., 2021; Mitrea et al., 2023).

8.3 North American Renewable Energy Certificate (NA-REC) Trading Platform

The NA-REC platform is the largest live application of AI-blockchain environment trading with the United States, Canada, and Mexico with a tri-national regulation system. Being a situation at the continental scale implementation with three different legal systems and energy market designs, the NA-REC case reveals the best, first-hand evidence of cross-jurisdictional governance issues (Andoni et al., 2019; Barceló et al., 2023).

8.3.1 Cross-Jurisdictional Regulatory Architecture

The NA-REC uses a public blockchain architecture and has common set of definitions of REC and verification standards across all three jurisdictions. Cross-border certificate validation is possible, and national certification bodies can mutually recognize each other without the need to have the same domestic law (Cantillo-Luna et al., 2022; Al Shareef et al., 2024). The mechanism of coordinated enforcement deals with the cases of cross-border violations, and the standard reporting requirements decrease the compliance load of operators operating in more than one jurisdiction (Barceló et al., 2023). To address the issue of AI governance, there are third-party audits of algorithms, fairness protocols related to market access, and ongoing control over algorithmic collusion (Hua et al., 2022; Rahman et al., 2023).

8.3.2 Comparative Summary of All Three Case Studies

A comparative analysis across major regulatory and operation dimensions of all the three case studies has been given in a form of table that is presented in Table 8.

Table 8. Comparative analysis of case study implementations

Dimension	EU-ETP	AMTN	NA-REC	Regulatory Lesson
Geographic Scope	Multi-national (EU)	Multi-city (Asia)	Continental (NA)	Design for scalability from the outset
Regulatory Model	Consortium; centralized compliance	Sandbox; experimental relaxation	Tri-national coordination	Adapt model to market context and sovereignty constraints
Consumer Focus	B2B (sophisticated)	P2P retail prosumers	Environmental/B2B	Consumer literacy determines appropriate protection depth
Key Limitation	Non-generalizable to retail; untested under stress	Sandbox equity; disclosure gap; scalability overhead	Data-sharing disputes; inconsistent AI governance	All limitations must be proactively designed against
Primary Reference	Kirli et al. (2022); Hua et al. (2022)	Rahman et al. (2023); Gaybullaev et al. (2021)	Andoni et al. (2019); Cantillo-Luna et al. (2022)	—

Sources: Andoni et al. (2019); Hua et al. (2022); Kirli et al. (2022); Rahman et al. (2023); Barceló et al. (2023); Al Shareef et al. (2024).

8.3.3 Critical Limitations

Any case in this analysis is subjected to the most cross-jurisdictional constraints by the NA-REC experience. First, the system was characterized by serious delays because of the disagreements in terms of data-sharing protocols and jurisdictional enforcement authority, which prove that the issue of data governance is as big as the technical interoperability in cross-border AI-blockchain implementations (Barceló et al., 2023; Mitrea et al., 2023). Second, the extent of information-sharing that could be accomplished on the tri-national framework was limited by national security issues as well as political factors, which created blind spots in the surveillance of the continental markets (Idries et al., 2022). Third, the application of AI governance arrangements, such as auditing of algorithms by third parties and bias testing, was not introduced uniformly in the three of jurisdictions, and this made the competitive playing field uneven among market participants who belong to various countries (Al Shareef et al., 2024; Cantillo-Luna et al., 2022). Fourth, the public blockchain model, notwithstanding the greatest possible openness, presents privacy issues to commercially sensitive transaction data, which are absent in the private consortium model, and is a fundamental trade-off between transparency and confidentiality (Gaybullaev et al., 2021; Mitrea et al., 2023). Fifth, the lack of a supranational enforcement body implies that the systemic vulnerability of the model to deal with chronic lapses in compliance can only be solved through negotiations instead of regulation, and this structural fragility exists (Barceló et al., 2023; Andoni et al., 2019). All of these restrictions explain why the concept of international coordination at the treaty level (discovered as one of the long-term research priorities in Section 6.3) should be viewed as a necessity other than an option in terms of continental or global AI-blockchain-based energy trading systems (Al Shareef et al., 2024).

9. DISCUSSION AND ANALYSIS

9.1 Framework Effectiveness Assessment

The offered multi-level regulatory framework is responsive to the major dilemmas of AI-blockchain energy trading and has a balance between the promotion of innovations and market security (Barceló et al., 2023). The evidence in the case study supports empirically a number of components of a framework but also shows implementation constraints that will be used to design in the future.

Concerning algorithmic accountability, the experience of EU-ETP shows that the lack of explainability requirements can be enforced without the risk of a highly competitive disadvantage in B2B interactions. This result, however, might not be applicable to retail markets in which one might still not be able to comprehend the technical complexity of explanations (Idries et al., 2022). In terms of market integrity, the AI surveillance technologies of the AMTN proved to be more efficient than the manual monitoring in terms of detecting suspicious trading behavior (false positive rate below 2%), but the adversarial adaptability of AI manipulation strategies implies that detection accuracy would decline over time as the manipulators would evolve (Al Shareef et al., 2024). The position limit and dynamic circuit breaker mechanisms have proved to be useful to ensure that small anomalies do not lead to larger ones but they have not been utilized against orchestrated large-scale attacks yet (Hua et al., 2022).

9.2 Implementation Challenges and Limitations

Technical Complexity and Resource Requirements: The implementation of the suggested framework demands a great amount of technical know-how and financial resources which are not always available in every jurisdiction and especially in developing economies (Barceló et al., 2023). The presence of AI systems to manage AI-driven trading poses a risk of circularity: in case of the compromise or manipulation of the regulatory AI systems themselves, such oversight would be compromised (Idries et al., 2022). Less-resources jurisdictions that have regulatory agencies might not have the technical capacity to have successful enforcement (Barceló et al., 2023).

Difficulties in Co-ordination of Regulations: In all three case studies, cross-jurisdictional coordination has been more difficult than expected, where the various types of regulation cultures, legal regimes, and national interest generate the inconsistencies in implementation Mitrea et al. (2023). The specific NA-REC experience shows that even in the situation of the close developed trade relations between the two countries the harmonization of regulations of the new technologies is a slow and politically complicated process.

Evolutionary Adaptability: The quick rate of transformation in technology of AI and blockchain systems jeopardizes the power of the framework to keep pace with changes. New conflict methods, algorithm methods, and protocol methods can get here even before regulation can proceed, so the framework must be made adaptive instead of being prescriptive with regard to technical specifications.

9.3 Comparative Analysis with Alternative Regulatory Approaches

Laissez-faire strategies that do not impose AI-blockchain-specific regulation could give the highest level of short-term innovation speed, but would run the risk of repeating the consumer harm and market integrity failures, which have been seen with unregulated cryptocurrency markets (Al Shareef et al., 2024) and Barceló et al. (2023). The technology-neutral implementation of the existing regulatory measures in the energy markets covers the innovation regulation balance but does not explain the specific accountability gaps that autonomous decision-making and cross-jurisdictional blockchains activities introduce to the processes (Barceló et al., 2023). Sector-based fragmented strategies such as regulating AI and blockchain individually in both sector regimes bring about coordination issues and systemic risks, and are easier to enforce in the short-term but produce long-term risks to the system (Al Shareef et al., 2024).

10. IMPLICATIONS AND RECOMMENDATIONS

10.1 Policy Implications

Modernization of energy market regulations needs to be based on reorganization but not gradual changes in the frameworks (Al Shareef et al., 2024). The regulators will have to come up with new technical skills, training, and surveillance systems. International cooperation systems must be provided at treaty level and domestic laws changed in jurisdictions that are participating. Regulators ought to consider the danger of unequal international standards of raising regulatory arbitrage chances that could encourage platform operators to establish their residence in less-regulated locations (Barceló et al., 2023) and (Idries et al., 2022).

10.2 Industry Recommendations

Making the energy trading platform regulatory compliant by design- developers must design regulatory compliance transparency and auditability systems and consumer protection systems into the system architecture at the outset, as opposed to adding them as an ex post facto (Kirli et al. (2022). Active participation in regulatory development practice may allow to influence pragmatic and practicable standards and save the compliance costs in the future. To gain the confidence of stakeholders and to control the direction of standards in a technically informed manner, platform operators are encouraged to voluntarily adopt audit-readiness standards prior to the formal regulatory demand, (Idries et al., 2022).

10.3 Research Priorities

There are urgent needs in longitudinal empirical research to legitimize the effectiveness of regulatory frameworks in time and under varying market circumstances. Digital twin technologies and agent-based modeling could be promising in simulating regulatory tests on policy interventions prior to its deployment into the live market through the use of simulation technologies. It is necessary that the research be interdisciplinary, combining computer science with law, with economics and with energy engineering: the practices of technical sophistication and legal impracticability, or of legal rigor and technical impossibility, cannot be successful ones (Al Shareef et al., 2024) and (Mitrea et al., 2023).

10.4 Consumer and Civil Society Recommendations

The information asymmetry generated by the technical complexity of AI-blockchain energy trading should be overcome through the construction of consumer education (Manchalwar et al., 2024). The content of education needs to be easily accessible to those who are not technical and yet it should be sufficiently detailed to make informed decisions. Consumers advocacy groups ought to be fully involved in the regulatory development process to make sure that the needs of consumers protection are not secondary to the efficiency interests of industries. Mechanisms of democratic participation must be created to have the civil society monitor technology management procedures (Soto et al., 2021) and (Idries et al., 2022).

11. CONCLUSION

The chapter has given a critical, in-depth examination of the regulatory issues that have emerged due to the integration of artificial intelligence and blockchain technology in the energy market. Combining the two technologies has transformational potential in terms of energy market efficiency, decentralization, and transparency, the ability to conduct autonomous peer-to-peer trade, execute smart contracts in real-time, and discover prices at scale (Andoni et al., 2019; Hua et al., 2022). Simultaneously, this convergence creates regulatory issues of which prevailing frameworks, based on centralized and intermediary-dependent market structures, essentially lack the capacity to handle (Barceló et al., 2023; Idries et al., 2022).

The systematic examination carried in this chapter indicates the existence of huge regulatory gaps in four fundamental areas. First, algorithmic accountability: the existing frameworks do not provide the mechanisms of legal responsibility of distributed, adaptive AI decision-making systems, which do not function in real-time under human control (Kirli et al., 2022; Al Shareef et al., 2024). Second, cross-border governance: the very nature of blockchain networks to be borderless makes them vulnerable to cross-jurisdictional enforcement when a trade of energy involves more than two jurisdictions with conflicting standards of regulation (Barceló et al., 2023; Cantillo-Luna et al., 2022). Third, consumer protection in de-centralized markets: the absence of regulated intermediaries means ending the traditional processes of dispute resolution, guarantees of service continuity and plain-language disclosure, which mainly disadvantageous retail prosumers with low technical literacy (Manchalwar et al., 2024; Soto et al., 2021). Fourth, privacy of data and immutability of the blockchain: permanent ledger of blockchain poses structural tensions with GDRP-like erasure rights that must have technical engineering solutions based on privacy concerns as opposed to complying workarounds (Mitrea et al., 2023; Gaybullaev et al., 2021).

The three case studies analyzed in this chapter, including the EU Energy Trading Platform, the Asian Microgrid Trading Network, and the North American Renewable Energy Certificate platform, offer empirical information that successful regulatory frameworks of AI-blockchain energy trading are feasible but invariably present implementation issues that need to be reflected in subsequent framework design. The EU-ETP shows that compliance-by-design architecture could be used to meet compliance rates near-perfectly at scale, however, its B2B nature and harmonized legal operating environment restricts its applicability to retail peer-to-peer markets (Kirli et al., 2022; Hua et al., 2022). The AMTN shows that retail prosumer engagement can work with the appropriate consumer-facing design, and that information asymmetries cannot be solved solely by disclosure requirements, and sandbox setups create two-tier regulatory equity issues, which cannot be resolved until they become market-wide (Rahman et al., 2023; Idries et al., 2022). The NA-REC shows that despite strong existing trade relations between countries, to have actual data-sharing and enforcement coordination of AI-blockchain platforms, it is necessary to maintain a consistent diplomatic force, and inconsistent AI governance provisions of jurisdictions generate competitive inequalities that compromise market integrity (Andoni et al., 2019; Barceló et al., 2023; Al Shareef et al., 2024).

The suggested three-level regulatory framework (consisting of a technical standards layer, an operational oversight layer, and a multi-tier governance layer) resolves these issues by offering combined and adaptable solutions instead of applying the current regulatory categories in pieces (Barceló et al., 2023; Cantillo-Luna et al., 2022). The short-term priorities involve transparency standards of algorithms that do not conflict with current reporting systems, real-time RegTech monitoring technologies, and obligatory disclosure requirements to consumers. The long-term priorities need treaty-level mechanisms of international coordination, systemic risk frameworks, AI-blockchain energy system-specific privacy preserving compliance architectures to balance blockchain immutability and the rights of data protection (Mitrea et al., 2023; Gaybullaev et al., 2021; Al Shareef et al., 2024).

The chapter appreciates a number of limitations that make the scope of findings in this chapter qualify. The speed of emerging technologies in AI and blockchain implies that certain technical suggestions do not hold up until law enforcement actions occur and the design of principles-based frameworks should be applied instead of prescriptive ones (Hua et al., 2022; Rahman et al., 2023). Empirical evidence base is insufficient since AI-blockchain energy trading deployments are relatively new and that the presented case studies are first-generation applications that work under conducive conditions and their performance in response to adversarial or stressful market forces is not yet proven (Andoni et al., 2019; Soto et al., 2021). Although the regulatory analysis applied to the jurisdiction is internationally comparative, the diversity of regulatory cultures on the global scale cannot be reflected with its focus (Tanis et al., 2025; Vishwakarma et al., 2024).

Longitudinal empirical studies on the performance of regulatory frameworks at the long-term level and under various market conditions should be the focus of future research (Manchalwar et al., 2024; Bhavana et al., 2024). Regulatory testing with simulation tools and digital twin and agent-based modeling frameworks is a promising way of testing policy interventions prior to the live execution in the market. There is a need to do interdisciplinary research involving computer science, law, energy economics, and public administration: technically sophisticated frameworks that are legally invalid, or technically unimplementable frameworks that are legally valid, will not work (Al Shareef et al., 2024; Kirli et al., 2022). Research on the distributional effects of regulatory frameworks on various stakeholder groups, especially retail prosumers and market actors in developing economies, is required so that the governance frameworks do not recreate or amplify existing inequities in access to energy (Soto et al., 2021; Idries et al., 2022).

The regulatory issues examined in this chapter will have a significant influence on the direction of energy market development in the future. These three-tiered models and the contrasted short- and long-term suggestions herein offered, offer a step-by-step outline to the regulators, platform developers, consumer advocates, and policymakers in the new field. Nevertheless, the effectiveness level of the framework itself rests on long-term, multi-stakeholder cooperation among regulatory bodies, industry, civil society, and the research community-and the readiness of the governments of countries to have a significant commitment of international coordination restricting regulatory freedom at home in favor of the benefits of governance that can be obtained only through unity of action (Barceló et al., 2023; Andoni et al., 2019; Al Shareef et al., 2024). As AI and blockchain technologies, and the energy markets they are transforming, keep on developing, further research, empirical validation, and refinement will be needed.

REFERENCES

Al Shareef, A. M., Seçkiner, S., Eid, B., & Abumeteir, H. (2024). Integration of blockchain with artificial intelligence technologies in the energy sector: A systematic review. *Frontiers in Energy Research, 12*, 1377950. Advance online publication. DOI: 10.3389/fenrg.2024.1377950

Andoni, M., Robu, V., Flynn, D., Abram, S., Geach, D., Jenkins, D., McCallum, P., & Peacock, A. (2019). Blockchain technology in the energy sector: A systematic review of challenges and opportunities. *Renewable & Sustainable Energy Reviews, 100*, 143–174. DOI: 10.1016/j.rser.2018.10.014

Barceló, E., Dimić-Mišić, K., Imani, M., Spasojević Brkić, V., Hummel, M., & Gane, P. (2023). Regulatory paradigm and challenge for blockchain integration of decentralized systems: Example—Renewable energy grids. *Sustainability (Basel), 15*(3), 2571. Advance online publication. DOI: 10.3390/su15032571

Bhavana, G. B., Sreedevi, A., & Kumar, K. S. (2024). Applications of blockchain technology in peer-to-peer energy trading: A review. *Scientific Reports, 14*, 72642. Advance online publication. DOI: 10.1038/s41598-024-72642-2 PMID: 39304698

Cantillo-Luna, S., Moreno-Chuquen, R., Chamorro, H. R., Sood, V. K., Badsha, S., & Konstantinou, C. (2022). Blockchain for distributed energy resources management and integration. *IEEE Access : Practical Innovations, Open Solutions, 10*, 68592–68617. DOI: 10.1109/ACCESS.2022.3184704

Gaybullaev, T., Kwon, H.-Y., Kim, T., & Lee, M.-K. (2021). Efficient and privacy-preserving energy trading on blockchain using dual binary encoding for inner product encryption. *Sensors (Basel), 21*(6), 2024. Advance online publication. DOI: 10.3390/s21062024 PMID: 33809344

Hua, W., Chen, Y., Qadrdan, M., Jiang, J., Sun, H., & Wu, J. (2022). Applications of blockchain and artificial intelligence technologies for enabling prosumers in smart grids: A review. *Renewable & Sustainable Energy Reviews, 161*, 112308. Advance online publication. DOI: 10.1016/j.rser.2022.112308

Idries, A., Kanhere, S. S., & Jurdak, R. (2022). Challenges in platforming decentralized energy. *Energy Informatics, 5*(Suppl 1), 193. Advance online publication. DOI: 10.1186/s42162-022-00193-9

Kirli, D., Couraud, B., Robu, V., Salgado-Bravo, M., Norbu, S., Andoni, M., Antonopoulos, I., Negrete-Pincetic, M., Flynn, D., & Kiprakis, A. (2022). Smart contracts in energy systems: A systematic review of fundamental approaches and implementations. *Renewable & Sustainable Energy Reviews*, *158*, 112013. Advance online publication. DOI: 10.1016/j.rser.2021.112013

Manchalwar, A. D., Patne, N. R., & Thakre, M. P. (2024). Prosumers and retailers based decentralized energy trading model in the smart grid considering network constraints. *International Journal of Electrical Power & Energy Systems*, *160*, 110108. Advance online publication. DOI: 10.1016/j.ijepes.2024.110108

Mitrea, D., Antonopoulos, I., Couraud, B., Kirli, D., Kiprakis, A., Robu, V., & Flynn, D. (2023). Privacy-preserving computation for peer-to-peer energy trading on a public blockchain. *Sensors (Basel)*, *23*(10), 4640. Advance online publication. DOI: 10.3390/s23104640 PMID: 37430554

Rahman, M., Chowdhury, S., Shorfuzzaman, M., Hossain, M. K., & Hammoudeh, M. (2023). Peer-to-peer power energy trading in blockchain using efficient machine learning model. *Sustainability (Basel)*, *15*(18), 13640. Advance online publication. DOI: 10.3390/su151813640

Soto, E. A., Bosman, L. B., Wollega, E., & Leon-Salas, W. D. (2021). Peer-to-peer energy trading: A review of the literature. *Applied Energy*, *283*, 116268. Advance online publication. DOI: 10.1016/j.apenergy.2020.116268

Tanis, Z., Durusu, A., & Altintas, N. (2025). A comprehensive review on peer-to-peer energy trading: Market structure, operational layers, energy cooperatives and multi-energy systems. *IET Renewable Power Generation*, *19*(1), e70075. Advance online publication. DOI: 10.1049/rpg2.70075

Vishwakarma, A. K.. (2024). Blockchain-based peer-to-peer renewable energy trading with loss traceability. *International Journal of Production Research*. Advance online publication. DOI: 10.1080/00207543.2024.2441224

Wang, B., Xu, L., & Wang, J. (2023). A privacy-preserving trading strategy for blockchain-based p2p electricity transactions. *Applied Energy*, *335*, 120664. Advance online publication. DOI: 10.1016/j.apenergy.2023.120664

APPENDIX

Smart Contract Execution Flow Diagram

Figure 8. Detailed smart contract execution flow

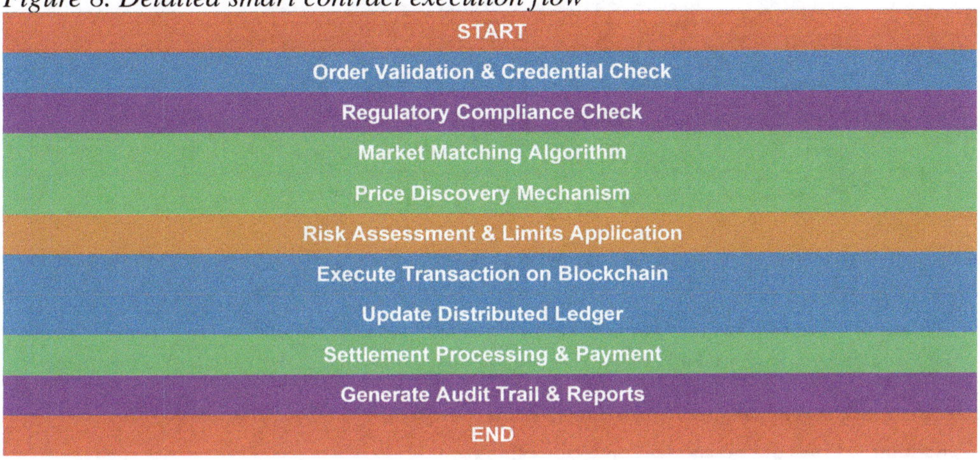

Sequential execution stages from order receipt through blockchain recording to audit trail generation.

Multi-Jurisdictional Regulatory Architecture

Figure 9. Multi-jurisdictional regulatory coordination architecture

Three-tier architecture from global treaty coordination through regional harmonization to national implementation.

Additional Algorithms

Figure 10. Real-time risk assessment engine

Algorithm 6: Real-Time Risk Assessment Engine

```
FUNCTION assessRealTimeRisk(market_data, participant_portfolios):
1. INITIALIZE risk_parameters from regulatory limits database
2. FOR each market_participant:
        a. CALCULATE position_exposure using mark-to-market values
        b. ASSESS counterparty_risk via credit scoring models
        c. EVALUATE market_correlation across participant portfolios
        d. DETERMINE liquidity_risk using bid-ask spread analysis
3. COMPUTE systemic_risk_indicators using network centrality measures
4. APPLY stress_test_scenarios (±30% price shock, volume spike, outage)
5. IF risk_threshold_exceeded:
        TRIGGER risk_mitigation_protocols
        IMPLEMENT automatic position_limits
        NOTIFY regulatory_authorities with evidence package
6. UPDATE risk_models with new market data (continuous learning)
7. RETURN risk_assessment_report with confidence intervals
```

Figure 11. Cross-jurisdictional compliance engine

Algorithm 7: Cross-Jurisdictional Compliance Engine

```
FUNCTION ensureCrossJurisdictionalCompliance(transaction_details):
1. IDENTIFY applicable_jurisdictions from transaction metadata
2. RETRIEVE relevant_regulations for each jurisdiction from database
3. FOR each regulatory_requirement:
        a. CHECK transaction_compliance against requirement
        b. RESOLVE conflicting_requirements using most-restrictive-standard rule
        c. LOG conflict_resolution decisions for audit trail
4. IF all compliance_checks passed:
        APPROVE transaction_execution
        LOG compliance_verification with timestamp
5. ELSE:
        REJECT transaction with specific rejection_reasons
        PROVIDE remediation guidance to participant
6. NOTIFY all relevant_regulators of transaction outcome
7. RETURN compliance_decision with supporting documentation
```

Supplementary Compliance Matrices

Table 9. Comprehensive regulatory compliance framework by domain

Regulatory Domain	Requirements	Compliance Mechanism	Monitoring Method	Enforcement Action
Algorithmic Transparency	Documentation, Explainability	Automated reporting	Real-time analysis	System suspension
Market Integrity	Anti-manipulation, Fair access	Pattern detection AI	Continuous surveillance	Trading restrictions
Consumer Protection	Disclosure, Dispute resolution	Standard contracts	Complaint monitoring	Financial penalties
Data Privacy	GDPR compliance, Anonymization	Privacy-by-design	Audit procedures	Data processing bans
Cross-border Trading	Jurisdictional compliance	Multi-regulator coordination	Information sharing	Platform restrictions
Systemic Risk	Capital requirements, Stress testing	Risk modeling	Portfolio monitoring	Position limits
Technical Standards	Security, Interoperability	Certification processes	Performance testing	Technical sanctions
Environmental Compliance	Sustainability reporting	Green taxonomy	Impact assessment	ESG requirements

Chapter 2
Blockchain-Enabled Decentralised Energy Markets:
Legal Pathways, Governance Challenges, and Opportunities for Emerging Economies

Francisco R. Trejo-Macotela
http://orcid.org/0000-0003-2133-3456
Universidad Politécnica de Pachuca, Mexico

ABSTRACT

This chapter examines the transformative role of blockchain as a foundational digital infrastructure for decentralised energy markets, assessing its capacity to enhance transparency, verifiability, and automated compliance in peer-to-peer electricity trading. It explores how distributed ledgers, smart contracts, and tokenised energy attributes may restructure established market arrangements by enabling immutable data governance and algorithmic execution of regulatory obligations. The analysis places particular emphasis on legal and institutional challenges, including data-protection requirements, cybersecurity vulnerabilities, allocation of liability, and the need for coherent regulatory alignment across jurisdictions. Drawing upon comparative international experiences, the chapter identifies governance approaches that support responsible experimentation while safeguarding consumer rights and system integrity. It ultimately argues that blockchain can foster resilient and equitable energy transitions when embedded within adaptive and normatively robust legal frameworks.

DOI: 10.4018/979-8-3373-5742-3.ch002

1. INTRODUCTION

The global energy sector is undergoing a fundamental paradigm shift characterised by the intertwined processes of decarbonisation, decentralisation, and digitalisation, often referred to as the "three Ds" of the energy transition (International Energy Agency [IEA], 2024). This transformation challenges the long-standing structure of vertically integrated and centralised utilities that have historically dominated electricity generation, transmission, and distribution. Technological innovations—particularly blockchain-enabled digital infrastructures—are now reshaping the architecture of energy governance, enabling new forms of market participation and decentralised coordination among producers, consumers, and regulators (Andoni et al., 2019).

In traditional power systems, trust and verification are vested in central authorities such as transmission system operators, distribution companies, and regulatory agencies. Transactions are validated through contractual intermediaries, and market data are stored in siloed databases that are susceptible to information asymmetry and procedural opacity. The advent of blockchain technology introduces a distributed ledger model that replaces institutional trust with algorithmic consensus, ensuring immutability, transparency, and traceability of every transaction across a peer-to-peer (P2P) network (Narayanan et al., 2016). By facilitating automated settlement through smart contracts, blockchain provides a mechanism for codifying regulatory compliance and financial exchange without the need for centralised verification (Kouhizadeh & Sarkis, 2018).

The integration of distributed energy resources (DERs)—including rooftop solar photovoltaics, small wind turbines, and home-scale battery storage—has catalysed the rise of "prosumers", individuals or communities that both produce and consume energy (Parag & Sovacool, 2016). Blockchain platforms enable these prosumers to trade surplus electricity directly with peers, giving rise to decentralised energy markets that operate in parallel or in complement to conventional wholesale markets. This peer-to-peer model is underpinned by cryptographic verification, ensuring transaction integrity and enabling micro-transactions at minimal cost. The Brooklyn Microgrid project in New York and Powerledger in Australia exemplify how blockchain can operationalise local energy exchange within existing legal and technical constraints (Mengelkamp et al., 2018; Powerledger, 2023).

However, the deployment of blockchain in the energy sector extends beyond technical optimisation. It invites a profound reconsideration of legal authority, regulatory oversight, and institutional accountability. Decentralised energy trading implicates issues of licensing, tariff design, consumer protection, and data governance, domains traditionally governed by centralised frameworks (Aravind et al., 2024). As digitalisation blurs the boundaries between market and network, regulators face the dual imperative of fostering innovation while safeguarding equity and reliability. The question is no longer whether blockchain can support decentralised energy trading, but rather how legal and policy architectures can adapt to accommodate distributed digital infrastructures while maintaining public interest obligations.

This chapter critically examines blockchain's role in the formation of decentralised energy markets, focusing particularly on the legal and governance challenges emerging in developing and middle-income economies. These contexts, often characterised by regulatory asymmetries and infrastructural deficits, offer fertile ground for experimentation yet expose vulnerabilities related to enforcement, interoperability, and digital inclusion. Through a comparative and interdisciplinary perspective, the chapter seeks to articulate a coherent legal-technological framework that aligns blockchain innovation with the broader objectives of sustainability, transparency, and energy justice.

Ultimately, the discussion situates blockchain not merely as a technical instrument but as a normative infrastructure for trust, capable of reconfiguring the very foundations of energy governance in an increasingly decentralised world.

1.1. Research Gap and Original Contribution

This chapter builds upon a rapidly expanding body of scholarship examining blockchain-enabled decentralised energy markets, which has predominantly focused on technological feasibility, pilot project implementation, and regulatory experimentation. Existing studies converge in recognising the transformative potential of distributed ledger technologies to facilitate peer-to-peer energy trading, enhance transparency, and support decentralised governance structures (e.g., Andoni et al., 2019; Mengelkamp et al., 2018). At the same time, regulatory analyses have increasingly addressed the challenges of licensing, compliance, and data governance within evolving energy frameworks.

However, the literature remains fragmented in three important respects. First, it often treats technological innovation and legal governance as parallel rather than deeply interdependent domains. Second, much of the existing work is descriptive in nature, with limited critical synthesis across jurisdictions and regulatory models. Third, there is insufficient attention to the normative implications of algorithmic governance, particularly in relation to accountability, equity, and institutional oversight.

Against this background, this chapter offers an integrated legal–technological analysis that systematically examines how blockchain-based energy trading systems interact with regulatory architectures across multiple jurisdictions. Its original contribution lies in bridging doctrinal legal analysis with socio-technical governance perspectives, while also critically evaluating the limits of automation and "rule by code" in complex energy systems. In doing so, the chapter advances a more nuanced understanding of decentralised energy governance beyond prevailing techno-optimistic narratives.

1.2. Methodological Approach

This chapter adopts a qualitative, interdisciplinary methodology combining doctrinal legal analysis with comparative regulatory assessment and socio-technical interpretation. The study examines selected jurisdictions based on their relevance to blockchain-enabled energy innovation, including the presence of regulatory sandboxes, pilot projects, or advanced digital energy policies. The analytical approach is primarily interpretive, focusing on legal frameworks, policy instruments, and governance mechanisms, while also incorporating insights from technological system design. This integrative methodology enables a comprehensive assessment of how decentralised energy markets operate within, and challenge, existing regulatory architectures.

2. FUNDAMENTALS OF BLOCKCHAIN TECHNOLOGY IN THE ENERGY CONTEXT

Blockchain technology represents a transformative paradigm in digital infrastructures, providing a decentralised, transparent, and tamper-evident mechanism for recording and verifying transactions. Conceptually derived from distributed ledger technologies (DLTs), blockchain enables the synchronisation of transactional data across multiple nodes in a network, eliminating the need for a single trusted intermediary (Narayanan et al., 2016). In the energy sector, this architecture offers a means of coordinating complex flows of electricity, information, and value between heterogeneous actors—generators, consumers, prosumers, aggregators, and regulators—without reliance on central clearing entities (Andoni et al., 2019).

By integrating blockchain with smart metering, Internet-of-Things (IoT) sensors, and digital identity systems, it becomes possible to construct verifiable records of energy production and consumption. These records underpin decentralised market mechanisms where energy can be traded peer-to-peer, settlements are executed automatically, and compliance with regulatory parameters is embedded within programmable logic. Such capabilities are critical for achieving transparency, accountability, and operational integrity within next-generation smart grids (Mengelkamp et al., 2018; Shamaseen et al., 2025).

2.1. Operational Mechanisms of Blockchain for Energy Transactions

A blockchain consists of a continuously growing sequence of data blocks linked through cryptographic hashes. Each block contains a timestamp, transaction data, and the hash of the previous block, forming an immutable chronological chain. When applied to energy trading, each transaction—representing, for instance, a kilowatt-hour of renewable generation or a tokenised energy credit—is verified through a consensus algorithm before inclusion in the ledger (Androulaki et al., 2018):

- **Transaction initiation: A** participant (prosumers, aggregators, or utilities) submits an energy trade request—typically measured in kilowatt-hours (kWh)—via a blockchain-enabled platform.
- **Validation:** Network nodes validate the authenticity of the transaction by verifying metering data, generation source, and digital signatures.
- **Consensus mechanism:** Agreement among nodes ensures that the transaction reflects a legitimate and non-duplicated energy exchange. Prominent consensus models include:
    – **Proof of Work (PoW)**—computationally intensive but secure; now uncommon in sustainable energy contexts due to its energy cost.
    – **Proof of Stake (PoS)**—validators are selected based on staked value; more energy-efficient and suitable for permissioned energy networks (Buterin, 2014).
    – **Proof of Authority (PoA)**—validation by approved authorities (e.g., distribution system operators or regulators), frequently used in energy-sector pilots such as the Energy Web Chain (EWF, 2019).
- **Block creation and distribution:** Validated transactions are bundled into a block, hashed, and appended to the ledger. All nodes update their copies of the ledger simultaneously, ensuring redundancy and resilience against data manipulation.

This process enables real-time settlement and verification of energy trades, substantially reducing transaction costs and counterparty risk compared to conventional market mechanisms (Aravind et al., 2024). Figure 1 below illustrates the adapted operational flow of blockchain within a decentralised energy market.

Figure 1. The functioning of blockchain

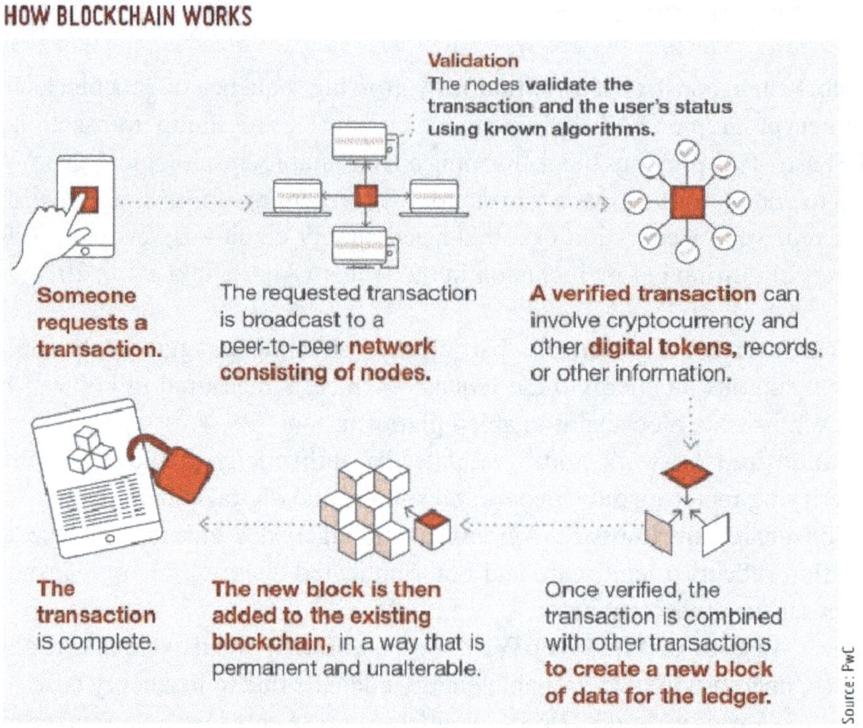

(How Blockchain Works, n.d.).

2.2. Defining Attributes and Their Relevance to Energy Systems

The defining technical attributes of blockchain—decentralisation, traceability, immutability, and transparency—carry distinct implications for the governance of energy markets (Table 1).

Table 1. Core attributes of blockchain and their functional value in decentralised energy governance

Attribute	Technical Definition	Application in Energy Markets	Regulatory/Operational Benefit
Decentralisation	Validation distributed across multiple nodes rather than a central server.	Enables peer-to-peer energy exchange without dependence on a single utility.	Reduces single points of failure; enhances market resilience.
Traceability	Every transaction recorded in an immutable chronological chain.	Allows end-to-end tracking of renewable generation, ownership, and consumption.	Supports verification of renewable certificates and compliance with carbon targets.
Immutability	Once validated, records cannot be altered without network consensus.	Protects integrity of smart-meter data and energy certificates.	Enhances trust in emission reporting and grid auditing.
Transparency	Ledger data viewable (in whole or part) by authorised stakeholders.	Provides regulators and consumers visibility over pricing, flows, and emissions.	Strengthens accountability and consumer confidence.

Transparency and immutability, when combined with selective privacy controls, are particularly critical in energy systems where data contain sensitive personal or commercial information. Permissioned or hybrid blockchains allow regulators to monitor transactions while maintaining confidentiality for end-users, aligning technological capabilities with privacy frameworks such as the EU General Data Protection Regulation (Finck, 2019).

Decentralisation also has profound implications for grid resilience. By distributing validation and data storage, blockchain minimises the vulnerability of market operations to cyberattacks or systemic outages. Each node retains a copy of the ledger, ensuring business continuity even in the event of localised network failures (Ahl et al., 2020). This decentralised redundancy complements ongoing efforts to create self-healing energy grids capable of autonomous recovery and adaptive load management.

Finally, traceability provides the foundation for verifiable green energy certification. Projects such as WePower and Energy Web Origin leverage blockchain to issue digital renewable energy certificates (RECs) that can be independently verified from generation to retirement, thereby preventing double counting and improving carbon-accounting accuracy (EWF, 2019; Invest in Estonia, 2018). These capabilities reinforce compliance with international sustainability reporting standards, including the Renewable Energy Directive (EU, 2018) and the Paris Agreement's transparency framework.

2.3. From Technical Infrastructure to Regulatory Enabler

Beyond its technical sophistication, blockchain functions as a regulatory enabler, embedding compliance mechanisms directly within digital code. Smart contracts can encode tariff structures, grid-balancing obligations, or subsidy eligibility criteria, executing them automatically when pre-specified conditions are met. In this sense, blockchain acts as a "rule of code" instrument that complements conventional legal enforcement (Bassan & Rabitti, 2024).

Nevertheless, the codification of regulatory logic introduces challenges. Translating the complexity of energy law—often marked by exceptions, human discretion, and interpretative flexibility—into deterministic smart-contract code requires rigorous interdisciplinary collaboration between engineers, lawyers, and policymakers. The following section (Section 3) will therefore examine how blockchain's technical affordances may foster trust and transparency while ensuring that automation remains compatible with existing legal and ethical norms.

3. BLOCKCHAIN AS AN INFRASTRUCTURE FOR ENERGY TRUST AND TRANSPARENCY

Trust has always constituted the cornerstone of modern energy systems. From regulated utilities to regional transmission operators, consumers have historically relied on institutional intermediaries to guarantee that electricity is generated, transmitted, and billed in a fair and reliable manner. However, the emergence of distributed energy resources and peer-to-peer trading models challenges this institutional monopoly of trust. As energy systems become increasingly decentralised, blockchain technology offers a novel infrastructure of algorithmic trust, substituting centralised verification with cryptographically secured and collectively validated consensus mechanisms (Andoni et al., 2019).

3.1. Redefining Trust in Decentralised Energy Systems

In conventional electricity markets, trust is produced and maintained through a combination of legal regulation, market oversight, and contractual enforcement. Consumers rely on regulators and utilities to ensure grid stability, billing accuracy, and non-discriminatory access. In contrast, decentralised energy systems require that participants—often acting simultaneously as producers and consumers—verify each other's actions without recourse to a central authority. This transition from institutional trust to technological trust necessitates mechanisms that can guarantee data integrity and market fairness autonomously.

Blockchain achieves this through the decentralised validation of transactions. Each energy exchange recorded on the ledger is timestamped, cryptographically signed, and approved by consensus among network nodes. Because no single actor can alter or delete entries unilaterally, the resulting record is both immutable and auditable. This immutability provides the foundation for algorithmic accountability—an emerging form of governance in which compliance and transparency are assured by code rather than discretionary oversight (Kouhizadeh & Sarkis, 2018).

3.2. Transparency and Data Verifiability in Energy Transactions

Transparency represents one of the most salient benefits of blockchain deployment in energy governance. By enabling traceability of energy origin, ownership, and consumption, distributed ledgers enhance both consumer confidence and regulatory observability (Aravind et al., 2024). For example, renewable-energy certificates (RECs) or guarantees of origin can be registered on a blockchain to prevent double counting, thereby satisfying international reporting standards under the Paris Agreement (European Commission, 2020).

The level of transparency, however, must be carefully calibrated. Full public disclosure of transaction data may conflict with data-protection obligations and the legitimate need for commercial confidentiality. Consequently, permissioned or hybrid blockchains—in which read/write access is granted selectively to authorised stakeholders such as regulators, distribution operators, or verified market participants—are often preferred in the energy sector. Table 2 summarises how blockchain's trust-building features translate into specific governance outcomes for decentralised energy markets.

Table 2. Blockchain trust mechanisms and their governance outcomes in decentralised energy systems

Trust Mechanism	Operational Feature	Energy-Sector Application	Governance Outcome
Immutable Ledger	Cryptographic linkage of blocks; tamper-evident history	Recording of generation and consumption data from smart meters	Prevents manipulation of billing and reporting records
Consensus Validation	Collective verification by distributed nodes	Authentication of energy trades and certificate issuance	Eliminates single-point failures; strengthens procedural legitimacy
Transparency with Controlled Access	Permissioned visibility of transaction data	Regulator access to live market information	Enhances oversight while protecting privacy
Programmable Smart Contracts	Automated execution of predefined rules	Dynamic pricing, settlement, and subsidy disbursement	Ensures rule compliance and reduces administrative cost

Such technological transparency has regulatory implications. It enables near real-time supervision of energy transactions, potentially replacing ex-post audits with continuous compliance monitoring. For instance, tariff deviations, network congestion, or unauthorised energy injections can be detected immediately, allowing regulators to intervene dynamically rather than retrospectively (Ahl et al., 2020).

3.3. Algorithmic Accountability and the Legitimacy of Digital Trust

While algorithmic trust enhances efficiency, it also redistributes responsibility. If compliance is automated through smart contracts, the locus of accountability shifts from human decision-makers to code developers and network validators. This transformation raises normative questions regarding the legitimacy of algorithmic governance in essential public infrastructures such as electricity (Niet et al., 2021).

Legal scholars have argued that technological systems must remain subject to democratic oversight and due-process guarantees, even when automated (Finck, 2019). Hence, blockchain should not be viewed as replacing institutional trust, but rather as complementing it through enhanced transparency and traceability. A hybrid governance model—combining regulatory supervision with verifiable technological enforcement—appears most compatible with the constitutional principles that underpin energy regulation across democratic jurisdictions.

The trust architecture of decentralised energy markets therefore rests upon three interrelated pillars:

- **Technical integrity**, guaranteed by cryptographic validation and decentralised consensus.

- **Regulatory legitimacy**, ensured through alignment with existing legal frameworks and procedural rights.
- **Social acceptability**, maintained through inclusiveness, accessibility, and clear allocation of responsibility.

In combination, these pillars position blockchain not merely as a disruptive technology, but as a normative infrastructure for trustworthy energy governance, capable of reconciling innovation with the enduring public values of transparency, accountability, and justice.

4. USE CASES IN DECENTRALISED ENERGY MARKETS

The practical implementation of blockchain in energy markets demonstrates how digital infrastructures can decentralise value exchange, enhance transparency, and reshape governance models. Numerous pilot projects and commercial initiatives across the globe illustrate blockchain's capacity to transform traditional electricity markets into peer-to-peer (P2P) ecosystems. These use cases provide empirical evidence for assessing the technical feasibility, regulatory compatibility, and institutional adaptability of decentralised trading mechanisms (Andoni et al., 2019). As illustrated in Figure 2, the blockchain-enabled decentralised energy market operates through a multi-layered architecture that connects producers, consumers, and prosumers via distributed ledgers and smart contracts. This architecture ensures system robustness (Nakamoto, 2008) (see Figure 2).

Figure 2. Decentralisation in blockchain networks

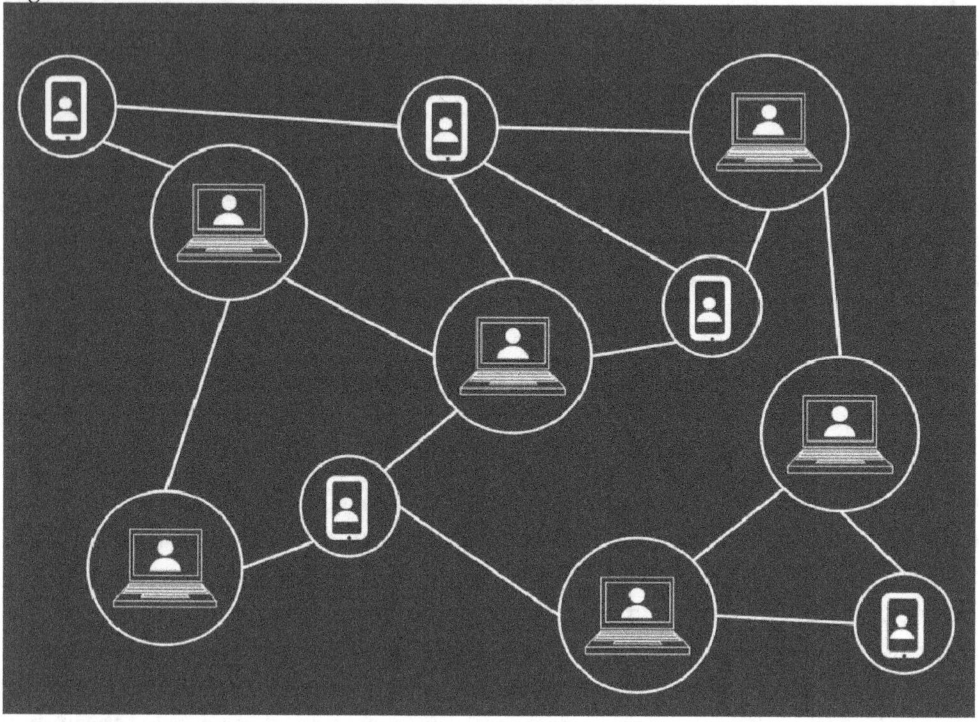

(TYN Magazine, 2023)

4.1. Peer-to-Peer Trading Platforms

Perhaps the most emblematic example is the Brooklyn Microgrid in New York City, initiated by LO3 Energy in 2016. The project allows households equipped with rooftop photovoltaic panels to sell excess generation directly to neighbouring consumers via a private blockchain network. Energy transactions are recorded and verified in real time, while pricing is established through local market auctions (Mengelkamp et al., 2018). Legally, the initiative operates under the framework of New York State's Reforming the Energy Vision (REV) programme, which authorises limited community-based energy exchanges. Despite its success, the project exposes the regulatory tension between retail competition laws and distributed market autonomy: the platform's peer-to-peer trades must still be cleared through licensed energy service companies (Aravind et al., 2024).

In Australia, Power Ledger Pty Ltd developed one of the first commercial blockchain-enabled trading systems. Built upon a dual-token architecture, it facilitates local energy trading, carbon-credit exchange, and renewable certificate issuance

(Powerledger, 2023). Power Ledger collaborates with distribution network operators and regulators through regulatory sandboxes sanctioned by the Australian Energy Regulator (AER), ensuring that experimental digital markets remain compliant with statutory metering and settlement standards (Australian Energy Regulator, 2023). This institutional flexibility exemplifies how adaptive legal frameworks can nurture innovation while preserving consumer protection and grid reliability.

4.2. Renewable-Certificate Traceability and Carbon-Credit Markets

The European-based WePower platform tokenises power-purchase agreements (PPAs) and renewable certificates through Ethereum smart contracts. Each token represents a defined volume of electricity from a specific renewable asset and is linked to verifiable generation data (Invest in Estonia, 2018). By encoding attributes such as time of generation, carbon intensity, and source verification, WePower creates digital guarantees of origin (GoOs) compliant with the EU Renewable Energy Directive (European Parliament & European Council, 2018).

A comparable initiative, the Energy Web Origin framework, operated by the Energy Web Foundation (EWF), provides open-source modules that enable energy suppliers and grid operators to issue, transfer, and retire renewable certificates on a permissioned blockchain (EWF, 2019). These mechanisms directly support compliance with corporate decarbonisation commitments and voluntary carbon markets. Legally, such systems raise complex questions regarding recognition of tokenised certificates as legal property, jurisdictional oversight of cross-border trading, and auditability under financial-market regulation (Church, 2022).

4.3. Smart Metering, Dynamic Billing, and Tariff Automation

Blockchain has also been integrated into smart metering infrastructures to enhance billing transparency and automate tariff adjustment. In a pilot project conducted by EDF Energy UK in collaboration with Electron Ltd, blockchain was used to reconcile metering data across multiple suppliers, thus enabling dynamic tariff calculation and consumer switching in near real time (Baillie, 2017).

Similarly, TenneT, the Dutch-German transmission system operator, has partnered with the EWF to employ blockchain for frequency-response balancing by aggregating distributed battery storage resources. Consumers who allow their batteries to participate in grid-stabilisation programmes are rewarded automatically through smart-contract settlements (TenneT, n.d.). These initiatives demonstrate how blockchain not only supports market transactions but also automates compliance with technical-grid codes and service obligations.

4.4. Community Energy Cooperatives and Local Microgrids

In Spain, the Pylon Network project seeks to empower community energy cooperatives by providing an open-source blockchain protocol for managing generation data, billing, and renewable certificates (Klenergy Tech, 2018). The system enables cooperatives to operate as micro-utilities under Spanish cooperative law while complying with EU data-protection and energy-licensing requirements. It exemplifies the potential for energy democratisation through decentralised digital infrastructures that respect national regulatory boundaries.

Other cooperative-oriented projects include SonnenCommunity in Germany and Piclo Flex in the UK, both of which apply blockchain-inspired registries to coordinate flexibility services among distributed participants (López et al., 2024). These initiatives foreground the socio-legal challenge of balancing local autonomy with system-level coordination—an issue increasingly central to the design of decentralised energy governance frameworks.

Table 3 shows a comparative overview of selected blockchain-based energy projects.

Table 3. Comparative analysis of selected blockchain-based energy-market initiatives and their governance implications

Project / Location	Core Function	Blockchain Type	Regulatory Interface	Key Legal or Governance Insight
Brooklyn Microgrid (USA)	P2P energy trading among households	Private Ethereum-based	Operates under New York REV pilot exemptions	Reveals limits of retail competition law and licensing constraints
Powerledger (Australia)	Energy & carbon trading with tokenised certificates	Permissioned Proof-of-Authority (Energy Web Chain)	Regulatory sandbox authorised by AER	Demonstrates adaptive regulation fostering innovation
WePower (EU)	Tokenised PPAs and renewable GoOs	Public Ethereum	Must comply with EU Renewable Directive	Illustrates need for legal recognition of digital certificates
Energy Web Origin (Global)	Open-source registry for renewable tracking	Permissioned (EWF)	Aligns with corporate & national decarbonisation frameworks	Highlights interoperability and data-audit obligations
Pylon Network (Spain)	Cooperative energy-data management	Open-source DLT hybrid	Compatible with EU data-protection law	Shows potential for community energy governance

Not all pilot initiatives have achieved their intended outcomes. For instance, certain blockchain-based energy trading projects have encountered regulatory resistance due to unclear licensing requirements and incompatibility with existing grid management structures. In some cases, initiatives were discontinued or scaled back following challenges related to user participation, interoperability, or economic viability. These experiences illustrate that technological feasibility does not necessarily translate into regulatory or market success, reinforcing the importance of adaptive governance and institutional alignment.

4.5. Synthesis: Lessons for Regulatory and Governance Design

Collectively, these projects illustrate that blockchain-enabled energy markets are technically viable yet legally contingent upon adaptive regulatory ecosystems. Jurisdictions employing experimental sandboxes—such as the United Kingdom, Australia, and Singapore—tend to achieve faster innovation diffusion because they allow limited-scope pilots within a controlled legal perimeter (Beckstedde et al., 2023). Conversely, regions lacking explicit frameworks for decentralised trading face legal uncertainty concerning data ownership, settlement finality, and consumer-protection duties.

From a governance standpoint, the alignment of digital protocols with statutory obligations remains essential. Smart-contract logic must embed tariff-compliance parameters, renewable-portfolio standards, and cybersecurity requirements. Moreover, given the internationalisation of energy data flows, interoperability between national blockchains becomes a precondition for transboundary electricity trade and carbon accounting (Aravind et al., 2024).

Thus, while blockchain offers an unprecedented infrastructure of trust, its institutional legitimacy depends on the parallel evolution of legal architecture, ensuring that decentralised trading reinforces, rather than undermines, the public values of equity, accountability, and reliability.

In summary, this section highlights that while blockchain technologies offer significant opportunities to enhance transparency and decentralisation in energy markets, their implementation remains contingent upon evolving regulatory frameworks and institutional capacities. The findings underscore the need for integrated governance approaches that reconcile technological innovation with legal certainty and public accountability.

5. SMART CONTRACTS AND AUTOMATED COMPLIANCE IN ENERGY REGULATION

Smart contracts—self-executing code deployed on distributed ledgers—are rapidly emerging as a legal-technological mechanism for automating regulatory compliance within decentralised energy markets. They enable the automatic execution of contractual clauses once predefined conditions are satisfied, thereby reducing transaction costs and mitigating counterparty risk (Szabo, 1997; Werbach & Cornell, 2017). When integrated into blockchain-based energy systems, smart contracts can implement tariff rules, renewable-certificate issuance, and settlement of power-purchase agreements (PPAs) without human intervention (Andoni et al., 2019).

From a legal perspective, however, the deployment of smart contracts in essential infrastructure such as electricity networks raises questions of enforceability, accountability, and due process. National legal systems vary widely in their recognition of code-based agreements, and the translation of statutory energy obligations into deterministic code presents interpretive and governance challenges (Niet et al., 2021).

5.1. Legal Enforceability and Contractual Validity

Traditional contract law relies upon the principles of offer, acceptance, consideration, and intention to create legal relations. A smart contract satisfies these elements when its digital execution embodies mutual consent and value exchange (Raskin, 2017). Jurisdictions such as the United Kingdom, Singapore, and certain U.S. states (e.g., Arizona, Tennessee) have explicitly recognised electronic records and signatures as legally binding under electronic-transactions legislation (UK Law Commission, 2021).

Nevertheless, questions persist regarding interpretability and remedies. Code operates deterministically: once deployed, it performs its logic irrespective of unforeseen contingencies or equitable considerations. For this reason, leading scholars advocate a dual-layered model—a natural-language contract expressing the legal obligations, and an accompanying code layer implementing operational terms (Werbach & Cornell, 2017; Finck, 2019). Such "Ricardian contracts" ensure that the legal narrative remains intelligible to courts and regulators while benefiting from automation.

In energy trading, smart-contract enforceability is often tested through pilot programmes within regulatory sandboxes (Beckstedde et al., 2023). These controlled environments allow innovators to assess whether automated settlement mechanisms comply with consumer-protection, billing-accuracy, and dispute-resolution standards.

5.2. Operational Applications in Energy Regulation

Smart contracts have been deployed across multiple regulatory dimensions of the energy sector:

- **Dynamic Tariff Adjustment:** Platforms such as *Electron* in the UK and *Powerledger* in Australia employ smart contracts to calculate tariffs based on real-time grid conditions. Tariff logic references external data ("**oracles**") such as wholesale prices, carbon intensity, or grid-congestion indicators (Baillie, 2017).
- **Power-Purchase Agreements (PPAs):** Blockchain-enabled PPAs automate payment flows between renewable generators and off-takers once generation data from certified meters is recorded. This reduces administrative overhead and enhances auditability for regulators overseeing renewable-portfolio standards (Invest in Estonia, 2018).
- **Grid-Service Compliance:** Transmission operators such as *TenneT* use smart contracts to verify participation of distributed storage assets in frequency-response markets. When a resource delivers balancing energy within predefined tolerances, compensation is triggered automatically (TenneT, n.d.).
- **Subsidy and Incentive Disbursement:** Smart contracts can encode eligibility criteria for feed-in tariffs or renewable-energy credits. Once verification oracles confirm compliance, payments are executed without manual processing, reducing the risk of fraud (Aravind et al., 2024).

These implementations illustrate the gradual evolution from ex-post regulatory enforcement towards embedded ex-ante compliance, where conformity to legal norms is ensured through code before a transaction occurs. Table 4 illustrate alignment between smart-contract functionality and energy-regulatory objectives.

Table 4. Smart-contract functions and corresponding regulatory interfaces

Regulatory Objective	Smart-Contract Function	Data/Oracle Input	Relevant Authority or Legal Basis	Expected Compliance Outcome
Dynamic pricing compliance	Algorithmic tariff adjustment	Wholesale price index, grid frequency	Energy regulator (e.g., OFGEM, AER)	Real-time tariff conformity
Renewable certification	Automatic REC/ GoO issuance	Smart-meter generation data	EU Directive 2018/2001; national RES laws	Verified certificate traceability
Grid-service remuneration	Performance-based settlement	Frequency-response telemetry	Transmission-system operator	Transparent, auditable balancing payments
Subsidy disbursement	Eligibility verification & payment	Policy parameters, consumption data	National energy-ministry schemes	Fraud reduction; procedural efficiency

5.3. Accountability, Liability, and the Role of Regulators

Automated execution complicates the attribution of responsibility when errors occur. If a smart contract executes incorrectly due to faulty code or malicious data input, identifying a liable party—developer, validator, or operator—becomes complex (Niet et al., 2021). Energy regulators therefore face the challenge of integrating algorithmic accountability into compliance frameworks. Potential solutions include:

- **Code auditability requirements**, ensuring that deployed smart contracts are subject to third-party verification before activation;
- **Mandated "human-in-the-loop" oversight** for high-risk operations affecting public grids;
- **Fallback clauses** enabling regulatory override or emergency suspension of code execution in cases of systemic risk (Finck, 2019).

Such hybrid models preserve automation's efficiency while upholding the rule of law and public-interest safeguards.

The emerging doctrine of "rule of code" (Bassan & Rabitti, 2024) envisions legislation that delegates certain procedural functions to programmable logic while retaining normative control in human institutions. In this model, regulators transition from ex-post arbiters to ex-ante code designers, collaborating with technologists to translate statutory requirements into verifiable computational conditions. The success of blockchain-based energy governance will thus depend on cultivating interdisciplinary capacity between law, computer science, and systems engineering.

Smart contracts demonstrate that compliance can be embedded directly within digital infrastructures, effectively turning legal rules into executable protocols. Yet, their legitimacy depends on maintaining transparency, reversibility, and recourse. Decentralised energy systems cannot rely solely on code; they require institutional guardianship to ensure justice and procedural fairness.

Consequently, the future of blockchain-enabled energy governance lies in institutional integration—a model in which smart contracts operate under explicit legal mandates, subject to regulatory auditing and public accountability. This balance between innovation and oversight ensures that automation enhances, rather than erodes, the foundational values of reliability, fairness, and consumer trust within the energy transition.

6. TOKENISATION OF ENERGY ASSETS AND MARKETS

The tokenisation of energy assets constitutes one of the most conceptually innovative applications of blockchain technology in the contemporary energy transition. Through tokenisation, a physical or contractual energy right—such as a kilowatt-hour (kWh), a renewable-energy certificate, or a carbon offset—is represented as a digital token recorded on a distributed ledger. This conversion enables fractional ownership, rapid transferability, and automated settlement, thereby introducing liquidity and transparency into traditionally illiquid energy markets (Kouhizadeh & Sarkis, 2018; Church, 2022).

Tokenisation thus bridges the physical and digital layers of the energy system. Each token acts as a cryptographically secured claim to an underlying asset or service, verifiable through smart-meter data and validated by consensus across network nodes. While the process enhances market efficiency, it also raises questions concerning legal recognition, financial regulation, and consumer protection, particularly when tokens acquire characteristics analogous to securities or derivatives (Finck, 2019; European Securities and Markets Authority [ESMA], 2025).

Figure 3 visualises the process of energy tokenisation, where renewable generation is represented through digital tokens that embody verified units of electricity or environmental attributes.

Figure 3. Tokenisation of renewable energy assets

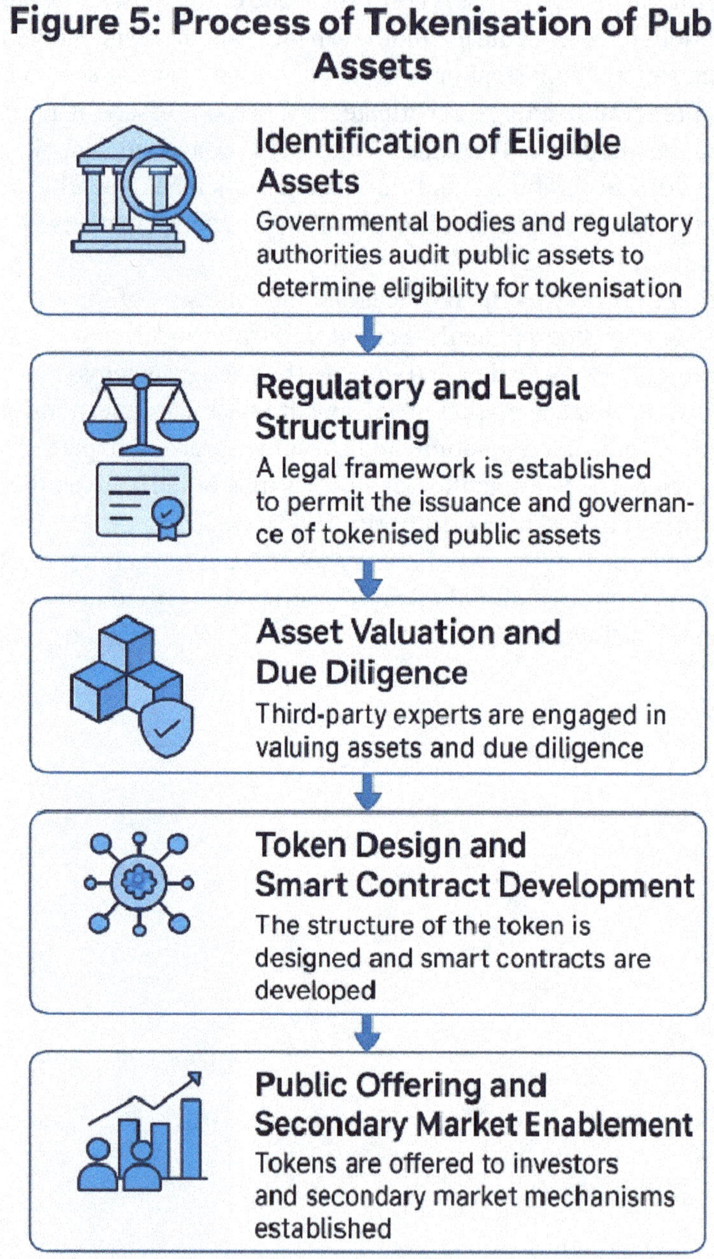

6.1. Conceptual and Technical Foundations

In blockchain-enabled energy systems, tokenisation typically follows one of two models:

- **Asset-backed tokens**, where each token corresponds to a measurable unit of energy or environmental value—such as 1 kWh of renewable electricity or 1 tonne of CO_2 avoided.
- **Utility tokens**, representing access rights or participation within a decentralised energy platform (e.g., governance voting, network fees, or storage usage).

The technological process of token creation involves the following sequential stages:

- **Measurement and verification** – Energy output is authenticated through smart-meter readings or supervisory control and data acquisition (SCADA) systems.
- **Data oracles** – Verified measurements are transmitted to the blockchain to trigger token minting.
- **Token issuance and smart-contract registration** – Each token carries metadata including timestamp, generation source, and geographical coordinates.
- **Trading and redemption** – Tokens circulate in secondary markets or are redeemed to claim the underlying energy value or environmental credit.

This architecture establishes a digital audit trail from generation to consumption, aligning technological traceability with the transparency obligations embedded in renewable-energy regulation (European Commission, 2020).

6.2. Legal Qualification and Regulatory Treatment

The legal status of tokenised energy assets remains unsettled and jurisdiction-dependent. In the European Union, the classification of tokens hinges on their economic function: if transferable and profit-oriented, they may fall within the definition of "financial instruments" under MiFID II or "crypto-assets" under the Markets in Crypto-Assets Regulation (MiCA) (Regulation (EU) 2023/1114).

Tokens representing renewable-energy certificates (RECs) or guarantees of origin (GoOs) are generally treated as environmental commodities, subject to energy-law oversight rather than financial regulation (European Parliament & European Council, 2018).

In Australia, the Australian Securities and Investments Commission (ASIC) has published guidance distinguishing between utility and investment tokens, while the Australian Energy Regulator recognises blockchain-issued energy certificates within regulatory sandboxes (Australian Energy Regulator, 2023).

In Singapore, the Monetary Authority of Singapore (MAS) permits tokenised PPAs under its FinTech Regulatory Sandbox Express, provided that settlement processes comply with existing energy-market rules (Monetary Authority of Singapore, 2024).

The challenge lies in reconciling token liquidity with the principles of energy-market integrity and consumer protection. If energy tokens are freely tradable on secondary exchanges, regulatory supervision must ensure that they do not constitute speculative financial products detached from their environmental purpose. Table 5 shows a comparative overview of selected regulatory frameworks applicable to energy-asset tokenisation.

Table 5. Comparative regulatory approaches to energy-asset tokenisation

Jurisdiction / Authority	Regulatory Instrument	Token Category	Supervisory Body	Key Legal Characteristic
European Union	MiCA (Reg. EU 2023/1114); Renewable Energy Directive 2018/2001	Crypto-asset / environmental commodity	ESMA; DG ENER	Dual regime distinguishing financial from environmental tokens
United Kingdom	Financial Services and Markets Act 2023; OFGEM Energy Data Strategy	Utility token	Financial Conduct Authority (FCA); OFGEM	Case-by-case classification; emphasis on consumer data rights
Australia	ASIC Info Sheet 225; AER Sandbox Framework 2023	Utility / asset-backed token	ASIC; AER	Sandbox-based authorisation enabling live pilots
Singapore	MAS FinTech Regulatory Sandbox Express 2023	Security / utility token	Monetary Authority of Singapore (MAS)	Conditional approval for tokenised PPAs
United States (New York)	Energy Vision REV programme; State Public Service Law	Energy-credit token	New York Public Service Commission	Treated as retail-energy instrument subject to state licensing

6.3. Economic and Governance Implications

Tokenisation introduces fractionalisation of energy ownership, allowing citizens or institutional investors to hold digital shares of renewable projects. Platforms such as WePower and Sun Exchange (South Africa) enable micro-investments in solar generation, democratising access to clean-energy financing (Invest in Estonia, 2018; Sun Exchange, 2023).

This financial inclusion must, however, be balanced against prudential oversight to avoid speculative volatility reminiscent of unregulated crypto-asset markets.

From a governance standpoint, tokenisation facilitates programmable compliance: smart contracts can embed emission caps, regional content requirements, or subsidy conditions. When transactions breach predefined sustainability thresholds, automated alerts can trigger regulatory intervention. Such mechanisms could enhance Environmental, Social and Governance (ESG) accountability, supporting international disclosure frameworks such as the Task Force on Climate-related Financial Disclosures (TCFD) (Financial Stability Board, 2021).

Moreover, tokenisation contributes to market interoperability. By representing energy attributes in standardised digital units, cross-border trading between heterogeneous systems becomes technically feasible. Initiatives like the Energy Web Chain propose universal token standards (EW Origin Token) that allow mutual recognition of renewable certificates across jurisdictions (EWF, 2019).

6.4. Risks, Consumer Protection, and Data Governance

Despite its promise, tokenisation introduces new categories of systemic risk. Cybersecurity breaches, private-key loss, and oracle manipulation could compromise market integrity. Furthermore, data-protection obligations under the EU General Data Protection Regulation (GDPR) require that personally identifiable energy-consumption data be processed lawfully and, where possible, anonymised before token issuance (Finck, 2019).

Regulators must also address market-conduct risks, including potential price manipulation and information asymmetry. The integration of blockchain analytics tools for transaction monitoring—analogous to anti-money-laundering (AML) regimes—could mitigate these concerns without undermining decentralisation (ESMA, 2025).

A balanced regulatory approach should thus combine technological safeguards (secure key management, oracle redundancy) with institutional safeguards (licensing, dispute resolution, consumer education). Only through such hybrid governance can tokenisation evolve from experimental novelty to legitimate infrastructure for the global energy transition.

6.5. Towards Legally-Embedded Digital Commodities

Tokenisation transforms energy into a legally cognisable digital commodity, enabling programmable trading and transparent certification. Yet, its durability depends on legal clarity. Without harmonised definitions of digital assets, tokenised energy instruments risk existing in a grey zone between property law, financial regulation, and energy governance.

To ensure legitimacy, regulators must recognise the dual nature of energy tokens: as technological artefacts that facilitate efficient trading and as normative instruments that embody sustainability commitments. Establishing this dual recognition will allow tokenisation to serve the twin objectives of decarbonisation and market democratisation within a stable legal order.

7. INTEGRATION OF BLOCKCHAIN WITH IOT, AI, AND BIG DATA IN ENERGY SYSTEMS

The convergence of blockchain, the Internet of Things (IoT), artificial intelligence (AI), and big data analytics is reshaping the operational and regulatory architecture of energy systems. Each of these technologies addresses a specific dimension of the energy transition: IoT enables pervasive measurement, AI delivers predictive optimisation, and big data provides systemic visibility. Blockchain, in turn, ensures data integrity, interoperability, and automated governance across these domains (Kouhizadeh & Sarkis, 2018; Shamaseen et al., 2025).

By combining these technologies, decentralised energy networks can achieve autonomous coordination and verifiable compliance, enhancing both system resilience and regulatory observability. This "cyber-physical fusion" is foundational to the digitalisation of energy governance, in which data streams generated by distributed assets become the operational backbone of decision-making and regulatory enforcement (IEA, 2024).

7.1. Internet of Things (IoT) and Real-Time Energy Data Integrity

The Internet of Things forms the sensory layer of decentralised energy systems. Smart meters, sensors, and control devices continuously record and communicate parameters such as voltage, frequency, and generation output. Integrating these data flows with blockchain creates a tamper-resistant audit trail, ensuring that energy transactions are verifiable from the moment of measurement to the point of settlement (Dorri et al., 2017).

For example, EDF Energy UK and Siemens Microgrid have piloted IoT–blockchain architectures that synchronise metering data with distributed ledgers to reduce data discrepancies and billing disputes (Siemens, 2020; EDF, n.d.). In such systems, IoT devices act as trusted oracles: they initiate smart-contract execution upon detecting predefined conditions—such as excess generation or grid imbalance. The resulting transactions are then validated by blockchain consensus, ensuring the authenticity of metered data.

This integration strengthens regulatory data governance, since it provides regulators and auditors with immutable datasets for compliance verification. Moreover, the elimination of manual data reconciliation reduces administrative overhead and the potential for fraud in energy trading schemes (Agarwal & Jain, 2023).

7.2. Artificial Intelligence for Predictive Load Management and Algorithmic Optimisation

Artificial intelligence augments blockchain systems by transforming historical and real-time data into actionable insights. Machine-learning models can forecast demand fluctuations, optimise dispatch from distributed generation units, and identify anomalies that indicate potential non-compliance or cyber threats (Biswas et al., 2024).

When coupled with blockchain, AI-generated decisions can be executed automatically via smart contracts. For instance, if predictive models detect imminent grid congestion, a smart contract may autonomously adjust tariff signals or trigger distributed storage dispatch to maintain equilibrium. This integration introduces a new paradigm of "autonomous energy governance", where regulatory objectives—such as reliability, sustainability, and equity—are enforced algorithmically (Niet et al., 2021).

However, embedding AI within regulatory frameworks raises profound legal and ethical considerations. Algorithmic bias, lack of explainability, and data-provenance uncertainty can undermine accountability. European energy regulators, following the EU Artificial Intelligence Act (COM/2021/206 final), now consider high-stakes energy applications as "high-risk systems," requiring human oversight and auditability (European Commission, 2020). Therefore, regulatory design must ensure that AI decisions remain transparent, contestable, and aligned with statutory mandates.

7.3. Big Data and Systemic Transparency

Big data analytics complements blockchain and AI by providing the macro-level visibility necessary for strategic governance. The exponential growth of smart-meter data enables granular analysis of consumption patterns, emissions profiles, and renewable-integration dynamics. Blockchain contributes by guaranteeing that these data remain authentic, timestamped, and tamper-evident, ensuring evidentiary reliability for audits and policymaking (Ahl et al., 2020).

Energy regulators such as OFGEM (UK) and the Australian Energy Market Operator (AEMO) increasingly employ data-analytics frameworks powered by blockchain to conduct near real-time compliance monitoring. These tools assist in dynamic tariff regulation, network-balancing verification, and fraud detection (Office of Gas and Electricity Markets, 2023; Australian Energy Market Operator, 2025).

Through open-data policies, anonymised blockchain datasets also contribute to energy research and innovation ecosystems, supporting the EU Open Data Directive (EU, 2019/1024) and comparable initiatives in the Asia–Pacific region. Table 6 illustrate the convergent technological functions enabling trustworthy, data-driven energy regulation.

Table 6. Synergistic functions of Blockchain, IoT, AI, and big data in decentralised energy governance

Technology Layer	Primary Function	Integration with Blockchain	Governance / Regulatory Implication
IoT	Real-time metering and device communication	Acts as data oracle for blockchain transactions	Ensures data authenticity and auditability
AI	Predictive optimisation and anomaly detection	Executes automated decisions via smart contracts	Enables proactive regulation and demand management
Big Data	Aggregation and macro-level analysis	Provides validated datasets for analytics	Facilitates evidence-based policy and transparency
Blockchain	Secure ledger and programmable logic	Integrates all layers within a verifiable ecosystem	Embeds compliance and trust into energy governance

Figure 4. Technological integration of blockchain, Internet of Things, and artificial intelligence for real-time energy data governance

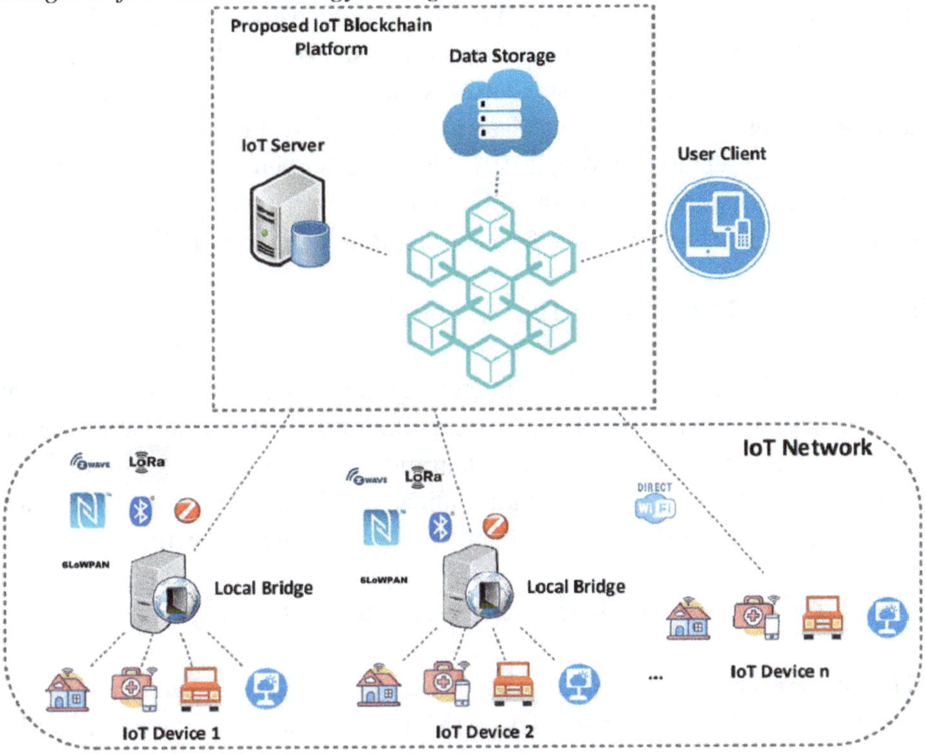

(Hang & Kim, 2019)

As shown in Figure 4, the integration of blockchain with IoT sensors and artificial-intelligence algorithms enables dynamic optimisation of distributed energy systems.

7.4. Case Studies: Cyber-Physical Convergence in Practice

Siemens Microgrid and Edge Blockchain Integration

The Siemens Microgrid Platform employs blockchain and AI at the grid edge to coordinate distributed resources. Sensors collect generation and load data, which AI algorithms use to optimise dispatch, while blockchain records and validates every control instruction (Siemens, 2018). This configuration enables self-balancing microgrids that comply autonomously with frequency and voltage constraints defined by national grid codes. Legally, Siemens collaborates with regulators to ensure that such systems remain auditable under standard technical-compliance procedures.

Energy Web Chain (EWC)

The Energy Web Chain, developed by the Energy Web Foundation, integrates IoT and big-data analytics within a permissioned blockchain designed specifically for energy-sector applications. It standardises identity management (via the Decentralised Operating System, D3A) and data interoperability, allowing certified devices to transact energy securely (EWF, 2019). The EWC's open-source framework has been adopted in pilot projects by Shell, Elia Group, and TenneT, demonstrating the feasibility of large-scale, cross-border coordination of decentralised energy assets.

EDF Energy's Smart Meter Blockchain Pilot

In the EDF–Electron pilot in the United Kingdom, IoT meters feed verified consumption data directly into blockchain smart contracts, enabling automated billing and tariff settlement (EDF, n.d.). The integration of AI algorithms further allows EDF to identify abnormal consumption indicative of energy poverty or technical losses, linking data analytics to social and regulatory policy objectives.

7.5. Governance and Ethical Considerations

The integration of these technologies introduces multi-layered governance complexity. Each data flow—from IoT device to blockchain ledger—traverses multiple regulatory regimes: data protection, cybersecurity, consumer rights, and energy-market law. Effective governance requires interoperable regulatory frameworks that harmonise these domains under a unified principle of data accountability (Finck, 2019).

Furthermore, the ethical dimension cannot be overlooked. The deployment of AI and big data in energy governance raises issues of privacy, algorithmic discrimination, and digital exclusion. Regulators must ensure that technological innovation supports, rather than compromises, energy justice—the equitable distribution of benefits and burdens of the energy transition (Sovacool et al., 2017).

7.6. Towards Autonomous but Accountable Energy Systems

The synergistic integration of blockchain with IoT, AI, and big data forms the technological core of the emerging autonomous energy system. In this paradigm, compliance, optimisation, and transparency are embedded within a single cyber-physical infrastructure. However, autonomy must not equate to opacity. Human oversight, legal accountability, and ethical alignment remain indispensable. The next generation of energy governance will therefore depend not only on computational efficiency but also on the normative capacity to align algorithmic processes with democratic oversight and public interest.

8. LEGAL AND REGULATORY ARCHITECTURE FOR DECENTRALISED ENERGY TRADING

The transition towards decentralised energy trading requires not only technological innovation but also the evolution of a coherent legal and regulatory architecture. While blockchain enables peer-to-peer (P2P) electricity exchange through distributed consensus, the absence of harmonised legal norms risks undermining market integrity, consumer protection, and system reliability (Aravind et al., 2024).

Existing electricity-market legislation—originally designed for vertically integrated utilities—must therefore adapt to accommodate digital, autonomous, and transnational trading platforms (Ahmed & Măgurean, 2024).

A well-designed regulatory framework must balance innovation with legal certainty, fostering experimentation without eroding the fundamental principles of fairness, transparency, and accountability. The following subsections outline the key components of such an architecture.

8.1. Legal Recognition of Prosumers and Energy Communities

The legal identity of prosumers—actors who both generate and consume electricity—is foundational to decentralised trading. Under the EU Clean Energy Package, Directive (EU) 2019/944 and Directive (EU) 2018/2001 formally recognise citizen energy communities and renewable energy communities as legitimate market participants entitled to sell, store, and share energy (European Parliament & European Council, 2019; 2018).

This recognition marks a paradigm shift from passive consumption to participatory governance. Prosumers are now permitted to engage in peer-to-peer transactions and collective self-consumption, subject to national regulatory transposition. For example, Spain's Royal Decree 244/2019 authorises shared self-consumption within buildings and local grids, while Germany's Energiewirtschaftsgesetz (EnWG) accommodates citizen cooperatives under specific licensing thresholds (López et al., 2024).

However, the definition of "market participant" varies across jurisdictions, creating uncertainty for blockchain platforms operating transnationally. Harmonisation efforts should clarify liability boundaries—particularly concerning metering accuracy, data ownership, and dispute resolution—to ensure that prosumers are treated consistently within energy-law frameworks.

8.2. Smart Contracts and Regulatory Certainty

Smart contracts introduce algorithmic enforcement into energy regulation, automating settlement and compliance processes. Yet, legal enforceability depends on jurisdictional recognition of electronic transactions and signatures (Finck, 2019). The UK Law Commission (2021) confirmed that smart contracts meeting contractual principles (offer, acceptance, consideration, intention) are enforceable under English law. Similarly, Singapore and the U.S. states of Arizona and Tennessee explicitly acknowledge smart-contract validity within their electronic-transactions legislation.

To ensure regulatory certainty, energy authorities are increasingly integrating code-based mechanisms into statutory frameworks. The Australian Energy Regulator (AER), for instance, allows blockchain pilots under its Regulatory Sandbox Framework (2022), granting conditional exemptions while ensuring compliance with consumer-protection standards.

Nonetheless, regulators must retain override authority in case of systemic malfunction or unforeseen harm—affirming that digital automation operates under, not above, the rule of law (Bassan & Rabitti, 2024).

8.3. Data Protection, Cybersecurity, and Privacy in Smart-Grid Environments

Decentralised trading generates massive volumes of granular consumption data, raising significant privacy and cybersecurity concerns. In the European context, the General Data Protection Regulation (GDPR) imposes strict obligations regarding lawful processing, purpose limitation, and data minimisation. Energy-consumption data, being personally identifiable, qualifies as protected personal data under Article 4 GDPR (Finck, 2019).

Blockchain's immutability poses a direct tension with the GDPR's "right to erasure." Legal scholars propose mitigation through off-chain storage and hash-referencing techniques, ensuring that only pseudonymised references—not raw personal data—are recorded on the ledger (Hewa et al., 2021).

Cybersecurity standards such as the NIS 2 Directive (EU 2022/2555) and ISO/IEC 27019 establish technical baselines for protecting networked energy infrastructure. Compliance monitoring may itself be enhanced through blockchain's audit trails, creating a reflexive loop between technological and regulatory assurance.

8.4. Cross-Border Energy Exchange and Jurisdictional Complexity

Because electrons and data both traverse borders, decentralised trading inevitably implicates conflict-of-laws issues. Questions arise concerning the governing law of smart contracts, the location of data processing, and the territorial scope of regulatory jurisdiction (Ahmed & Măgurean, 2024).

The European Network of Transmission System Operators for Electricity (ENTSO-E) has proposed blockchain-based mechanisms for cross-border certificate exchange, yet jurisdictional fragmentation persists.

In regions such as the Nordic Electricity Market, interoperability between national regulatory systems has been achieved through mutual-recognition agreements for guarantees of origin. A similar model could be adopted for blockchain certificates, whereby digital proofs recognised in one jurisdiction maintain validity across others. This approach would require international standardisation, potentially through the International Renewable Energy Certificate Standard (I-REC Standard Foundation) and collaboration under UN Energy Compact (2022) frameworks.

Cross-border trading also triggers taxation and consumer-law obligations, particularly under value-added-tax (VAT) regimes for digital services. Regulators should develop explicit guidance on when blockchain-based energy exchanges constitute taxable events or fall within existing exemptions for renewable generation.

8.5. Regulatory Sandboxes and Adaptive Governance Models

Given the rapid pace of technological change, traditional regulatory instruments often prove too rigid. Regulatory sandboxes—controlled environments that allow experimentation under regulatory supervision—have become pivotal in balancing innovation with risk management (Beckstedde et al., 2023).

The OFGEM Regulatory Sandbox (UK), the AER Sandbox (Australia), and the Energy Market Authority Sandbox (Singapore) exemplify adaptive governance mechanisms that provide temporary waivers for blockchain pilots while ensuring oversight through data reporting and impact assessment.

Adaptive governance also entails iterative policy learning: regulators collect empirical evidence from sandbox trials to refine future rules. This feedback loop fosters anticipatory regulation, where policy evolves alongside technology rather than lagging behind it (Ahmed & Măgurean, 2024).

In emerging economies, similar frameworks could facilitate experimentation with microgrids, community energy trading, and renewable-certificate tokenisation without prematurely codifying restrictive norms.

8.6. Towards an Integrated Legal Ecosystem

A coherent legal architecture for decentralised energy trading must integrate the following principles:

- **Legal Recognition** – Formal acknowledgement of prosumers and digital-contract validity.
- **Regulatory Certainty** – Clear criteria for code-based enforcement and oversight.
- **Data Accountability** – Harmonisation of privacy and cybersecurity standards.
- **Cross-Border Interoperability** – Mutual recognition of digital certificates and transactions.
- **Adaptive Governance** – Continuous policy learning through sandbox experimentation.

Such an architecture will ensure that blockchain innovation operates within a normatively stable framework, aligning technological autonomy with democratic legitimacy and the overarching objective of a just, sustainable energy transition.

Beyond ex-ante regulatory design, the effectiveness of decentralised energy markets depends significantly on ex-post enforcement mechanisms and dispute resolution frameworks. Smart contracts, while self-executing, do not eliminate the need for legal recourse in cases of error, fraud, or unforeseen contingencies. Courts and regulatory authorities continue to play a crucial role in interpreting contractual obligations, resolving disputes, and ensuring compliance with broader legal principles. Hybrid approaches combining automated execution with institutional oversight are therefore essential for maintaining legal certainty.

9. COMPARATIVE EXPERIENCES AND EMERGING BEST PRACTICES

The diffusion of blockchain-enabled decentralised energy trading has produced diverse regulatory responses worldwide. Comparative examination reveals that jurisdictions differ not only in their technological maturity but also in their regulatory philosophies—ranging from proactive experimentation to cautious incrementalism (Andoni et al., 2019; Beckstedde et al., 2023).

Identifying emerging best practices is essential to inform future harmonisation and governance strategies. Comparative insights from the European Union (EU), Australia, Singapore, United Kingdom, and selected Latin American jurisdictions illustrate how legal systems can enable innovation while safeguarding public interests such as reliability, transparency, and consumer protection.

9.1. European Union: Legal Integration and Renewable Traceability

The European Union represents the most comprehensive legal environment for decentralised energy governance. Through the Clean Energy Package—including Directive (EU) 2018/2001 and Directive (EU) 2019/944—the EU formally recognises renewable energy communities and peer-to-peer trading as legitimate market activities (European Parliament & European Council, 2018; 2019).

Several EU member states have implemented blockchain pilots consistent with this legislative framework. For instance, France and Netherlands employ blockchain for guarantees of origin (GoOs) under the Association of Issuing Bodies (AIB), while Spain's Pylon Network and Germany's SonnenCommunity support local energy sharing compliant with national transpositions.

A key best practice emerging from the EU model is the integration of digital innovation into existing legal regimes rather than the creation of parallel frameworks. The adoption of the Markets in Crypto-Assets Regulation (MiCA, Regulation EU 2023/1114) further strengthens this approach by clarifying the classification of energy-related tokens as crypto-assets or environmental commodities, depending on function (ESMA, 2025).

9.2. United Kingdom: Regulatory Sandboxes and Data-Centric Governance

Following its departure from the EU, the United Kingdom has developed an independent, data-driven regulatory strategy. The OFGEM Regulatory Sandbox and Energy Data Taskforce initiatives allow innovators to test blockchain-based projects—such as Electron's local flexibility markets—under supervised conditions (OFGEM, 2023).

The UK's legal architecture is characterised by a principles-based approach, emphasising proportionality and outcome-oriented compliance. Rather than specifying technologies, regulators define desired objectives (e.g., transparency, consumer benefit) and permit diverse technical pathways to achieve them.

This flexibility has made the UK a reference point for adaptive energy governance, especially regarding algorithmic compliance and interoperability between energy and financial regulators.

9.3. Australia: Sandbox Governance and Decentralised Market Integration

Australia stands at the forefront of sandbox-based regulatory governance in the energy sector. The Australian Energy Regulator (AER) and Australian Securities and Investments Commission (ASIC) jointly administer sandbox environments where blockchain energy pilots operate under controlled exemptions (Australian Energy Regulator, 2023).

Projects such as Powerledger and Transactive Grid Australia demonstrate how blockchain can support renewable-certificate trading, demand-response optimisation, and P2P energy exchange (Power Ledger, 2022).

Australia's key regulatory innovation lies in the creation of inter-agency coordination between energy and financial regulators, ensuring that digital tokens and settlement processes remain consistent with both energy-market and financial-law requirements.

Moreover, Australia's Consumer Data Right (Energy) initiative exemplifies data-empowerment policy: consumers own their energy data and can authorise its use in blockchain-based applications. This reinforces data sovereignty as a core principle of decentralised energy governance (ACCC, n.d.).

9.4. Singapore: FinTech Regulation Applied to Energy Innovation

Singapore applies its FinTech expertise to energy digitalisation through the Monetary Authority of Singapore's (MAS) FinTech Regulatory Sandbox Express (Monetary Authority of Singapore, 2024). Energy-trading platforms employing tokenised power-purchase agreements (PPAs) are tested under the same principles governing financial innovation.

This cross-sectoral regulatory model ensures legal coherence between energy, data, and finance. Singapore's blockchain pilots—such as the SP Group's renewable-certificate marketplace—operate under mandatory cybersecurity and Know-Your-Customer (KYC) requirements, combining technological flexibility with strict compliance oversight (SP Group, 2018).

Best practices identified include the integration of blockchain with national digital-identity systems and the alignment of smart-contract logic with legal-contract enforceability—both of which enhance trust and traceability in decentralised trading.

9.5. Latin America: Experimental Governance and Energy Inclusion

In Latin America, blockchain initiatives are emerging within contexts of regulatory experimentation and energy-access challenges. Chile, Mexico, and Brazil have initiated blockchain pilots for renewable-certificate management and community solar projects (IRENA, 2019).

For instance, Powerledger and the Bloxberg network collaborated with Mercado Libre de Energía in Brazil to test blockchain-based green-energy certificates (Yamaguchi, et al., 2021).

While regulatory frameworks remain less mature than in Europe or Oceania, Latin America's experimentation highlights two critical dimensions of best practice:

(1) leveraging blockchain to enhance transparency and reduce corruption in public renewable programmes, and
(2) using tokenisation to enable community ownership and micro-investment in distributed generation, promoting energy inclusion.

To fully realise these benefits, Latin American jurisdictions require regulatory sandboxes, data-protection reforms, and public–private partnerships that align digital innovation with socio-economic objectives. Table 7 shows a comparative regulatory strategies and best practices for blockchain-enabled decentralised energy trading.

Table 7. Comparative overview of regulatory experiences and emerging best practices

Jurisdiction / Region	Regulatory Strategy	Key Instruments / Authorities	Notable Projects	Emerging Best Practice
European Union	Legal integration within Clean Energy Package and MiCA	European Commission, ESMA, AIB	WePower, SonnenCommunity, Pylon Network	Harmonised digital–energy governance and traceability of renewable assets
United Kingdom	Principles-based adaptive regulation and sandboxing	OFGEM, BEIS, UK Law Commission	Electron, Piclo Flex	Regulatory learning and data-centric oversight
Australia	Cross-agency sandbox governance and consumer data rights	AER, ASIC, ACCC	Powerledger, Transactive Grid	Inter-agency coordination and data sovereignty

continued on following page

Table 7. Continued

Jurisdiction / Region	Regulatory Strategy	Key Instruments / Authorities	Notable Projects	Emerging Best Practice
Singapore	FinTech regulatory framework applied to energy	MAS, EMA	SP Group Renewable Marketplace	Integration of digital identity and smart-contract legality
Latin America (Chile, Brazil, Mexico)	Experimental, hybrid approaches	National energy ministries, IRENA	Mercado Libre de Energía, Bloxberg pilots	Transparency, inclusion, and community tokenisation

9.6. Synthesis: Converging Principles of Good Energy Governance

Despite diverse contexts, comparative analysis reveals several converging principles across successful jurisdictions:

- **Regulatory Experimentation** – Sandbox models enabling safe innovation.
- **Interdisciplinary Coordination** – Collaboration between energy, finance, and data authorities.
- **Legal Clarity on Token Classification** – Differentiation between financial and environmental tokens.
- **Consumer Data Empowerment** – Legal recognition of data ownership and consent mechanisms.
- **Transparency and Accountability** – Mandatory reporting, open data, and traceability mechanisms.

Together, these principles form the blueprint for an international governance regime where decentralised energy markets operate under the shared ethos of trust, inclusivity, and legal predictability.

Such a framework not only promotes technological advancement but also anchors it within the rule of law—ensuring that decentralisation contributes to a sustainable and equitable energy future.

Figure 5. Conceptual framework of blockchain-enabled decentralised energy governance

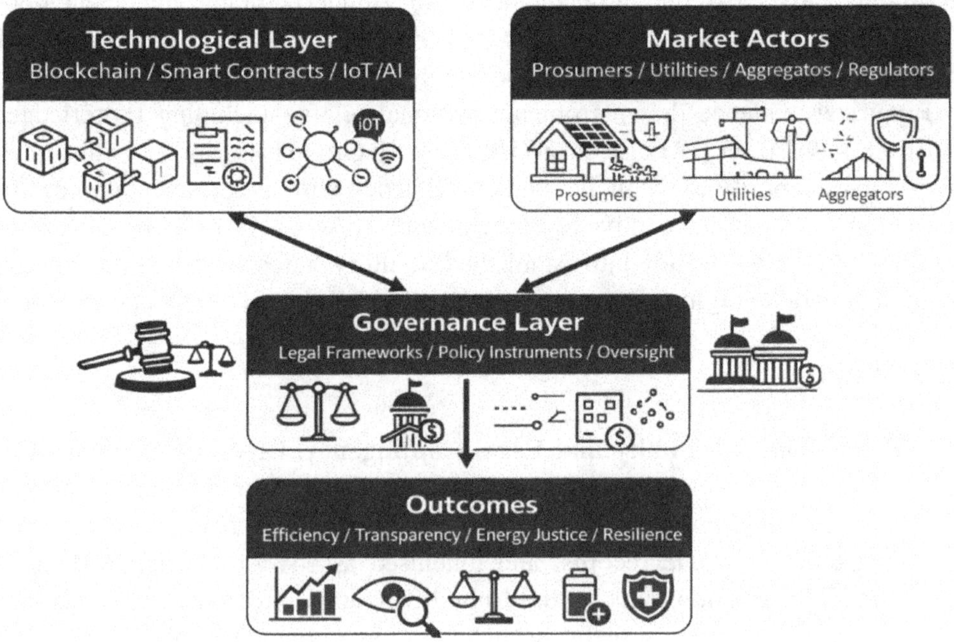

The conceptual framework (Figure 5) illustrates the interaction between technological components (blockchain, smart contracts, IoT), market actors (prosumers, utilities, regulators), and governance structures (legal frameworks, policy instruments, institutional oversight). These elements collectively shape outcomes related to efficiency, transparency, and energy justice.

10. CONCLUSIONS AND POLICY RECOMMENDATIONS

This chapter has explored the intersection between blockchain technology and the legal-regulatory frameworks governing decentralised energy markets. The analysis demonstrates that while blockchain facilitates transparency, traceability, and automation, its legitimacy and long-term viability depend upon legal institutionalisation. Without coherent rules, decentralised trading may undermine consumer protection, grid stability, and equitable access—objectives that remain fundamental to public energy law (Ahmed & Măgurean, 2024; Aravind et al., 2024).

Although much of the current development of blockchain-enabled energy systems is concentrated in advanced economies, their implications for developing and emerging contexts are equally significant. In such settings, decentralised energy trading may offer opportunities to expand energy access and enhance system resilience. However, structural challenges—including limited infrastructure, regulatory capacity constraints, and socio-economic inequalities—may hinder equitable implementation. Consequently, a context-sensitive approach is required to ensure that decentralised energy innovations contribute to inclusive and sustainable development.

Empirical evidence drawn from international pilots—including Powerledger (Australia), WePower (EU), Electron (UK), and SP Group (Singapore)—illustrates blockchain's capacity to reconcile market efficiency with environmental integrity when embedded within adaptive legal architectures. Across jurisdictions, the integration of digital innovation into established statutory frameworks, rather than the creation of parallel regimes, emerges as a defining feature of effective governance.

Key Findings

- **Technological Viability and Legal Contingency**. Blockchain systems can deliver trust by design through immutable ledgers and programmable compliance. Yet, their enforceability depends on jurisdictional recognition of smart contracts, electronic records, and tokenised assets. As Section 8 demonstrated, jurisdictions such as the United Kingdom, Australia, and Singapore provide clear statutory pathways, whereas others remain legally ambiguous. Hence, technological reliability must be complemented by legal certainty.
- **Data Governance as Regulatory Infrastructure**. The convergence of blockchain with IoT, AI, and big data (Section 7) redefines data not merely as operational input but as a regulatory asset. Immutable audit trails enable real-time oversight; however, they also necessitate rigorous compliance with privacy and cybersecurity standards under frameworks such as the GDPR and NIS 2 Directive. Effective governance thus requires technological and legal interoperability.
- **Tokenisation and Market Democratisation**. Tokenisation transforms energy and environmental values into legally cognisable digital commodities (Section 6). Properly regulated, it facilitates fractional ownership and inclusive financing of renewable projects. Nonetheless, classification clarity—distinguishing environmental from financial tokens—is essential to avoid speculative volatility. The EU's MiCA Regulation and Australia's joint energy–financial oversight exemplify best practice.
- **Regulatory Sandboxes and Adaptive Governance**. As evidenced in Sections 8 and 9, sandbox frameworks foster learning-based regulation that accommodates rapid innovation while controlling systemic risk. They exemplify adaptive governance, wherein regulators and innovators co-produce

policy knowledge through empirical experimentation. Such mechanisms should be institutionalised globally to manage technological uncertainty.

- **Converging Principles of Global Good Governance.** Comparative insights (Section 9) reveal convergence around five guiding principles:
  - Legally recognised prosumers and communities;
  - Algorithmic accountability in smart contracts;
  - Data sovereignty and consumer empowerment;
  - Cross-sectoral coordination between energy, finance, and data regulators;
  - Transparency and inclusivity as normative anchors.

These principles define the emergent paradigm of lawful decentralisation.

Policy Recommendations

- **Harmonisation of Legal Definitions.** International standard-setting bodies—such as the International Renewable Energy Certificate (I-REC) Foundation, ISO/TC 307 on Blockchain Technologies, and the UN Energy Compact—should collaborate to establish uniform terminology and classification of digital energy assets. This would reduce jurisdictional ambiguity and support cross-border certificate exchange.
- **Embedding Legal Compliance in Code.** Energy regulators should require that smart contracts incorporate statutory compliance parameters—covering tariff caps, grid-service obligations, and consumer-protection clauses—prior to deployment. Independent code audits and regulatory certification of algorithms would operationalise the principle of regulation by design (Bassan & Rabitti, 2024).
- **Data-Protection Alignment.** Regulators must reconcile blockchain's immutability with data-protection rights by encouraging hybrid architectures (on-chain hashes with off-chain storage). Supervisory authorities should issue specific guidance on privacy-preserving distributed ledgers within the energy sector (Finck, 2019).
- **Institutionalising Regulatory Sandboxes.** Governments should formalise sandbox regimes through legislation that specifies eligibility, duration, and evaluation criteria. This would enable continuous innovation under controlled legal conditions while ensuring that public-interest outcomes—sustainability, affordability, and equity—remain paramount (Beckstedde et al., 2023).
- **Ensuring Algorithmic Transparency and Ethics.** Given the integration of AI and blockchain, transparency and explainability must be legally mandated for automated decision-making affecting energy access or pricing. Ethical-

impact assessments should accompany algorithmic deployment to uphold the principle of energy justice (Sovacool et al., 2017).

- **Capacity-Building and Multilevel Coordination.** International cooperation and professional education are essential. Policymakers, engineers, and legal practitioners must develop shared competencies in energy informatics and digital regulation. Multilevel coordination—between national regulators, municipalities, and regional agencies—should ensure coherence across decentralised governance scales.

Risks, Limitations, and Governance Challenges

While blockchain-enabled decentralised energy markets present considerable opportunities, they also entail significant risks and limitations that warrant careful consideration. Algorithmic governance mechanisms, particularly smart contracts, may introduce rigidity and reduce the scope for discretionary intervention, thereby complicating dispute resolution and regulatory oversight. Moreover, issues of algorithmic bias and data asymmetry may lead to unequal outcomes among market participants.

From a socio-economic perspective, the deployment of decentralised energy systems may inadvertently exacerbate existing inequalities, particularly where access to digital infrastructure and technical expertise is uneven. Furthermore, reliance on automated systems raises concerns regarding accountability, especially in cases of system failure or unintended contractual execution.

These considerations highlight the limits of a purely techno-centric approach and underscore the continued importance of human oversight, institutional capacity, and adaptive regulatory frameworks in ensuring equitable and resilient energy systems.

Final Reflections

Blockchain and allied technologies are catalysing a transition from centralised control to distributed coordination in the energy sector. Yet decentralisation does not imply deregulation. Rather, it necessitates a reconfiguration of governance—from reactive enforcement to proactive, code-based compliance grounded in legal legitimacy.

The ultimate challenge lies in achieving a symbiosis between technological autonomy and institutional accountability. If successfully realised, blockchain-driven decentralised energy markets can advance not only efficiency and transparency but also the normative goals of sustainability, inclusivity, and justice that underpin modern energy law.

Future research should further investigate the long-term regulatory implications of decentralised energy markets, particularly in relation to cross-border interoperability, data governance, and institutional adaptation. Greater empirical analysis of implemented systems, including both successful and unsuccessful cases, would provide valuable insights into real-world performance. Additionally, interdisciplinary approaches integrating legal, technical, and socio-economic perspectives will be essential for advancing the governance of emerging energy systems.

REFERENCES

Agarwal, S., & Jain, A. (2023). Blockchain applications in smart grid: Challenges and opportunities. In *Proceedings of the 2023 IEEE International Conference on Blockchain and Distributed Systems Security (ICBDS)* (pp. 1-7). IEEE. DOI: 10.1109/ICBDS58040.2023.10346363

Ahl, A., Yarime, M., Goto, M., Chopra, S. S., Kumar, N. M., Tanaka, K., & Sagawa, D. (2020). Exploring blockchain for the energy transition: Opportunities and challenges based on a case study in Japan. *Renewable & Sustainable Energy Reviews, 117*, 109488. DOI: 10.1016/j.rser.2019.109488

Ahmed, S., & Măgurean, A. M. (2024). Renewable energy communities: Towards a new sustainable model of energy production and sharing. *Energy Strategy Reviews, 55*, 101522. Advance online publication. DOI: 10.1016/j.esr.2024.101522

Andoni, M., Robu, V., Flynn, D., Abram, S., Geach, D., Jenkins, D., McCallum, P., & Peacock, A. (2019). Blockchain technology in the energy sector: A systematic review of challenges and opportunities. *Renewable & Sustainable Energy Reviews, 100*, 143–174. DOI: 10.1016/j.rser.2018.10.014

Androulaki, E., Barger, A., Bortnikov, V., Cachin, C., Christidis, K., De Caro, A., & Yellick, J. (2018, April). Hyperledger fabric: a distributed operating system for permissioned blockchains. In *Proceedings of the thirteenth EuroSys conference* (pp. 1-15). DOI: 10.1145/3190508.3190538

Aravind, A., Santhi, G., Patil, S. A., Selvakumar, S., Sharma, G., Dhamone, J., & Nathan, S. R. (2024). Blockchain Technology in Energy Markets: Enabling Peer-to-Peer Energy Trading. E3S Web of Conferences, 591, 06002. DOI: 10.1051/e3sconf/202459106002

Australian Competition and Consumer Commission (ACCC). (n.d.). The Consumer Data Right. Retrieved November 3, 2025, from https://www.accc.gov.au/by-industry/banking-and-finance/the-consumer-data-right

Australian Energy Market Operator. (2025). Strategic corporate plan. https://www.aemo.com.au/about/corporate-governance/corporate-plan

Australian Energy Regulator. (2023). Trial projects guidelines – Regulatory sandboxing. Retrieved November 1, 2025, from https://www.aer.gov.au/industry/registers/resources/guidelines/trial-projects-guidelines-regulatory-sandboxing

Baillie, K. (2017). Blockchain: talking utility switching and more with Electron. Retrieved November 13, 2025, from https://www.power-technology.com/features/featureblockchain-talking-utility-switching-and-more-with-electron-5904299/?cf-view

Bassan, F., & Rabitti, M. (2024). From smart legal contracts to contracts on blockchain: An empirical investigation. *Computer Law & Security Review*, *55*, 106035. Advance online publication. DOI: 10.1016/j.clsr.2024.106035

Beckstedde, E., Correa Ramírez, M., Cossent, R., Vanschoenwinkel, J., & Meeus, L. (2023). Regulatory sandboxes: Do they speed up innovation in energy? *Energy Policy*, *180*, 112672. Advance online publication. DOI: 10.1016/j.enpol.2023.113656

Biswas, P., Rashid, A., Biswas, A., Al Nasim, A., Datta Gupta, K., & George, R. (2024). AI-driven approaches for optimizing power consumption: A comprehensive survey (arXiv preprint arXiv:2406.15732v1). https://arxiv.org/abs/2406.15732

Buterin, V. (2014). A next-generation smart contract and decentralized application platform. *white paper, 3*(37), 2-1.

Church, S. (2022). Tokenizing Renewable Energy Certificates (RECs)—A Blockchain Approach for REC Issuance and Trading. *IEEE Access : Practical Innovations, Open Solutions*, *10*, 134477–134490. DOI: 10.1109/ACCESS.2022.3230937

Dorri, A., Kanhere, S. S., Jurdak, R., & Gauravaram, P. (2017). Blockchain for IoT security and privacy: The case study of a smart home. 2017 IEEE International Conference on Pervasive Computing and Communications Workshops (PerCom Workshops), Kona, HI, USA, 2017, pp. 618-623. DOI: 10.1109/PERCOMW.2017.7917634

Energy, E. D. F. (n.d.). Smart meters help you save energy. Retrieved October 13, 2025, from https://www.edfenergy.com/smart-meters?_%24ja=tsid%3A44771%7Ccgn%3ARedirect%7Ckw%3Asmart

Energy Web Foundation (EWF). (2019). The Energy Web Chain: Accelerating the energy transition with an open-source, decentralized blockchain platform (Version 2.0). https://cdn.prod.website-files.com/65e5b22cf7f0480a007420fc/67adf26e027740e64f98b5df_EWF-Paper-TheEnergyWebChain-v2-201907.pdf

European Commission. (2020). Energy communities in the clean energy package: Best practices and recommendations for implementation (Publication No. MJ-01-20-760-EN-N; ISBN 978-92-76-27320-2; DOI). Publications Office of the European Union. https://op.europa.eu/en/publication-detail/-/publication/4b7d5144-91c9-11eb-b85c-01aa75ed71a1/language-enDOI: 10.2833/51076

European Parliament & European Council. (2018). Directive (EU) 2018/2001 of the European Parliament and of the Council of 11 December 2018 on the promotion of the use of energy from renewable sources (recast). *Official Journal of the European Union, L, 328*, 82–209. https://eur-lex.europa.eu/eli/dir/2018/2001/oj/eng

European Parliament & European Council. (2019). Directive (EU) 2019/944 of the European Parliament and of the Council of 5 June 2019 on common rules for the internal market for electricity and amending Directive 2012/27/EU (recast). *Official Journal of the European Union, L, 158*, 125–199. https://eur-lex.europa.eu/eli/dir/2019/944/oj/eng

European Securities and Markets Authority (ESMA). (2025). Markets in Crypto-Assets Regulation (MiCA). Retrieved October 18, 2025, from https://www.esma.europa.eu/esmas-activities/digital-finance-and-innovation/markets-crypto-assets-regulation-mica

Financial Stability Board. (2021). 2021 status report: Task Force on Climate-related Financial Disclosures. https://www.fsb.org/2021/10/2021-status-report-task-force-on-climate-related-financial-disclosures/

Finck, M. (2019). Blockchain and the general data protection regulation: Can distributed ledgers be squared with European data protection law? Max Planck Institute for Innovation & Competition. DOI: 10.2139/ssrn.3080322

Group, S. P. (2018). SP Unveils Blockchain-Powered Renewable Energy Certificates Marketplace. SP Group. https://www.spgroup.com.sg/about-us/media-resources/energy-hub/innovation/sp-unveils-blockchain-powered-renewable-energy-certificates-marketplace

Hewa, T., Ylianttila, M., & Liyanage, M. (2021). Survey on blockchain based smart contracts: Applications, opportunities and challenges. *Journal of Network and Computer Applications, 177*, 102857. DOI: 10.1016/j.jnca.2020.102857

How Blockchain Works. (n.d.). Blockchain: Empowering manufacturing sector. MM India. https://www.mmindia.co.in/article/533/blockchain-empowering-manufacturing-sector

International Energy Agency (IEA). (2024). World Energy Outlook 2024 [Report]. https://www.iea.org/reports/world-energy-outlook-2024

International Renewable Energy Agency (IRENA). (2019). Innovation landscape brief: Blockchain (Innovation Landscape Brief). International Renewable Energy Agency. https://www.irena.org/-/media/Files/IRENA/Agency/Publication/2019/Feb/IRENA_Landscape_Blockchain_2019.pdf

Invest in Estonia. (2018). Digital revolution in Estonia's energy sector: WePower is the first blockchain firm to tokenize an entire grid. https://investinestonia.com/wepower-is-the-first-blockchain-firm-to-tokenize-an-entire-grid/

Klenergy Tech. (2018). Pylon Network: The energy blockchain platform — Token Paper [White paper]. https://www.allcryptowhitepapers.com/wp-content/uploads/2018/05/Pylon-Network.pdf

Kouhizadeh, M., & Sarkis, J. (2018). Blockchain Practices, Potentials, and Perspectives in Greening Supply Chains. *Sustainability (Basel)*, *10*(10), 3652. DOI: 10.3390/su10103652

Law Commission of England and Wales (UK Law Commission). (2021). Smart legal contracts: Advice to government. London: Law Commission. https://www.lawcom.gov.uk

López, I., Goitia-Zabaleta, N., Milo, A., Gómez-Cornejo, J., Aranzabal, I., Gaztañaga, H., & Fernández, E. (2024). European energy communities: Characteristics, trends, business models and legal framework. *Renewable & Sustainable Energy Reviews*, *197*, 113046. Advance online publication. DOI: 10.1016/j.rser.2024.114403

Magazine, T. Y. N. (2023, August 2). Redes descentralizadas: el entorno perfecto para el crecimiento del blockchain. https://tynmagazine.com/redes-descentralizadas-el-entorno-perfecto-para-el-crecimiento-del-blockchain/

Mengelkamp, E., Gärttner, J., Rock, K., Kessler, S., Orsini, L., & Weinhardt, C. (2018). Designing microgrid energy markets: A case study of the Brooklyn Microgrid. *Applied Energy*, *210*, 870–880. DOI: 10.1016/j.apenergy.2017.06.054

Monetary Authority of Singapore. (2024). Regulatory Sandbox. Retrieved November 14, 2025, from https://www.mas.gov.sg/development/fintech/regulatory-sandbox

Narayanan, A., Bonneau, J., Felten, E., Miller, A., & Goldfeder, S. (2016). *Bitcoin and cryptocurrency technologies: a comprehensive introduction*. Princeton University Press.

Niet, I., van Est, R., & Veraart, F. (2021). Governing AI in Electricity Systems. *Frontiers in Artificial Intelligence*, *4*, 690237. DOI: 10.3389/frai.2021.690237 PMID: 34396090

Office of Gas and Electricity Markets. (2023). Consultation on updates to Data Best Practice Guidance and Digitalisation Strategy and Action Plan Guidance (Consultation OFG1163) [PDF]. https://www.ofgem.gov.uk/sites/default/files/2023-02/Consultation%20on%20updates%20to%20Data%20Best%20Practice%20Guidance%20and%20Digitalisation%20Strategy%20and%20Action%20Plan%20Guidance.pdf

Parag, Y., & Sovacool, B. K. (2016). Electricity market design for the prosumer era. *Nature Energy*, *1*(4), 16032. DOI: 10.1038/nenergy.2016.32

Powerledger. (2023). Global Blockchain Business Council – 2022 Annual Report [Case study]. https://powerledger.io/media/global-blockchain-business-council-2022-annual-report/

Raskin, M. (2017). The law and legality of smart contracts. Georgetown Law Technology Review, 1(2), 305–341. https://georgetownlawtechreview.org/the-law-and-legality-of-smart-contracts

Shamaseen, A., Qatawneh, M., & Elshqeirat, B. (2025). Blockchain for future smart grid: A comprehensive survey. *Bulletin of Electrical Engineering and Informatics*, *14*(4), 2497–2513. DOI: 10.11591/eei.v14i4.8956

Siemens. (2018). A microgrid grows in Brooklyn. Siemens. Retrieved October 14, 2025, from https://www.siemens.com/global/en/company/stories/research-technologies/energytransition/a-microgrid-grows-in-brooklyn.html

Siemens. (2020). Microgrid project in Vienna: Small grid, major impact. Siemens. https://www.siemens.com/global/en/company/stories/infrastructure/2020/microgrid-project-in-vienna.html

Sovacool, B. K., Burke, M., Baker, L., Kotikalapudi, C. K., & Wlokas, H. (2017). New frontiers and conceptual frameworks for energy justice. *Energy Policy*, *105*, 677–691. DOI: 10.1016/j.enpol.2017.03.005

Sun Exchange. (2023). Solar power for the people: Crowd-funded energy projects in emerging markets. Sun Exchange Ltd. https://thesunexchange.com/

Szabo, N. (1997). Formalizing and securing relationships on public networks. *First Monday*, *2*(9). Advance online publication. DOI: 10.5210/fm.v2i9.548

Tenne, T. (n.d.). Innovations. Retrieved November 14, 2025, from https://www.tennet.eu/innovations-tennet

Werbach, K., & Cornell, N. (2017). Contracts Ex Machina. *Duke Law Journal*, *67*(2), 313–382. https://scholarship.law.duke.edu/dlj/vol67/iss2/2/

Yamaguchi, J. A. R., Teresa Rachael Santos, T. R., Vargas, G., & Pereira, A. de C. (2021). Blockchain technology in renewable energy certificates: An exploratory study. Business & Accounting Review, 20(1), 934-958. https://doi.org/DOI: 10.1590/1807-7692bar2021200069

KEY TERMS AND DEFINITION

Algorithmic Governance: A mode of governance in which decision-making processes are partially or fully automated through algorithms, raising important questions regarding accountability, transparency, and legal oversight.

Blockchain: A distributed digital ledger technology that records transactions across a decentralised network of nodes, ensuring transparency, immutability, and cryptographic security without reliance on a central authority.

Decentralised Energy Markets: Energy systems in which electricity generation, distribution, and exchange occur through distributed actors—often including prosumers—facilitated by digital platforms that enable peer-to-peer transactions.

Energy Justice: A normative framework concerned with the fair distribution of energy benefits and burdens, equitable access to energy resources, and inclusive participation in energy decision-making processes.

Ex-Post Enforcement: Legal and regulatory mechanisms applied after the execution of transactions or contracts, including dispute resolution, judicial interpretation, and compliance monitoring.

Interoperability: The capacity of different technological systems, platforms, or regulatory frameworks to exchange data and operate effectively together, which is essential for scaling decentralised energy markets.

Prosumer: An energy user who both produces and consumes electricity, typically through distributed generation technologies such as solar photovoltaic systems, and who may participate actively in local energy markets.

Regulatory Sandbox: A controlled environment established by regulatory authorities that allows innovators to test new technologies or business models under relaxed regulatory conditions while maintaining oversight and risk mitigation.

Smart Contracts: Self-executing digital agreements encoded on a blockchain, in which contractual terms are automatically enforced through predefined conditions, reducing the need for intermediaries.

Tokenisation: The process of representing physical or digital assets—such as units of energy—through blockchain-based tokens that can be traded, transferred, or programmed within decentralised systems.

Chapter 3 Decentralized Energy Markets: Blockchain Frameworks, Regulatory Pathways, and Compliance Models

Bhavya Bhasuran
http://orcid.org/0009-0000-2299-1405
Sathyabama Institute of Science and Technology, India

G. Ashwin Prabhu
http://orcid.org/0000-0001-5673-647X
St. Joseph's College of Engineering, India

G. Subash
Sathyabama Institute of Science and Technology, India

E. Raviendiran
Sathyabama Institute of Science and Technology, India

M. Ashok Kumar
CVR College of Engineering, India

Karthika P. Devan
Vishwakarma University, India

Vela N. Yendeesh
St. Joseph's College of Engineering, India

Pandilla Nithin
St. Joseph's College of Engineering, India

ABSTRACT

Decentralized energy markets are transforming electricity generation, distribution, and consumption by enabling peer-to-peer trading, active prosumer participation, and localized energy governance. Blockchain technology underpins these markets by delivering transparent, tamper-resistant, and automated transaction infrastruc-

DOI: 10.4018/979-8-3373-5742-3.ch003

tures through smart contracts and distributed ledgers. The discussion analyzes key blockchain frameworks for decentralized energy trading, focusing on platform architectures, consensus mechanisms, and interoperability models suited to energy systems. It also examines evolving regulatory pathways across jurisdictions, emphasizing interactions between decentralized trading models and existing energy laws, market rules, and grid codes. Emphasis is placed on compliance models that reconcile innovation with legal certainty, consumer protection, data privacy, and grid stability, while outlining challenges, best practices, and policy directions for scalable and compliant blockchain-enabled energy trading ecosystems.

1. INTRODUCTION

The global energy sector is undergoing a structural transformation driven by decarbonization goals, rapid deployment of renewable energy sources, and the increasing digitalization of electricity systems. Traditional electricity markets were designed around centralized generation, one-directional power flows, and hierarchical governance structures. While this model enabled large-scale electrification and operational stability for decades, it is increasingly misaligned with contemporary energy realities characterized by distributed generation, active consumer participation, and localized energy exchanges. Decentralized energy markets have emerged as a response to these shifts, offering new mechanisms for electricity trading, coordination, and governance at the distribution and community levels (Burger et al., 2019). Centralized electricity markets face several structural and regulatory limitations. Market concentration, high entry barriers, and reliance on centralized intermediaries often restrict consumer choice and innovation. Moreover, conventional regulatory frameworks are primarily designed to manage large utilities and wholesale transactions, leaving limited flexibility to accommodate small-scale producers, rooftop solar owners, battery storage operators, and community energy initiatives (Littlechild, 2021). The integration of intermittent renewable energy sources such as solar and wind further stresses centralized systems, creating challenges related to grid congestion, price volatility, and balancing mechanisms. These constraints have prompted policymakers and market participants to explore alternative market architectures that can better align economic incentives with technological and environmental objectives. Within this evolving context, peer-to-peer energy trading has gained prominence as a decentralized market model that enables direct electricity transactions between participants without relying exclusively on centralized utilities or retailers. Similar decentralized frameworks are being explored in agriculture, where AI-powered agrobots enable localized resource management and peer-to-peer exchanges of crop health data and energy inputs, reinforcing the cross-sectoral relevance of decentralized trading (Dhanushya et al., 2025). IoT-integrated smart vertical crop cultivation further exemplifies how decentralized digital ecosystems can optimize resource flows, underscoring parallels between agriculture and energy market decentralization (Gowthaman et al., 2025). In peer-to-peer systems, consumers can become "prosumers" by generating electricity locally and selling surplus energy to neighbors or community members. This shift redefines traditional roles within electricity markets, transforming passive consumers into active market participants and fostering localized energy ecosystems (Parag & Sovacool, 2016). Such models promise improved economic efficiency, enhanced renewable energy utilization, and greater consumer empowerment, but they also raise complex coordination, trust, and regulatory challenges.

Blockchain technology has emerged as a key enabler of decentralized energy markets by addressing many of the coordination and trust issues inherent in peer-to-peer trading environments. As a distributed ledger technology, blockchain allows transactions to be recorded in a transparent, tamper-resistant, and verifiable manner without requiring a central authority. Smart contracts further automate market processes such as energy settlement, pricing, and compliance enforcement, reducing transaction costs and administrative complexity (Mengelkamp et al., 2018). These features make blockchain particularly attractive for managing high-frequency, low-volume energy transactions that characterize decentralized energy trading systems. Beyond technical efficiency, blockchain frameworks introduce new governance possibilities for energy markets. Decentralized consensus mechanisms can facilitate collective decision-making, while tokenization models enable innovative incentive structures linked to renewable generation, flexibility services, and grid support. However, the deployment of blockchain in energy markets also introduces legal and regulatory uncertainties. Questions related to contractual enforceability, data protection, consumer rights, market licensing, and grid responsibility remain insufficiently addressed within existing legal frameworks (De Filippi & Wright, 2018). As a result, technological innovation in decentralized energy trading often advances faster than regulatory adaptation, creating a gap between operational feasibility and legal compliance. Against this background, the primary objective of this chapter is to provide a structured analysis of decentralized energy markets through the combined lenses of blockchain frameworks, regulatory pathways, and compliance models. The chapter seeks to examine how blockchain-based architectures support peer-to-peer energy trading, how existing energy laws and regulations interact with decentralized market designs, and what compliance mechanisms are necessary to ensure legal certainty, consumer protection, and grid stability. Rather than treating technology and regulation as separate domains, the chapter adopts an integrated legal-technical perspective that recognizes their mutual interdependence.

The scope of the chapter encompasses both conceptual and practical dimensions of decentralized energy markets. It explores foundational market concepts, technical design considerations, and governance models, while also engaging with regulatory developments across multiple jurisdictions. Particular attention is given to compliance challenges arising from smart contracts, data governance, and cross-sector regulation at the intersection of energy and digital markets. By situating decentralized energy trading within broader legal and policy debates, the chapter aims to contribute to ongoing discussions on how innovation can be aligned with regulatory objectives. The chapter is structured as follows. Following this introduction, subsequent sections examine the conceptual foundations of decentralized energy markets, blockchain frameworks supporting energy trading, regulatory pathways shaping market participation, and compliance models addressing legal and operational risks. The concluding sections synthesize key findings and outline future directions for policy and governance. Through this structure, the chapter seeks to provide researchers, policymakers, and practitioners with a comprehensive understanding of the legal architecture underpinning blockchain-driven decentralized energy markets.

2. CONCEPTUAL FOUNDATIONS OF DECENTRALIZED ENERGY MARKETS

Decentralized energy markets represent a structural shift in the organization and governance of electricity systems, moving away from centrally controlled generation and top-down market clearing toward distributed, participatory, and digitally mediated exchange mechanisms. At their core, decentralized energy markets enable electricity generation, consumption, and trading to occur closer to the point of use, often at the distribution or community level, rather than through large centralized power plants and wholesale exchanges. These markets are characterized by localized decision-making, distributed ownership of energy assets, and the use of digital platforms to coordinate transactions among multiple small-scale participants (Burger et al., 2019; Mengelkamp et al., 2018). Key characteristics of decentralized energy markets include bidirectional energy flows, high penetration of distributed energy resources such as rooftop solar photovoltaics and battery storage, and the active participation of end users in market activities. Unlike conventional electricity markets, where consumers are largely passive price-takers, decentralized markets enable participants to simultaneously consume, produce, and trade electricity. Digitalization, automation, and data-driven coordination are essential enablers, allowing real-time measurement, pricing, and settlement of energy transactions at a granular level (Parag & Sovacool, 2016). Similar IoT-enabled wearable health monitoring systems demonstrate how decentralized digital infrastructures can enhance trust and responsiveness, offering transferable lessons for energy market coordination (Sibiya et al., 2025). Practical demonstrations, such as AI-enabled rover control systems and smart inspection technologies, highlight how decentralized digital infrastructures can be applied to both energy and non-energy domains, strengthening resilience and trust in distributed systems. These features collectively challenge traditional assumptions about market boundaries, regulatory oversight, and the role of intermediaries in electricity systems.

Peer-to-peer (P2P) energy trading is a foundational mechanism within decentralized energy markets. P2P models allow individual participants to directly exchange electricity with one another based on predefined rules, often supported by digital platforms or blockchain-based infrastructures. Transactions may occur within neighborhoods, apartment complexes, or virtual communities, enabling localized balancing of supply and demand (Tushar et al., 2020). Closely related to P2P trading are local energy communities, which are organized groups of participants who collectively generate, consume, and manage energy resources. These communities often pursue broader objectives beyond economic efficiency, such as environmental sustainability, energy autonomy, and social inclusion (Lowitzsch et al., 2020). Microgrids further operationalize decentralization by providing a physical and operational framework for localized energy systems that can operate in coordination with or independently from the main grid, enhancing resilience and reliability. The functioning of decentralized energy markets depends on the evolving roles of key actors, particularly prosumers, aggregators, and distribution system operators (DSOs). Prosumers occupy a central position, as they own and operate distributed generation assets and actively participate in market transactions. Their dual role blurs the traditional distinction between producers and consumers, raising new questions about pricing, responsibility, and regulatory treatment (Parag & Sovacool, 2016). Aggregators act as intermediaries that bundle the capacities of multiple small participants to enable efficient market participation, risk management, and grid coordination. By pooling resources, aggregators reduce transaction costs and facilitate compliance with technical and regulatory requirements that individual prosumers may be unable to meet independently (Burger et al., 2019). DSOs, traditionally responsible for operating and maintaining distribution networks, increasingly assume a market-facilitation role, ensuring grid stability while accommodating decentralized trading activities. This transition requires DSOs to balance neutrality, transparency, and system reliability within more dynamic and decentralized market environments.

Market design principles play a critical role in shaping the effectiveness and legitimacy of decentralized energy markets. Transparency is essential to build trust among participants and ensure that pricing, settlement, and allocation rules are clearly defined and accessible. Digital platforms enable enhanced transparency through real-time data availability and auditable transaction records (Mengelkamp et al., 2018). Efficiency is pursued by reducing transaction costs, minimizing losses, and enabling localized balancing, which can lower system-wide costs and congestion. Resilience is another core principle, as decentralized systems can improve the ability of energy networks to withstand disruptions by distributing generation and control across multiple nodes. Fairness, both procedural and distributive, is increasingly emphasized to ensure equitable access to market participation, prevent discrimination, and avoid the exclusion of vulnerable consumers from the benefits of decentralization (Lowitzsch et al., 2020). A comparison with traditional wholesale and retail electricity market structures highlights the transformative implications of decentralized energy markets. Wholesale markets are typically centralized, with large generators competing to supply electricity through organized exchanges, while retail markets focus on supply contracts between utilities and end consumers under regulated or semi-liberalized conditions. These models rely on hierarchical control, standardized products, and centralized clearing mechanisms. In contrast, decentralized markets operate at smaller scales, involve heterogeneous participants, and rely on continuous, data-driven coordination rather than periodic market clearing. While wholesale and retail markets prioritize system-wide optimization and regulatory control, decentralized markets emphasize local value creation, participant autonomy, and flexible governance arrangements (Burger et al., 2019; Tushar et al., 2020). Overall, decentralized energy markets represent not merely a technological innovation but a reconfiguration of economic, social, and institutional relationships within the energy sector. Understanding their conceptual foundations is essential for designing regulatory pathways and compliance models that can accommodate innovation while preserving system reliability, consumer protection, and public policy objectives.

3. BLOCKCHAIN FRAMEWORKS FOR DECENTRALIZED ENERGY TRADING

Blockchain technology has emerged as a critical digital infrastructure for enabling decentralized energy markets by providing secure, transparent, and automated mechanisms for coordinating transactions among distributed participants. In contrast to traditional centralized databases, blockchain operates as a distributed ledger maintained collectively by network participants, ensuring that records of energy generation, consumption, and exchange are tamper-resistant and verifiable. These characteristics are particularly valuable in decentralized energy trading environments, where trust must be established among heterogeneous actors without relying on a single central authority (Andoni et al., 2019). At a structural level, blockchain frameworks for energy trading consist of three interrelated layers: the physical energy layer, the digital data and control layer, and the transaction and settlement layer. The physical layer includes distributed energy resources such as solar panels, wind turbines, batteries, and smart meters. The digital layer captures real-time data related to energy production and consumption, often through Internet of Things (IoT) devices. The transaction layer, implemented via blockchain, records energy exchanges, executes smart contracts, and facilitates settlement processes. The integration of these layers enables near real-time matching of supply and demand while maintaining a secure and auditable transaction history (Mengelkamp et al., 2018). Advances in high-performance memory architectures and AI-driven cybersecurity approaches provide the computational backbone for secure blockchain execution, ensuring resilience against denial-of-service attacks and efficient transaction validation (Kameswaran et al., 2017; Krithika et al., 2025). Different blockchain architectures are employed in decentralized energy markets, each with distinct governance and performance implications. Public blockchains offer high transparency and openness but may face scalability and energy-efficiency challenges. Private blockchains provide greater control and faster transaction speeds but rely on trusted operators, which may reintroduce elements of centralization. Consortium or permissioned blockchains are increasingly favored in energy applications, as they balance decentralization with regulatory oversight by limiting participation to authorized stakeholders such as utilities, aggregators, and community members (Andoni et al., 2019). The choice of architecture significantly influences compliance, accountability, and system scalability.

Smart contracts form the functional core of blockchain-based energy trading frameworks. These self-executing programs encode predefined rules for pricing, allocation, settlement, and enforcement of energy transactions. Once conditions specified in the contract are met—such as the delivery of a certain amount of electricity—execution occurs automatically without the need for manual intervention. In decentralized energy markets, smart contracts can automate peer-to-peer trading, dynamic pricing based on local supply and demand, and the distribution of payments among participants (Tushar et al., 2020). From a governance perspective, smart contracts reduce transaction costs and enhance predictability, but they also raise legal questions regarding enforceability, liability, and dispute resolution when automated outcomes conflict with regulatory or contractual expectations. Tokenization is another important feature of blockchain frameworks in decentralized energy trading. Energy tokens can represent units of electricity, grid access rights, or environmental attributes such as renewable energy certificates. By converting physical energy flows into digital assets, tokenization enables fractional ownership, real-time trading, and secondary market activities. This approach enhances liquidity and facilitates participation by small-scale actors who might otherwise be excluded from energy markets (Peters & Panayi, 2016). However, tokenized energy assets also intersect with financial regulation, as they may be classified as securities or financial instruments in certain jurisdictions, thereby triggering additional compliance requirements. Consensus mechanisms are central to maintaining the integrity of blockchain-based energy trading platforms. Comparable IoT-based smart traffic control systems highlight how decentralized consensus and automation can optimize complex, real-time flows, reinforcing blockchain's role in managing distributed energy transactions (Poovithaa et al., 2025). These mechanisms determine how transactions are validated and added to the ledger. Energy applications typically avoid energy-intensive consensus models and instead favor more efficient alternatives that align with sustainability objectives. The selection of consensus mechanisms affects transaction throughput, latency, and trust assumptions, making it a critical design consideration for real-time or near real-time energy markets (Andoni et al., 2019). Achieving an appropriate balance between decentralization, performance, and regulatory acceptability remains an ongoing challenge.

Scalability and interoperability are persistent concerns in blockchain-enabled decentralized energy markets. Energy systems generate large volumes of data at high frequency, particularly when smart meters and IoT devices are involved. Blockchain frameworks must therefore be designed to handle high transaction volumes without compromising speed or security. Layered architectures, off-chain processing, and hybrid solutions that combine blockchain with conventional databases are commonly proposed to address these limitations (Mengelkamp et al., 2018). Interoperability between different blockchain platforms and with existing energy market systems is equally important to prevent fragmentation and ensure seamless integration with wholesale markets, grid management systems, and regulatory reporting tools. Cybersecurity and data governance considerations further shape the design of blockchain frameworks for energy trading. While blockchain enhances data integrity through cryptographic techniques, vulnerabilities may still arise at the interface between physical devices and digital platforms. Secure identity management, access controls, and data privacy mechanisms are essential to protect sensitive consumption and generation data. These concerns are particularly salient in jurisdictions with stringent data protection regulations, where blockchain's immutability must be reconciled with rights such as data correction and erasure (Finck, 2018). Overall, blockchain frameworks provide a robust technological foundation for decentralized energy trading by enabling trusted coordination among distributed actors. Their effectiveness, however, depends on careful architectural choices that account for technical constraints, market design objectives, and regulatory requirements. As decentralized energy markets evolve, blockchain frameworks are likely to adopt hybrid and adaptive models that combine automation with institutional oversight, supporting innovation while ensuring compliance and system stability.

4. TECHNICAL INTEGRATION WITH ENERGY INFRASTRUCTURE

The successful operation of decentralized energy markets depends not only on market design and digital platforms but also on their seamless technical integration with existing energy infrastructure. Unlike conventional electricity markets, where transactions are largely abstracted from physical grid operations, decentralized markets are tightly coupled with real-time generation, consumption, and network constraints at the distribution level. This integration requires coordination between digital trading platforms, physical energy assets, and grid management systems to ensure reliability, efficiency, and regulatory compliance (Burger et al., 2019). A foundational component of technical integration is the deployment of smart meters and Internet of Things (IoT) devices. Smart meters provide high-resolution, near real-time data on electricity consumption and generation, enabling accurate measurement, verification, and settlement of decentralized energy transactions. IoT-enabled sensors extend this capability by monitoring voltage, frequency, and power quality across the distribution network. These data streams form the empirical basis for peer-to-peer trading, dynamic pricing, and automated settlement mechanisms, allowing decentralized markets to function with a level of granularity that is not feasible in traditional retail markets (Gungor et al., 2013). Data acquisition and validation are critical challenges in decentralized energy systems. Energy transactions must be based on trusted and verifiable measurements to prevent fraud, double counting, or manipulation. Technical integration therefore requires secure communication protocols, standardized data formats, and validation mechanisms that reconcile physical energy flows with digital transaction records. In blockchain-enabled systems, smart meters often act as trusted data oracles, transmitting verified measurements to the trading platform. However, ensuring the integrity of data at the point of collection remains a key vulnerability, as compromised devices can undermine the reliability of the entire market system (Andoni et al., 2019).

Interoperability with distribution networks is another central aspect of technical integration. Decentralized energy trading typically occurs at the distribution level, where networks were historically designed for unidirectional power flows from centralized generators to consumers. High penetration of distributed energy resources introduces bidirectional flows, localized congestion, and voltage fluctuations. Trading platforms must therefore incorporate network constraints into market operations, either by embedding grid-aware pricing mechanisms or by coordinating closely with distribution system operators (DSOs). This integration ensures that market outcomes remain physically feasible and do not compromise grid stability (Mengelkamp et al., 2018). The role of DSOs evolves significantly in technically integrated decentralized markets. Rather than acting solely as passive infrastructure managers, DSOs increasingly function as active coordinators that provide network data, enforce technical constraints, and facilitate local balancing. Advanced distribution management systems (ADMS) and digital twins are often employed to simulate network conditions and assess the impact of decentralized trading on grid performance. Similar FPGA-based adaptive systems have been successfully applied in biomedical signal processing, demonstrating the potential of hardware-accelerated solutions to enhance real-time responsiveness in energy networks (Shanmugaraja et al., 2023). These tools enable DSOs to support market-based coordination while maintaining their core responsibilities for safety, reliability, and quality of supply (Burger et al., 2019). Energy forecasting, demand response, and load balancing mechanisms further enhance the integration of decentralized markets with physical infrastructure. Accurate forecasting of local generation and demand is essential for efficient market clearing and congestion management. Machine learning and data analytics are increasingly used to predict consumption patterns and renewable generation variability. Comparable predictive analytics frameworks in wind turbine maintenance illustrate how AI can enhance reliability and optimize decentralized energy systems, reinforcing the importance of intelligent forecasting in energy markets (Cloudin et al., 2026). Demand response programs, enabled through automated control systems, allow flexible loads to adjust consumption in response to price signals or grid conditions, thereby aligning market incentives with system needs (Parag & Sovacool, 2016). These mechanisms reduce the need for costly network reinforcements and enhance the overall efficiency of decentralized energy systems. Latency and real-time responsiveness pose additional technical challenges. Electricity systems operate on sub-second timescales, whereas many digital transaction platforms are designed for slower financial settlements. Decentralized energy markets must therefore carefully define the temporal scope of trading activities, often separating real-time grid control from near real-time or periodic market settlements. Hybrid architectures, in which blockchain or market platforms handle financial and contractual settlement while conventional control systems manage real-time grid operations, are commonly proposed to address this mismatch (Andoni et al., 2019).

Cybersecurity and system reliability are integral to technical integration. The increased connectivity and digitalization of energy infrastructure expand the attack surface for cyber threats. Secure identity management, encryption, and access control mechanisms are essential to protect both operational systems and market platforms. Moreover, resilience strategies such as redundancy, fail-safe mechanisms, and islanding capabilities in microgrids enhance the ability of decentralized systems to withstand technical failures or external disruptions (Gungor et al., 2013). Overall, technical integration represents a critical bridge between decentralized market concepts and real-world energy systems. Effective integration requires aligning digital trading mechanisms with physical grid constraints, operational practices, and reliability standards. As decentralized energy markets scale, continued innovation in smart grid technologies, data management, and system coordination will be essential to ensure that market-based decentralization enhances, rather than undermines, the stability and performance of electricity infrastructure.

5. REGULATORY PATHWAYS FOR DECENTRALIZED ENERGY MARKETS

The proliferation of decentralized energy markets has introduced complex regulatory challenges that require a reassessment of traditional electricity governance frameworks. Conventional regulatory models are designed for centralized electricity systems, where market participants are largely segregated into producers, distributors, and consumers, and where market rules, grid codes, and licensing regimes are tailored to large-scale generation. In contrast, decentralized markets involve heterogeneous participants, bidirectional energy flows, and dynamic transactions that blur the boundaries between producers and consumers, necessitating new regulatory pathways to ensure legal certainty, grid reliability, and consumer protection (Lowitzsch et al., 2020; Parag & Sovacool, 2016). A fundamental regulatory consideration is the legal classification of peer-to-peer energy trading activities. In many jurisdictions, small-scale prosumers are not traditionally recognized as market participants, and their ability to sell or trade electricity is often limited by licensing requirements. Regulatory frameworks must therefore define the conditions under which prosumers can engage in market transactions, the permissible scale of such activities, and the rights and responsibilities associated with generation, consumption, and trading (Burger et al., 2019). Insights from crisis-driven healthcare equity frameworks highlight the importance of inclusive governance, suggesting that decentralized energy regulations must also prioritize accessibility and resilience for vulnerable communities (Ponkarthika et al., 2026). Clarifying these legal roles is essential for both market efficiency and compliance, as ambiguity can lead to disputes, market distortions, and barriers to participation. Licensing, market access, and participation requirements are critical elements of regulatory design. Regulators may implement tiered licensing schemes or exemptions for small-scale prosumers to reduce administrative burdens while maintaining oversight. Market participation rules often specify technical qualifications, data reporting obligations, and adherence to grid codes to ensure that decentralized trading does not compromise system stability (Tushar et al., 2020). Additionally, regulatory mechanisms such as caps on transaction volumes or geographic restrictions can be used to mitigate risks associated with network congestion or market fragmentation.

Regulatory sandboxes and experimental frameworks have emerged as practical tools to foster innovation in decentralized energy markets. Sandboxes allow startups, community energy projects, and technology providers to test new trading platforms, business models, and blockchain applications under regulatory supervision without being fully subject to existing market rules. Such approaches facilitate iterative learning, reduce entry barriers, and generate empirical insights for policy development, while protecting consumers and the grid from potential risks (Andoni et al., 2019). Examples include pilot programs in Europe and North America where blockchain-enabled P2P trading has been trialed under controlled conditions. Comparative analysis of regulatory approaches across jurisdictions reveals significant diversity in how decentralized energy markets are governed. European Union directives, such as the Clean Energy Package, explicitly recognize energy communities and prosumer rights, emphasizing social inclusion, environmental objectives, and participatory governance (Lowitzsch et al., 2020). In contrast, the regulatory landscape in many U.S. states varies widely, with some jurisdictions enabling community solar programs and P2P trading, while others maintain restrictive licensing and interconnection rules. Understanding these variations is critical for designing interoperable platforms and for anticipating legal compliance requirements in cross-border energy trading scenarios. The role of regulators and standard-setting bodies extends beyond market access and licensing to encompass grid stability, data governance, and financial accountability. Regulators must ensure that decentralized energy trading does not compromise reliability by requiring adherence to technical standards, coordination with distribution system operators, and contingency planning. Emerging scholarship also emphasizes the role of intercultural communication and organizational agility in shaping regulatory adaptation, with conversational AI and digital transformation frameworks offering new pathways for compliance and stakeholder engagement (Santhini et al., 2025; Ruba et al., 2026). Data governance regulations, including privacy and cybersecurity requirements, impose additional compliance obligations on platform operators and prosumers, particularly when blockchain or IoT technologies are deployed (Finck, 2018). Furthermore, financial oversight mechanisms, including reporting and settlement standards, are necessary to prevent fraud, double counting, and market manipulation. In addition to formal regulation, self-regulatory mechanisms and industry standards play a complementary role. Industry consortia, professional associations, and certification schemes can provide frameworks for technical interoperability, smart contract auditing, and cybersecurity compliance. Such initiatives are particularly relevant in emerging markets where statutory regulation may lag technological innovation. By combining formal and informal regulatory approaches, decentralized energy markets can achieve a balance between innovation, consumer protection, and system reliability. Overall, regulatory pathways for decentralized energy markets require a nuanced approach that reconciles technological innovation with legal, technical, and social objectives. Effective regulation should clarify the roles of prosumers, aggregators, and distribution system operators, facilitate market participation, ensure adherence to technical and data standards, and provide mechanisms for innovation under controlled conditions. Comparative insights and adaptive regulatory strategies are essential for creating a legally compliant and operationally efficient decentralized energy ecosystem.

6. COMPLIANCE MODELS AND LEGAL RISK MANAGEMENT

Compliance and legal risk management are central to the sustainable operation of decentralized energy markets. Unlike traditional electricity markets, decentralized systems involve a diverse set of actors—prosumers, aggregators, and technology providers—operating within dynamically evolving regulatory and technological environments. The complexity of these interactions necessitates robust compliance frameworks that ensure adherence to legal, technical, and financial obligations while enabling innovation (Finck, 2018; Andoni et al., 2019). A primary compliance concern arises from the enforceability of smart contracts. Smart contracts automate transactions and settlements in decentralized energy markets, executing predefined rules once specified conditions are met. However, the legal recognition of these digital contracts varies across jurisdictions. Issues such as liability for contract failures, dispute resolution, and alignment with statutory consumer protection provisions must be addressed to ensure that automated mechanisms do not operate outside the bounds of law (Peters & Panayi, 2016). Regulatory guidance or standardization of smart contract formats can mitigate these risks and provide legal clarity for all participants. Consumer protection represents another critical dimension of compliance. Decentralized energy markets expose participants to risks including misinformation, pricing volatility, and technical failures. Regulations must ensure transparent pricing, accurate billing, and recourse mechanisms for consumers in case of disputes. Provisions for vulnerable or energy-poor households are particularly important to prevent exclusion and inequitable outcomes. Digital platforms can support compliance by providing real-time transaction visibility, audit trails, and automated alerts for irregularities (Lowitzsch et al., 2020).

Data privacy and cybersecurity are increasingly central to compliance in decentralized energy markets. The collection, transmission, and storage of granular consumption and generation data raise significant privacy concerns. Blockchain and IoT technologies, while enhancing transparency and automation, introduce potential risks for unauthorized access, identity theft, or misuse of sensitive information. Legal frameworks such as the General Data Protection Regulation (GDPR) in Europe impose strict requirements on data processing, necessitating technical measures like encryption, pseudonymization, and controlled access (Finck, 2018). Cybersecurity compliance also involves regular audits, penetration testing, and adherence to industry standards to ensure system resilience against cyber threats. Grid security and operational compliance constitute additional layers of risk management. Decentralized trading may lead to local congestion, voltage fluctuations, or imbalance between supply and demand if not properly coordinated with distribution system operators (DSOs). Compliance frameworks require adherence to grid codes, operational standards, and reporting obligations, often enforced through monitoring systems and automated control mechanisms. Aggregators often play a pivotal role in ensuring that individual prosumer activities collectively comply with network constraints, acting as intermediaries between market operations and physical grid stability (Burger et al., 2019). Financial compliance is also significant, particularly in tokenized energy markets. Energy tokens representing electricity, renewable certificates, or other tradable assets may fall under securities, commodity, or financial regulations depending on jurisdiction. Platforms must ensure accurate accounting, prevent double counting, and comply with anti-money laundering (AML) and know-your-customer (KYC) obligations. Failure to comply can result in legal sanctions, reputational damage, and systemic market risks (Peters & Panayi, 2016).

Self-regulatory initiatives complement formal compliance frameworks. Industry standards, certification schemes, and best practices for blockchain operations, cybersecurity, and smart contract auditing help reduce regulatory uncertainty and provide a foundation for trust among participants. Collaborative efforts among regulators, industry actors, and technology providers are essential to harmonize compliance approaches and develop adaptive governance structures capable of responding to emerging risks. In summary, compliance models in decentralized energy markets must address legal, technical, financial, and operational risks. A multi-layered approach that integrates smart contract enforceability, consumer protection, data privacy, grid security, and financial regulation is essential to enable sustainable market participation. By aligning technological innovation with robust compliance frameworks, decentralized energy markets can achieve both operational efficiency and regulatory legitimacy, fostering broader adoption and long-term stability.

7. ECONOMIC, SOCIAL, AND ENVIRONMENTAL IMPLICATIONS

Decentralized energy markets carry profound economic, social, and environmental implications, reflecting their potential to reshape traditional energy systems while promoting sustainability and inclusivity. Economically, these markets introduce novel value creation mechanisms by enabling prosumers to monetize their distributed energy resources through peer-to-peer trading, local energy communities, and microgrids. By reducing dependency on centralized generators and facilitating localized balancing of supply and demand, decentralized markets can lower overall system costs, minimize transmission losses, and enhance energy efficiency (Burger et al., 2019; Mengelkamp et al., 2018). Tokenization of energy assets and smart contract-enabled automated settlements also reduce transaction costs, improve liquidity, and support participation by smaller actors who were traditionally excluded from energy markets (Peters & Panayi, 2016). From a social perspective, decentralized energy markets empower communities and individual participants, promoting inclusivity, autonomy, and energy literacy. Local energy communities and microgrids enable collective decision-making, giving consumers and prosumers a direct voice in energy management. Such arrangements can strengthen community cohesion and support energy access in remote or underserved areas. Moreover, decentralized markets can facilitate equitable participation by reducing the scale and cost barriers associated with traditional energy investments, ensuring that benefits are distributed across a broader population (Lowitzsch et al., 2020). These social benefits are closely tied to regulatory frameworks that prioritize fairness, consumer protection, and participatory governance.

Environmentally, decentralized energy markets offer significant opportunities for decarbonization and renewable energy integration. By promoting local generation from solar, wind, and other low-carbon sources, these markets reduce reliance on fossil-fuel-based centralized generation and decrease greenhouse gas emissions. Dynamic pricing mechanisms, demand response, and load balancing can further optimize renewable energy utilization, mitigating curtailment and aligning consumption with periods of high renewable output (Parag & Sovacool, 2016). The ability to measure, verify, and trade renewable energy certificates via blockchain enhances transparency and accountability in environmental claims, supporting corporate sustainability goals and policy compliance. However, economic and environmental benefits are accompanied by potential challenges. Market fragmentation, inconsistent regulatory frameworks, and technical limitations in grid integration may limit the scalability and efficiency of decentralized energy systems. Furthermore, social risks such as the digital divide, technological literacy gaps, and inequitable access to distributed energy resources may exacerbate existing inequalities if not adequately addressed (Tushar et al., 2020). Policymakers and platform designers must therefore integrate social equity considerations into market design, ensuring that technological innovation does not lead to unintended exclusion or disadvantage for vulnerable populations. Overall, the economic, social, and environmental implications of decentralized energy markets are interdependent. By fostering efficient resource allocation, community participation, and renewable integration, these markets can contribute to sustainable energy transitions. The realization of these benefits depends on harmonized technical design, robust regulatory frameworks, and inclusive governance models that collectively support scalability, reliability, and equitable access.

8. CHALLENGES, LIMITATIONS, AND FUTURE DIRECTIONS

Decentralized energy markets represent a transformative paradigm in energy systems, promising enhanced efficiency, resilience, and consumer empowerment. By enabling peer-to-peer (P2P) trading, integrating distributed energy resources (DERs), and leveraging blockchain-based platforms, these markets can fundamentally reshape electricity generation, distribution, and consumption. However, despite their significant potential, the broad-scale implementation of decentralized energy markets faces multifaceted challenges and limitations spanning technical, regulatory, operational, and socio-economic dimensions. Addressing these barriers is essential to ensure that such markets are not only technically feasible but also legally compliant, economically viable, and socially inclusive. A primary challenge lies in the technical scalability of decentralized energy systems. Unlike traditional centralized power grids, decentralized markets involve large numbers of DERs, such as rooftop solar panels, small-scale wind turbines, and battery storage units, all connected through smart meters, sensors, and Internet of Things (IoT) devices. These components collectively generate vast volumes of data in real time, which must be efficiently processed to ensure accurate settlement, load balancing, and demand response. While blockchain technology offers a secure and transparent mechanism for recording transactions and automating market operations via smart contracts, it faces inherent limitations in terms of throughput, latency, and energy consumption, particularly in public or large-scale networks. For example, proof-of-work-based blockchains may struggle to handle thousands of micro-transactions per second, leading to performance bottlenecks. To address these constraints, researchers and industry stakeholders are exploring off-chain solutions, layer-two scaling mechanisms, and hybrid blockchain architectures that combine the security of distributed ledgers with the efficiency of conventional databases. Optimized consensus protocols, including proof-of-stake and delegated proof-of-stake models, further enhance scalability while reducing computational overhead (Andoni et al., 2019; Mengelkamp et al., 2018). Another significant limitation pertains to regulatory and legal uncertainty. Existing energy regulations in most countries were designed around centralized electricity generation and wholesale markets, leaving little clarity regarding the participation of prosumers, tokenized energy assets, or peer-to-peer trading schemes. Ambiguities in licensing requirements, grid interconnection standards, taxation policies, and liability frameworks create barriers for market entry, innovation, and investment. In some regions, prosumers may be legally restricted from selling excess energy or may face cumbersome approval processes that reduce the attractiveness of decentralized participation. While regulatory sandboxes, pilot programs, and adaptive policy frameworks can provide temporary relief, comprehensive and harmonized legal clarity is necessary to foster long-term investor confidence and ensure the sustainability of decentralized market models (Lowitzsch et al., 2020; Finck, 2018). To achieve equitable outcomes, policymakers and platform designers must adopt inclusive strategies that lower entry barriers, provide educational support, and ensure affordable access to necessary technologies (Tushar et al., 2020). Looking forward, future directions for decentralized energy markets emphasize technological innovation, regulatory evolution, and hybrid governance approaches. On the technological front, integrating artificial intelligence (AI), machine learning, and predictive analytics with blockchain platforms can enhance demand forecasting, optimize energy dispatch, and enable automated decision-making. Such advancements can also support real-time pricing, dynamic load management, and enhanced energy efficiency. Regulatory evolution is expected to focus on harmonizing rules across jurisdictions, standardizing smart contract frameworks, and establishing clear definitions of prosumer rights and obligations. Regulatory clarity will be critical to attract private investment and ensure consistent operational practices across different markets. Hybrid governance models, which combine automated, blockchain-driven coordination with institutional oversight, offer a pathway to balance efficiency, accountability, and consumer protection. By integrating algorithmic decision-making with regulatory supervision, such models can support innovation while ensuring grid reliability and compliance. Emerging trends also point toward greater integration of renewable energy communities, flexibility markets, and cross-border energy trading, enabled by interoperable blockchain networks. Pilot programs, field trials, and empirical studies will play an essential role in validating technical solutions, assessing socio-economic impacts, and informing future policy and regulatory frameworks. In conclusion, while decentralized energy markets present transformative opportunities for clean, efficient, and participatory energy systems, their success depends on overcoming complex challenges related to scalability, regulation, grid integration, and equity. By advancing technological capabilities, clarifying legal frameworks, and adopting inclusive governance strategies, these markets can evolve into scalable, sustainable, and legally compliant systems that contribute significantly to the global energy transition and democratization of electricity markets. Strategic collaboration among technology developers, policymakers, utilities, and communities will be vital to realizing the full potential of this emerging paradigm (Parag & Sovacool, 2016; Andoni et al., 2019).

9. CONCLUSION

Decentralized energy markets, supported by blockchain and advanced digital technologies, represent a paradigm shift in how electricity is generated, distributed, and consumed. By enabling peer-to-peer trading, local energy communities, and active prosumer participation, these markets challenge traditional centralized electricity models, offering novel opportunities for increased efficiency, resilience, and sustainability. Unlike conventional energy systems, where power generation and consumption are largely unidirectional and centrally managed, decentralized markets allow consumers to become active participants, generating, storing, and trading energy within local and regional networks. This transformation empowers communities, enhances energy accessibility, and fosters engagement in sustainable energy practices. The chapter has examined the conceptual foundations, technological frameworks, regulatory pathways, compliance mechanisms, and socio-economic and environmental implications of decentralized energy trading. At the technological core, blockchain frameworks provide a secure, transparent, and automated infrastructure that enables reliable energy transactions, real-time settlements, and smart contract execution. Tokenization of energy assets and automated settlements facilitate efficient microtransactions and innovative market designs, while integration with smart meters, IoT devices, and distribution networks ensures operational alignment with physical grid constraints and reliability requirements. These technological advancements not only enhance operational efficiency but also increase trust among market participants by providing auditable records and reducing reliance on intermediaries. Regulatory pathways for decentralized energy markets are evolving to balance innovation with system stability and consumer protection. Adaptive policies, regulatory sandboxes, and legal clarifications are gradually accommodating new market actors such as prosumers, aggregators, and community energy organizations. Compliance frameworks are critical in this context, addressing smart contract enforceability, data privacy, cybersecurity, grid security, and financial regulations. These measures ensure that decentralized trading is conducted within a robust legal and operational environment, fostering investor confidence and promoting wider adoption. The economic, social, and environmental impacts of decentralized energy markets are substantial. Economically, they can enhance local value creation, reduce energy costs, and incentivize renewable energy deployment. Socially, they encourage community engagement, inclusivity, and energy literacy, although attention must be given to prevent digital and financial inequalities that may exclude vulnerable populations. Environmentally, decentralized trading promotes higher integration of renewable sources, contributing to carbon emission reduction and sustainable energy transitions. Nonetheless, challenges such as market fragmentation, scalability limitations, regulatory uncertainty, and operational complexities must be addressed to realize these benefits fully. Future directions for decentralized energy markets emphasize the integration of emerging technologies, including artificial intelligence, machine learning, and predictive analytics, with blockchain-enabled platforms to optimize demand response, forecasting, and automated decision-making. Hybrid governance models, which combine decentralized, automated coordination with institutional oversight, offer a pathway to balance efficiency, accountability, and reliability. Cross-border energy trading, flexibility markets, and renewable energy communities are likely to become more prominent as interoperable frameworks and standardized protocols mature. In conclusion, decentralized energy markets exemplify the convergence of technological innovation, regulatory adaptation, and socio-environmental objectives. Their long-term success depends on harmonizing market design, technical infrastructure, legal compliance, and social equity considerations. With coordinated policy support, inclusive participation, and robust governance mechanisms, blockchain-enabled decentralized energy markets have the potential to become scalable, sustainable, and legally compliant systems, significantly advancing the global energy transition and shaping the future of electricity systems worldwide.

REFERENCES

Andoni, M., Robu, V., Flynn, D., Abram, S., Geach, D., Jenkins, D., McCallum, P., & Peacock, A. (2019). Blockchain technology in the energy sector: A systematic review of challenges and opportunities. *Renewable & Sustainable Energy Reviews, 100*, 143–174. DOI: 10.1016/j.rser.2018.10.014

Burger, S. P., Jenkins, J. D., Batlle, C., & Pérez-Arriaga, I. J. (2019). Restructuring revisited: Competition and coordination in electricity distribution systems. *Energy Journal, 40*(3), 1–33. DOI: 10.5547/01956574.40.3.sbur

Cloudin, S., Priya, R. S., Prabhu, G. A., Ruba, M., Joel, R., & Vanitha, R. (2026). AI-Enhanced Predictive Analytics and Reliability Management for Wind Turbine Maintenance. In AI-Powered Analysis, Modeling, and Monitoring of Wind Energy Systems (pp. 191-218). IGI Global Scientific Publishing.

De Filippi, P., & Wright, A. (2018). *Blockchain and the law: The rule of code.* Harvard University Press.

Dhanushya, J., Ruba, M., Ponkarthika, M., Soundarya, B., Sakthi Priya, V., Dhanushri, S., & Harikrishnan, S. (2025, April). AI-Powered Multipurpose Agrobot for Precision Farming: Crop Health Management, Fertilizer Recommendation and Yield Prediction. In *2025 International Conference on Recent Advances in Electrical, Electronics, Ubiquitous Communication, and Computational Intelligence (RAEEUCCI)* (pp. 1-6). IEEE.

Finck, M. (2018). *Blockchain and the General Data Protection Regulation: Can distributed ledgers be squared with European data protection law?* European Parliamentary Research Service. DOI: 10.1017/9781108609708.004

Gowthaman, N., Ruba, M., Aswath, S., Hemavarshni, S. N., Bragadeshwaran, R., & Nithish, S. J. (2025, November). Iot Integrated Solution For Smart Vertical Crop Cultivation. In 2025 IEEE 7th International Conference on Computing, Communication and Automation (ICCCA) (pp. 1-6). IEEE.

Gungor, V. C., Sahin, D., Kocak, T., Ergut, S., Buccella, C., Cecati, C., & Hancke, G. P. (2013). Smart grid technologies: Communication technologies and standards. *IEEE Transactions on Industrial Informatics, 7*(4), 529–539. DOI: 10.1109/TII.2011.2166794

Kameswaran, A., Nanthakumar, K., & Ruba, M. (2017). High performance precomputation based self-controlled precharge-free content-addressable memory. *Journal of Electronics and Communication Systems, 2*, 1–9.

Krithika, S., Jayachandran, T., Raghapriya, N. R., Ruba, M., Janani, C., & Sathiyavathi, S. (2025, April). An Effective Denial of Service Texture Features from Malware Grayscale Image. In 2025 3rd International Conference on Artificial Intelligence and Machine Learning Applications Theme: Healthcare and Internet of Things (AIMLA) (pp. 1-7). IEEE. DOI: 10.1109/AIMLA63829.2025.11041501

Littlechild, S. (2021). Decentralization, competition, and innovation in electricity markets. *Energy Policy*, *148*, 111968. DOI: 10.1016/j.enpol.2020.111968

Lowitzsch, J., Hoicka, C. E., & van Tulder, F. J. (2020). Renewable energy communities under the 2019 European Clean Energy Package. *Energy Policy*, *139*, 111212. DOI: 10.1016/j.enpol.2020.111212

Mengelkamp, E., Gärttner, J., Rock, K., Kessler, S., Orsini, L., & Weinhardt, C. (2018). Designing microgrid energy markets: A case study: The Brooklyn Microgrid. *Applied Energy*, *210*, 870–880. DOI: 10.1016/j.apenergy.2017.06.054

Parag, Y., & Sovacool, B. K. (2016). Electricity market design for the prosumer era. *Nature Energy*, *1*(4), 16032. DOI: 10.1038/nenergy.2016.32

Peters, G. W., & Panayi, E. (2016). Understanding modern banking ledgers through blockchain technologies: Future of transaction processing and smart contracts on the internet of money. In *Banking beyond banks and money* (pp. 239–278). Springer., DOI: 10.1007/978-3-319-42448-4_13

Ponkarthika, M., Prabhu, A., Ruba, M., Kishore, Y. R., & GV, M. (2026). Bridging Health Equity in Times of Crisis: Ensuring Inclusive Care During Emergencies and Disasters. In Advancing Equity in Global Healthcare Systems (pp. 27-54). IGI Global Scientific Publishing.

Poovithaa, M., Ruba, M., Ponkarthika, M., Ragul, N., Ragul, V., & Sangeetha, K. (2025, November). IoT Smart Traffic Control System for Hill Roads with Emergency Vehicle Priority. In 2025 IEEE 7th International Conference on Computing, Communication and Automation (ICCCA) (pp. 1-6). IEEE.

Ruba, M., Ashwin Prabhu, G., Robinson Joel, M., Vishnu Priya, K., Dharshani, L., & Dharani, R. (2026). Integrating emerging technologies for organizational agility: An interdisciplinary approach to digital transformation. In Roache, D. (Ed.), *High performance leadership for organizational excellence: Innovation, technology, and resilience* (pp. 63–92). IGI Global Scientific Publishing., DOI: 10.4018/979-8-3373-6207-6.ch003

Santhini, C., Ruba, M., Sakthi Priya, V., & Hemalatha, R. (2025). Intercultural communication and assessment practices in the integration of AI with education. In *Generative AI Approaches to Sustainable Development in Higher Education*. IGI Global. DOI: 10.4018/979-8-3693-5623-4.ch010

Shanmugaraja, T., Ruba, M., Priya, V. S., Venkatesh, T., Supriya, M., & Nallamuthu, M. (2023, September). ECG noise elimination using an FPGA version of a modified error normalized LMS adaptive filter. In AIP Conference Proceedings (Vol. 2764, No. 1, p. 060033). AIP Publishing LLC. DOI: 10.1063/5.0144367

Sibiya, S., Ruba, M., Ponkarthika, M., Vijayasabariswari, M., Sheba, S., Vignesh, S., . . . Sankar, S. S. (2025, October). Integrated Wearable Health Monitoring and Emergency Response System With IOT Sensors, AED and Mobile App for Enhanced Patient Outcomes. In 2025 IEEE 2nd International Conference on Green Industrial Electronics and Sustainable Technologies (GIEST) (pp. 1-6). IEEE.

Tushar, W., Yuen, C., Mohsenian-Rad, H., Saha, T. K., Wood, K. L., & Poor, H. V. (2020). Transforming energy networks via peer-to-peer energy trading: The potential of game-theoretic approaches. *IEEE Signal Processing Magazine*, *35*(4), 90–111. DOI: 10.1109/MSP.2018.2818327

Chapter 4
Regulatory Architecture for Blockchain Energy Trading:
Navigating Legal Frameworks for Decentralized Energy Markets

Manoj Govindaraj
http://orcid.org/0000-0003-2830-7875
Vel Tech Rangarajan Dr. Sagunthala R&D Institute of Science and Technology, India

Jayendira P. Sankar
http://orcid.org/0000-0001-8435-2123
University of Technology Bahrain, Salmabad, Bahrain

N. Mari Anand
http://orcid.org/0000-0002-6443-249X
University of Technology and Applied Sciences, Salalah, Oman

Chandramowleeswaran Gnanasekaran
http://orcid.org/0000-0002-1293-1043
Vel Tech Rangarajan Dr. Sagunthala R&D Institute of Science and Technology, India

Jenifer Lawrence
http://orcid.org/0000-0002-4115-1521
Woldia University, Ethiopia

DOI: 10.4018/979-8-3373-5742-3.ch004

ABSTRACT

Blockchain technology is rapidly transforming the energy sector by enabling decentralized, peer-to-peer energy trading systems that challenge traditional regulatory models. This chapter critically examines the existing legal frameworks governing energy markets and explores how they must evolve to accommodate the unique demands of blockchain-based energy trading. It analyzes the key legal and policy challenges posed by decentralization, including jurisdictional complexities, consumer protection, data privacy, liability, and smart contract enforceability. With the emergence of prosumers—individuals who both produce and consume energy—the chapter delves into questions of legal status, licensing, and equitable grid access. It also discusses the implications of tokenized energy assets, the interoperability of technical and regulatory standards, and the legal recognition of consensus mechanisms used in blockchain systems.

INTRODUCTION

The world energy scene has now entered into a sea change because of the twin forces of decarbonization and digitalization. With the fast-rising renewable energy resources and the close involvement of consumers in the production of energy, the centralized model of energy production and distribution is gradually making way to the decentralized, flexible and participatory models. At the center of this transition is the blockchain technology, which provides a secure, decentral, transparent infrastructure of peer-to-peer (P2P) energy trading. Using blockchain-based platforms, individual energy consumers who now are often termed prosumers because they can create as well as consume electricity can directly exchange surplus electricity with each other without going through the conventional energy intermediaries. This innovation does not only democratize the access to the energy, but it creates efficiency and sustainability and resilience of local energy.

Nevertheless, the opportunity to disrupt the energy trade with the help of blockchain creates new challenges to software and hardware. Many of the traditional regimes do not have the capacity to adapt distributed ledger technologies (DLTs), tokenized energy assets, and autonomous smart contracts, which do not respect jurisdictional boundaries. There is consequently a mounting urgency to reengineer regulatory fabrications that can effectively deal with the multifaceted nature of systems to decentralized energy as well as with enshrining fundamental standards, like consumer protection, system reliability, equity, and legal certainty.

The distribution of the legal responsibility and liability of different stakeholders can be seen as one of the central regulatory problems in blockchain-based energy trading. Energy markets that are prevalent are the traditional markets that are guided by the black and white relationships that exist between the producers, suppliers, distributors, and consumers. Conversely, blockchain systems frequently entail a dynamic and fluid net of participants, an automated decision-making process, and smart contracts with algorithms, as it clouds the boundaries of accountability. The question of the persons owning responsibility in case of the contractual failure, system malfunction, or cyber-attacks in the decentralized environment is the issue that legal systems have to address.

In addition, the advent of prosumers also creates another legal subject whose activity is not properly regulated in the current licensing and regulatory framework. Can prosumers be regarded as a utility, a micro-operator and end-users? Are they the same as larger energy suppliers in that they should face the same level of compliance burdens or alternate regulatory practice be considered? These questions are not just academics as they play a major role in market entry, costs structures and fair grid participation. Offering legal status to prosumers and the establishment of proportional or scaled down licensing platforms is therefore necessary in promoting inclusion in the blockchain-enabled energy system. A different field of legal ambiguity touches upon the validity of smart contracts self-executing digital contracts coded on the blockchain. Whereas smart contracts are efficient in operations and trusted transactions, they have been known to lack subtle interpretation and regulation by the law that is evident with normal contracts. The relationship between such digitally nowable agreements and the established legal doctrines of contracting, dispute resolution and consumer protection is only starting to be considered by courts and regulators.

Moreover, confidentiality and cybersecurity is especially important in blockchain energy trading where consumption data and real-time transactions are continuously shared and stored. The immutable and decentralized nature of blockchain makes it complicated to adhere to data protection regulations, including the General Data Protection Regulation (GDPR) that is used in the EU. The regulatory paradox between blockchain vibrancy and the right to be forgotten begs a serious consideration to guarantee that regulations should adequately be resolved. To address these complex issues, jurisdictions across the world are becoming experimental with adaptive modalities of regulatory frameworks like the regulatory sandboxes that enable a test-and-learn approach where blockchain applications can be tested in a conditional environment with a controlling regulatory oversight. The comparative analysis demonstrates the variety of the approaches that are applied within the EU, the US, and Asia-Pacific and that is caused by different legal tradition, the structure of energy market, and priorities in policy topics. The global experiments can help in giving answers to the design of regulatory frameworks that are both innovation-friendly and fits into the objectives of the public interest.

The purpose of this chapter is to critically analyze how the existing energy legislation would have to be amended to allow blockchain energy trade. It proposes an agile, adaptive, principles-oriented framework of regulation, which promotes innovation and protects legal responsibility, energy fairness and climate compliance. The chapter fills the gap between technological possibilities and normative legal practices to understand in a better way the navigability to the future of decentralized energy markets.

Although the potential of energy trading based on blockchain has a significant technological prospect, a regulatory paradox is created at the same time. The decentralised platforms are redistributing market power among prosumers and automate transactions using smart contracts, but extant energy law is still on the model of centralised licensing regimes, territorially oriented governance and well-known intermediaries. This structural incongruity results in confusion around the issue of liability distribution, the enforceability of smart-contracts, the type of tokens, and the issue of data-protection regulation. Specifically, the security of the blockchain records is a complication to privacy regulations like the GDPR, and tokenised kilowatt-hour transactions blur the line between energy legislation and financial regulation.

The main issue, accordingly, is not whether blockchain technologies are viable to design peer-to-peer energy markets; it is their ability to make the existing jurisprudential frameworks re-calibrate - or, indeed, re-engineer wholesale - in order to effectively govern the emerging decentralized, automated, and transnational energy systems without undermining the foundations of accountability, consumer protection, and the quality of the electric grid. This problem requires a delicate interpretation of the interplay between regulatory regimes and one that is coming up with new technologies. This chapter argues that the paradox can be resolved through the need to have a flexible, principle-driven regulatory framework that can be used to harmonize the energy law, digital governance and financial regulation in a harmonized institutional framework. This kind of architecture should be knowledgeable of the multidisciplinary scholarship and sensitive to the realities of the dynamic nature of decentralised energy markets.

BACKGROUND OF THE STUDY

They conduct a systematical review on P2P energy trading literature, market structure development, bidding strategy collocation, technology facilitation. It groups peer-to-peer schemes i.e. one-to-one, community, and hybrid schemes and assesses the techniques to a solution i.e. game theory, optimization and machine learning. The paper is concluded with the determination of existing limitations and proposes the further research. This review is extensive in that it summarizes the contexts of the whole research landscape defining P2P market mechanisms as well as their underlying computations. It is especially useful to regulatory architecture research, which can expose the interaction of trading strategies on the prosumer behavior and market design. The argument about hybridised models and method of optimisation reveals how the system of law could accommodate various interactions between participants, and ensure fairness in the market. It provides the threshold of knowledge of technical and economic motivation of decentralized energy systems (Islam, 2024).

They examines the recent developments in the P2P power systems in the setting of the community which include the technological developments, market dynamics as well as challenges like scalability, trust, and regulatory hurdles. It summarizes both scholarly and pilot case studies, and designates operational and policy gaps due to which large-scale adoption is not possible. This paper identifies technological as well as regulatory challenges on decentralized community energy models. Legal-framework research, its findings with regard to issues of trust, interoperability, and governance are of interest. It focuses on scalability and inter-jurisdictional issues, which reflect the legal issues of the access of the grid, prosumer licenses, and data compliance. Comparative context shown by the pilot helps to see how the alternative designs align under the current rules, which will be useful in proposing flexible, sandbox-enabled law (Tushar et al., 2021).

It provides an editorial overview of blockchain-based integration of multi-energy systems (electricity, heat, gas). They talk about how the use of smart contracts, consensus protocols and IoT-integration performs automation, possibilities of multi-vector prosumers and the use of AI to optimise the scheduling and trading of energy vectors. In this paper, blockchain energy trading is embedded in the context of multi-energy systems, where the issues of legal and regulatory relevance are taking center stage particularly in intersecting areas (electricity, gas, and heat). It offers an intellectual point of support about the potential prosumer conceptualization within the context of legally explicit and the idea that technical standards-based energy interoperability might have within the establishment of regulatory standards. Privacy and data-protection requirements forming a key focus of legal architecture design involve the AI and IoT (Hua, Luo, Du, & others, 2022).

Inefficiency, high cost of transactions, privacy/security issue in blockchain-based energy trading is the result of such survey. They categorise the schemes that exist into transaction design-oriented categories, consensus mechanism oriented, and system optimised categories. The article summarizes technical suggestions to address those drawbacks and proposes directions of future research. Despite being presented as preprint, the survey continues to be influential as a source of essential technological constraints that overlap with legal feasibility (latency, privacy and cost trade-offs). Become aware of these limitations to determine whether current regulation is appropriate or must be adjusted. As an example, feasibility of tokenized micro-transactions could be limited by the high transaction cost problem. It assists in knowing where regulation is to be backed by technical development (Wang et al., 2019).

It suggests a smart-contract energy trading scheme that will be applicable in a smart grid to electric vehicles (EVs). They propose an incentive mechanism of credit points tied with a smart contracts and employ a type of double auction to determine the participants of the trading. The results of simulation suggest better security, fairness and efficiency of the trading process that is in line with node reputation. The paper offers a realistic example of tokenized energy trading of EVs based on smart-contract logic, where the mechanisms of an incentive and auction system are activated. Regulatory viewpoints an example of the way legal frameworks must factor in dynamic eligibility, fraud prevention, and liability distribution is the credit based system of operation. It mediates between the enforceability of conventional technical contracts and trust-based systems and will influence legal architecture design in terms of automated bilateral transactions and participants screening (Xue et al., 2022).

They suggest a P2P energy trading model ruled with a decentralized system that highly employs agent reputation, electro-distance pricing, and an Anonymous Proof-of-Location algorithm to maintain privacy. Partner selection and market settlements are carried out proximity-wise and reputation-sensitive. Feasibility is shown by analysis and case studies of security and privacy. This is an example of a technical architecture, which shows how location restraint and identity protection may be incorporated into trading systems which are essential to the legal implications relating to jurisdictional authority, information privacy and grid parity. Geofencing, local licensing and anonymized participation may be one of the rules that regulators would require. The A-PoL algorithm begs legal issues of; Identity verification vs right to privacy. The present paper will give a practical design guide to regulatory adaptation to location-based decentralized trading (Khorasany, Dorri, Razzaghi, & Jurdak, 2020).

This paper shows how Ethereum-smart-contract can be used to implement a double auction mechanism of P2P energy trading and spinning reserves. Otherwise, sellers and buyers place bids/offers through Solidity contracts; spinning reserves may be ordered automatically, as well. Extensive testing is achieved by simulating a scenario to prove the auction logic, payments and transparent or efficient trades of energy and decentralized trading. The paper is an actual and viable parade of actuality of smart-contract in energy markets in how intelligent contracts can be deployed to automate decentralized auction, reserve mechanism and management. It can also help in regulatory analysis to indicate how auction and payment settlement logic can be encoded into code-increasing issues of enforceability, dispute resolution and liability. To the regulators, an appreciation of the nature of such implementations would be used in determining areas where legal supervision needs to be exerted and the application of traditional market regulation to code (Energies, 2022).

A solar blockchain-ML electric architecture P2P energy auctions. They embed weather informed ML forecasting into smart contracts in order to optimize the trading opportunity. The system has been tested on real-generation and weather data and perform well in terms of better auction efficiency and satisfaction of the participants. The paper presents the idea of how technical stacks can raise decentralized trading when it comes to hybrid technical stacks. Policy wise, it emphasizes the need of transparency and auditability of algorithms when using automated markets. The regulators ought to think about protocols with regard to ML-themed smart contract, inclination, fairness, and supervision so as to reinforce the relevance of the market integrity and guarantee consumer security in tokenized energy models (Rahman, Chowdhury, Shorfuzzaman, Hossain, & Hammoudeh, 2023).

They devises a game of dynamic pricing under a P2P microgrid based on a block chain. Their model of double auction will balance out the supply-demand ratios as well as satisfaction with users. Enhanced pricing efficiency, fairness in the market and the system performance, as achieved by the simulations, is confirmed. The regulatory concerns of the transparent pricing models in the decentralized energy markets are directly affecting the subject matter of this article. It aids regulatory regimes that enforce use of game theory-based pricing algorithms and non-obscure audit trails. Regulators can use such techniques to determine the sufficiency of pricing mechanisms to establish fair access and consumer welfare within independent energy systems (Dong, Ning, Zhao, Song, et al., 2025).

Blockchain trilemma The trilemma consists of three elements: scalability security and decentralization. Their empirical model takes Ethereum-based data and introduces a layer two scaling solution to increase throughput. Going straight to the legal issues, the work shows that technical scalability affects the ability to comply (speed of transactions), grid stability, and consumer protection. It highlights the necessity to regulate to necessitate performance benchmark of blockchain infrastructure in order to protect the reliability of the system, and confidence of the participants (Wongthongtham et al., 2021).

Incorporates a loss-aware, Lyapunov-based scheme into smart contracts so as to reduce the transmission loss of grid due to P2P trades among microgrids. Their Ethereum has implemented 25 per cent less power wastage and 50 per cent more profits to prosumers. The optimisation of a physical layer grid integrated into block chain is the business that has direct regulatory impact in terms of energy justice, efficiency, and tariff. It implies that the regulator ought to acknowledge and assist algorithms that contain network-cost restoration and the encouragement of effective local trading an engagement between technical excellence and legal frameworks (Zhu, Salam, et al., 2024).

They suggested a P2P energy trading system that operates locally and preserves privacy based on a consensus approach based on IOTA and anonymous proof-of-place. They are a combination of proximity pricing and reputation scales and display good privacy and low overhead in testing. This research contributes to debates over state enacted local licensing and privacy legislation. As its design clarifies, it is the balancing of geofenced participation and data rights which is important to grid equity and GDPR/adoption. It implies that location constraints and the reputation control in decentralizing energy networks can and should be formalized into policy frameworks (Rahimi, Khorasany, Dorri, & Jurdak, 2022).

They proposes an IBM Hyperledger-based solution to two phase crowdsourced energy management and real-time smart contracts within microgrids, which allows day-ahead scheduling, as well as realtime trading among prosumers. Their business model expresses how regulators can employ blockchain in determining schedules in grid routine. It gives a clue as to how the balance between centralized infrastructure control with decentralized crowd-sensitive transactions can be achieved: contributing to the development of hybrid regulatory frameworks and licensing regulations regarding prosumer operators (Wang, Taher, Wang, Kvaternik, & Hahn, 2019).

The authors introduce a credit based reputation-based system of enacting a smart-contract charging/trading of an EV based on double auctions. They are even fairer, more secure and more efficient in their trading in integrated rewards simulations. The paper clearly demonstrates the working of tokenized EV energy markets, and highlights some legal aspects of enforcement, legal liability, and credit governance. It references regulatory uncertainty of automatized rights to transaction under automated disputing in the IT-focused decentralized markets which are EV-centered (Xue et al., 2022).

They provide an overview of real-time, prosumer involvement and market flexibility, and renewables integration by blockchain-AI composites, including a variety of use cases, such as P2P trading, energy forecasting, and management. This work interlinks various stream of technologies: GDPR/policy interfaces, grid resilience, and data management by prosumers. Regulators may use it to build cross-sector systems that deal with data interoperability, algorithmic regulation, and cross-vector energy sharing (Hua, Luo, Du, & Qadrdan, 2022). They develop a short term of P2P market trading of wind power with a smart contract, reputation-based incentives and bi-level programming optimization indicating 9.6% increase in producer revenue and 5.6% consumer savings. The focus on volatility and storage economics achieves two things that interconnect technical design and legal issues such as transparency and grid reliability. It facilitates the regulatory framework that incorporates goals in accordance with incentives in decentralized markets and formal trading windows (Iglesias & Guerrero, 2022).

They are incredibly critical about the blockchain P2P trading and the GDPR relationships between immutability and right to erasure, anonymity, governance functions, and its regulatory compliance. This exploration of legal-tech is also directly relevant to your theme of regulatory architecture, when it comes to the need to resolve the co-existence between immutable ledgers and the right to privacy. It has a tendency to trigger legal amendments- it proposes the introduction of the specific data governance, consent, and a hybrid ledger framework in decentralized energy systems (Romano & Marino, 2022).

It offer a P2P energy trading framework of an Ethereum-based decentralized, transparent P2P energy trading (DT-P2PET) framework built on Ethereum and IPFS. They have a dynamic pricing scheme, smart contracts, and recommender system to foster more participation of prosumers. Measurement on performance indicates enhanced profit both to the prosumer and the consumer, saved storage costs and bandwidth consumption in contrast to the centralized options. The inquiry provides a practical technical implementation of P2P trading in the presence of off-chain storage and smart contracts. As a regulatory aspect, it induces legal doubts regarding trust, auditability, and license of decentralized platforms. It shows how to design cost-effective infrastructure to enable transparent energy markets, which are essential to regulators who want to find the frameworks to reconcile innovation with consumer protection, the affordability of prices, and the management of platforms (Javadi et al., 2022).

It proposed a game-theoretic dynamic pricing mechanism with a double auction based on a blockchain in microgrids. The resultant market equilibrium prices and high efficiency are obtained with their approach. Fairness of allocation is proven, better participant satisfaction and stable market in random demand-supply situations are confirmed with simulations. The current paper responds to the issue of how to design and regulate the pricing devices in decentralized markets directly. It favours legal systems whereby pricing algorithms are transparent and auditable to strike dynamic price balance between demand and supply. Regulators may look up to this model to break even pricing, track algorithm fairness, and price-discovery principle in self-executing trading systems (Dong, Ning, Zhao, & Song, 2025).

Introduces a two-phased crowdsourcing energy architecture to perform the P2P trading of the prosumers based on the use of IBM Hyperledger Fabric. Phase I does day-ahead and phase II does real-time trading. Coordinated operation of the microgrid through the use of smart contracts is provided by the system that allows smooth P2P transactions, supervised by the utility. This architecture is a middle ground between the centralized grid operation and decentralized prosumerdom markets providing an example of hybrid regulatory frameworks. One of the ideas is licensing schemes of prosumer aggregators, on-demand control mechanisms, and verification of compliance with blockchain based systems. Such a design gives regulators an opportunity to strike the appropriate balance between network reliability, participant autonomy and grid stability within the context of decentralized trading schemes (Wang, Taher, Wang, Kvaternik, & Hahn, 2019).

RESEARCH QUESTIONS

1. How do existing legal and regulatory frameworks address the emergence of blockchain-based peer-to-peer energy trading systems?
2. What are the primary legal and policy challenges associated with decentralization in blockchain-enabled energy markets, particularly concerning consumer protection, data privacy, and liability?
3. How are different jurisdictions—such as the EU, US, and Asia-Pacific—responding to regulatory challenges posed by tokenized energy assets and prosumer participation?
4. What strategic legal innovations and institutional mechanisms can ensure a flexible, climate-aligned, and inclusive regulatory environment for decentralized energy markets?

OBJECTIVES

To critically examine current energy law and regulatory provisions in relation to blockchain-enabled decentralized energy trading.

To analyze the legal implications of decentralization, including issues related to smart contract enforceability, jurisdictional authority, liability, and data governance.

To conduct a comparative analysis of regulatory responses and experimentation (e.g., sandboxes) across leading jurisdictions in managing blockchain energy disruptions.

To propose strategic legal reforms and policy frameworks that foster innovation, transparency, and inclusivity in peer-to-peer energy trading systems.

METHODOLOGY

The research uses a qualitative research approach in the form of secondary data analysis (including both scholarly literature and regulatory, legal, and policy reports) of the issues related to blockchain-based energy trading. Through a systematic review and doctrinal legal analysis, the study critically analyzes current legal structures, innovations in the regulators, and policy responses regarding the situation around the world. The paper is a synthesis of the peer-reviewed journals, government reports, and international guidelines to create an identification of the legal gap, obstacles, and a road map to strategic reforms in decentralized energy markets without case studies or questionnaires.

Adaptation of Legal Frameworks to Blockchain-Based Energy Trading

With blockchain-enabled peer-to-peer (P2P) energy-trading, the underlying assumptions of current energy law and regulatory systems are called into question since this is a model originally aimed at centralized, vertically integrated systems. The majority of the current laws are unready to address decentralized energy purchases and sales, programmed self-executing smart contracts, and tokenized energy resources. Tushar et al. (2021) claim that the traditional regulatory systems are based on a method that uses centralized operators of the market, licensed suppliers, and top-down grid governance which do not suit blockchain-based P2P trading where the prosumers act independently and frequently in multiple jurisdictions. The present legal structures are majorly concentrated on the large energy producers and utilities, which involves stringent licensing, consumer protection and tariff regulations in force. These mechanisms however fail to take into consideration individual prosumers, or microgrids using distributed ledger technologies. Dong et al. (2025) and Wang et al. (2019) note that the decentralization of consensus algorithms and cryptographic trust systems associated with its blockchain is a threat to the centralized systems of authority used in current energy regulations. As an example, smart contracts are not arms of the law in most regulations, which creates problems with dispute settlement and the distribution of liability in blockchain-based energy systems.

One of the biggest one concerns lack of legal recognition of prosumers. Prosumers as stated by Islam (2024) seemingly break down the conventional boundaries between the producers and consumers regarding the rights and responsibilities they enjoy and the accessibility to markets, thereby creating an uncertain regulatory situation regarding their rights and responsibilities. Indeed, existing frameworks are also opaque regarding the nature of energy tokens, both in matters of how they are to be taxed, priced, or governed by the securities or commodities laws (Romano & Marino, 2022). A second concern is information privacy and online security. Distribution of transaction data on blockchain cannot be altered, and it frequently consists of information on location or consumption (real time). Such practices are potentially incompatible with data protection regulations, including that of the EU (GDPR), notably in respect of the so-called right to be forgotten. A permanent record implied by use of blockchain, according to Rahimi et al. (2022), is anathema to this requirement and thus calls upon hybrid systems that incorporate both off-chain data storage and on-chain validation.

However certain regulatory experimentation is starting to appear. Innovators in the UK, Australia, and Singapore can test blockchain-based energy solutions in regulatory sandboxes in designated controlled environments under more lenient regulations. Iglesias and Guerrero (2022) also note that such sandboxes may be an effective site of experimentation with alternative market designs that could assist policymakers in digesting how decentralized trading systems may impact theoretically and practically. The sandbox is however in most cases jurisdiction-specific and limitly temporal and thus cannot give long-term stability in the way the courts should be feeling good about.

Certainly, in European case, the Clean Energy Package (CEP) of the EU has advanced in the direction of sustaining the active involvement of consumers and local energy communities, which implicitly coincide with P2P trading. Nevertheless, a majority of national legislations in the EU countries remain to convert these directives into elaborate policies of blockchain-based energy systems. In its turn, the U.S. portrays a disconnected regulatory strategy since its electricity markets are governed by individual states. Such states as New York and California introduced pilot projects to promote distributed energy resources, but the Federal government still does not present a unified approach to blockchain integration.

In short, although several regulatory indicators of flexibility are enough, the majority of the existing legal frameworks are not entirely prepared to properly support energy trading through blockchains. The most important issues are prosumer legal recognition, smart contracts, legislative treatment of tokenized energy assets, data protection legislation, and the lack of any standardization at different jurisdictions. According to the literature, there is an dire necessity of adaptive and principles-based regulation that will be able to accommodate the decentralized innovation and protect the transparency, equity, and reliability of the system at the same time.

Regulatory and Policy Challenges in Decentralized Energy Markets

On the one hand, the decentralization of the energy markets with the help of blockchain technology offers non-trivial changes in energy markets, and on the other hand, it creates a range of legal and policy issues that current regulatory frameworks have not yet considered. Topmost among are issues on consumer protection, data privacy and liability which are all further complexed in a decentralized, peer-to-peer (P2P) energy trading scenario. There is the big issue of consumer protection in decentralized energy markets. Conventional energy policies assure the constraints of quality assurance, pricing, and reliability by the licensed renderers including coherent dispute resolution arrangements and regulatory discharge. As opposed to this, blockchain-based energy systems are based on the autonomous smart contracts between prosumers that are not focused on the controlled parties without intermediaries (Mahmoud et al., 2017; Tushar et al., 2021). This transition eliminates the protection of the consumers. According to Rahman et al. (2023), absence of central authority presents a challenge to consumers to take redress in the event of fraud, non-performance, or system failure. In addition, P2P systems can employ dynamic pricing, which can endear the less-informed participants to unjust outcomes since it is usually based on algorithms, questioning both transparency and equity considerations (Dong et al., 2025).

Moreover, efficient as it is, smart contracts are associated with further risks. When these contracts are deployed in the block chain they become irreversible and automatic, and coding bugs in them can cause unintended effects without a remedy (Xue et al., 2022). The status and enforceability of such contracts (particularly across countries) remained unclear and have not been considered fully by courts and regulators. The existing legal framework fails to give a clear picture of the attribution of responsibility in situations where the autonomous digital contracts break Working or conduct themselves in a non-regulated manner (Wongthongtham et al., 2021). The other major challenge is data privacy. The key feature of blockchain, its immutability, conflicts with data protection regulation including EU General Data Protection Regulation (GDPR). Romano and Marino (2022) cited the fact that the right to be forgotten established by GDPR can hardly have any application on a public blockchain where all transactions are immutable. In addition to this, transaction contains sensitive information about personal use and location which may be used against a user in the case of insufficient protection. Some solutions like off-chain storage, pseudonymization have been proposed, though they also tend to appear incompatible with transparency and essentially create new laws that are legally ambiguous (Rahimi et al., 2022).

To compound the matter, matters of jurisdiction arise when there is a decentralized market that is not pegged on fixed geography. Blockchain networks do not have national boundaries, but rather the legal regulation does. Such questions, therefore, emerge regarding the jurisdiction to which the laws of transactions, disputes, and enforcement may be applied (Wang et al., 2019). To take one such example, a prosumer in Germany selling their electricity to an electricity prosumer in France through a blockchain intermediary might encounter conflicting national regulations and thus regulatory arbitrage, and enforcement problems.

Probability of liability is diffused in decentralized systems too. In traditional structures, the utilities and suppliers are conspicuously the agents behind power delivery and breakdown of the system. In the case of the decentralized platforms, though, the responsibility is shared by different actors: the developers of the blockchain platforms, the smart contract developers, their users, and potential third-party verifications (Javadi et al., 2022). Finding who is at fault when there has been an overbilling or when service or data damages has occurred, is thus problematic. In legal works, such risks are discussed as the necessity to organize distinct accountability schemes and schemes of insurance to prevent their occurrence (Iglesias & Guerrero, 2022).

Finally, there is a problem of regulatory interoperability. As the blockchain-based energy systems evolve side by side with different jurisdictions, insufficient legal standards of equivalence interfere with scalability by country and trust between countries. Regulatory sandboxes have been relatively accommodating when applied (e.g., in Singapore and the UK) yet they are experimental and have no legal force to them (Romano & Marino, 2022). In sum, decentralization raises legal and policy issues of critical importance including consumer protection and privacy, as well as the definition of liability. Current legal systems do not serve well in matters of these complexities. Within the current rise of blockchain energy trading, effective regulators should be out in front with technologically agnostic frameworks that are enabled to be more adaptive and focused on accountability, data sovereignty, fairness, and cross-border enforceability. The literature is heavy in favor of a synchronized, worldwide strategy of these new legal mandates.

Despite the fact that smart contracts are characterized by a positive reputation, which concerns the effectiveness of operations, they raise substantive issues related to enforceability and liability. The traditional definition of contract law assumes that a contract exists on the fact that parties were identified, the consent was deemed through different evidences and there should be a court where it can be interpreted. Contracts negotiated using blockchains, by comparison, are coded automatically and often lack any discretionary human interpretation. The critical issue posed by such a paradigm is whether or not the code-based consent provides adequate proof of the legal conditions under which a contractual formation and enforcement is possible.

This is complicated even more when there are malfunctions, errors in coding, or unanticipated consequences. Existing regulatory policies do not explicitly define the responsibility in such distributed ecosystems. In addition, cross-border transactions also complicate in establishing the governing law, as well as the choice of relevant mechanisms of resolving disputes. This leads to a need to regulate the legal soundness of smart contracts, establishment of liability due to the misuse of algorithms and legal jurisdiction of peer-to-peer energy trades across countries.

Prosumers' Legal Status and Participation in Blockchain Energy Systems

Countries worldwide are adopting different regulatory responses to the blockchain-unleashed energy trading potential: adapting regulation, experimenting, and transforming institutions. A handful of leading nations are taking an active approach to regulatory innovation but the majority of the current legal frameworks are still based on centralized models of energy. The three major questions, which are seeking legal identification of prosumers, treatment of tokenized energy assets, and regulatory match of blockchain-based platforms, are not receiving equal attention in the European Union (EU), the United States (US), and the Asia-Pacific region. Regulation Regulatory parts to decentralized energy trading in the European Union are in the development phase in terms of its larger climate and energy instructions. The so-called Clean Energy for all Europeans Package encourages the active involvement of energy consumers and the establishment of energy communities and prosumers. As much as this change in legislation offers solid ground on P2P trading, there is a wide gap in its application among the member states. To give an example, such countries as Germany and the Netherlands have already started to experiment with regulatory sandboxes and local energy cooperatives, and as a result, it is possible to test blockchain systems in real-market environments to some extent (Islam, 2024; Tushar et al., 2021). Nevertheless, the financial regulation status of the proposed tokenized energy products, i.e., blockchain-based representations of kilowatt-hours or grid rights, is unclear having regard to the financial regulation regime including MiFID II and the Markets in Crypto-Assets Regulation (MiCA). According to Romano and Marino (2022), such frameworks fail to clarify the nature of energy tokens into commodities, securities, or as a novel asset class that makes taxation, compliance and investor safeguards challenging.

However, contrary to this, the United States has a more decentralized environment because of its federal system of governance, as the individual states control their electricity markets. California, New York—and others are actively encouraging distributed power innovation by using pilot projects and regulatory promotions. Blockchain-based microgrid projects have been supported by the Energy Commission of California and the creation of decentralized energy markets has been made possible under the REV (Reforming the Energy Vision) initiative of New York (Dong et al., 2025). Other states however, have tight licensing conditions, restricting the entry of prosumers. There is still regulatory ambiguity on blockchain and tokenized energy market (Wang et al., 2019) since there has not been any clear statement released by the Federal Energy Regulatory Commission (FERC) to regulate electricity trade across states are under scrutiny until now (Zhou, 2018).

In addition, at the U.S. Securities and Exchange Commission (SEC) the approach towards tokenization has been conservative. The energy tokens may amount to securities depending on the manner in which they are designed, which implies disclosure requirements and restrictions over trading. This can cause immense regulatory burdens to decentralized platforms and can discourage small-scale prosumers/developers entering the market without legal provisions or institutional backing (Rahimi et al., 2022). Within the Asia-Pacific region, Singapore and Japan, as well as Australia, have exhibited leadership with respect to experimental regulation. The city-state of Singapore in particular has developed a strong regulatory sandbox framework in its Energy Market Authority (EMA) and Monetary Authority of Singapore (MAS). Such a sandbox enables developers to trial blockchain P2P trading platforms in the context of conditional regulatory relief to stimulate innovation, allowing regulators to keep an eye on risk (Iglesias & Guerrero, 2022). Singapore is also a supporter of tokenized energy markets combined with IoT, Artificial Intelligence as well as the development of regulatory clarity in terms of data processing across sectors and digital identity.

Australia has moved ahead actively with its Power Ledger system that facilitates real-time blockchain trading of energy among households. Regulators in Australia (Australian Energy Regulator or AER) and the government energy agencies collaborate with blockchain traders to harmonize trading in the blockchain with the grid operations and consumer protection laws (Wongthongtham et al., 2021). Nevertheless, as in the EU, the status of tokenized energy and smart contracts fulfilment should be clarified through law. In the meantime, Japan is ensuring blockchain energy projects lead to its carbon neutrality objectives whereby it supports tokenized renewable energy credits (RECs) and other community energy endeavors. Regulatory regulators have considered the potential of blockchain to monitor renewably generated power and selling surplus to others (Javadi et al., 2022), nonetheless development of grid access and liability regulation is pending (Javadi et al., 2022).

Overall, international regulatory prioritization differs by amount of readiness, creativity and prudence. The EU is harmonizing decentralized trading to its objectives of energy democracy however it has implementation gaps. The United States is based on powerful local experimentation, but lacks national leadership. The countries in the Asia-Pacific region are testing sandboxes and tokenization, but they are continue to face problems with classification and enforceability. All in all, the literature recommends an urgent need of harmonised regulation level that facilitates electricity cross-border P2P trading keeping a balance between innovation and consumer rights, data protection and legal certainty.

The legal standing of prosumers is not that well-articulated, which is the critical gap in the current regulation. The current licensing systems have been skewed towards large-scale utilities and therefore put the small-scale energy players in a risky legal grey zone. Therefore, an individualized regulatory approach seems to be inevitable. A three-party system would be able to provide commensurate regulation:

Tier 1- Residential Prosumers: These are people whose edges are the boundary of power production and they produce small surplus to sell in the intra-community markets. These investors are also exposed to streamlined registration filing, low compliance requirements and underlying consumer protection frameworks.

Tier II Community or Microgrid Aggregators Localised coalitions which facilitate collective trading arrangements. They must have intermediate licensing, make periodic reporting; and meet grid-coordination standards as provided by oversight bodies.

Tier III -Commercial Prosumers: Large-scale distributed generation participants which participate in the energy markets on a repeated basis. Their responsibilities resemble those of the more conventional suppliers that include complete licensing, taxation and wholesome market-compliance schemata.

This type of controlled regulation will allow healthy market participation and stability of the grid, consumer protection, and regulatory accountability.

Comparative Regulatory Approaches to Blockchain Energy Trading

In order to maximize the potential of blockchain-based decentralized energy markets, legal, and institutional environments would require restructuring toward encouraging scalability, inclusivity, and climate-compatible development. It is indicated in literature that such transformation needs a multi-dimensional framework with regulatory flexibility, legal acceptability of new players in the market, digital infrastructure legal regulation, and environmentally, social and technological needs in regulation and law and policy. Blockchain energy trading requires operationalizable laws based on interoperability, technologically neutral and innovation-friendly. According to Tushar et al. (2021), the majority of the current energy laws are applicable to centralized and linear electricity supply chains, instead of decentralized prosumer-based networks. To scale blockchain-based systems across the regions and the borders, there needs to be legal harmonization. It will involve establishing universal definitions and standards in the energy token, smart contracts, and prosumer rights. According to Romano and Marino (2022), there is a need to find a common ground of transnational legal cooperation, specifically in defining the nature of a digital energy asset which might be a commodity or a security depending on the relevant jurisdictional regimes.

Institutionally, regulatory sandboxes provide a mechanism of transition so as to use new models without full compliance with regulation as in the cases of Singapore, the UK, and Australia. Nevertheless, scalability must entail the transition to the long-term regulatory frameworks letting sandboxes morph into flexibility, digital resilience and automation of energy statutory texts. Iglesias and Guerrero (2022) promote the regulation based on principles, which emphasizes the outcomes, such as affordability, reliability, and sustainability, rather than a strict adherence to the procedures. This kind of regulation enables the regulator to meet new technologies such as blockchain, AI, and IoT without undermining the safeguards of the advice to the people.

With decentralized energy markets, inclusivity would mean the legal recognition and empowerment of prosumers, energy communities, energy communities and marginal users. Due to the current licensing, tariff, and grid access regulations, small-scale players are usually left out of the fair play of the market, as Islam (2024) and Rahman et al. (2023) claim. The key legal changes that should be made are simpler registration procedures of microgenerators, the possibility of net metering and dynamic prices and an obligation of a nondiscriminatory access to the grid. Also, laws are needed to enable the creation of energy cooperatives and local energy communities and, their cooperation, which are able to combine forces and democratize clean energy for all.

It is also necessary to speak about digital inclusion. According to Javadi et al. (2022), unless special legal instruments are found, a blockchain energy system can duplicate or increase the number of digital divisions that already exist. Thus, each framework ought to guarantee the availability of empowering technologies, digital literacy training, and inexpensive internet and metering infrastructure in underserved regions. This is particularly important in the Global South because here energy poverty is a burning issue. The adherence to climate will involve the integration of sustainability requirements into the legal framework of blockchain trading in energy. Systems based on tokenized energy need to be combined with carbon accounting, renewable energy certification, and climate finance systems. As an example, Wang et al. (2019) address the idea of blockchain-based transparent tracking of renewable energy certificate (REC) and reductions in emissions. Environmental disclosures should be required by the regulatory frameworks of the platforms, and it is recommended that low-energy blockchain consensus mechanisms, which include Proof-of-Stake, should be used to decrease the carbon footprint of the technology (Rahimi et al., 2022).

An important role in the governance is that played by public institutions. There is a need to have independent regulatory bodies, data protection, and environment agencies who interpolate and view blockchain platforms. There has to be a cross-sectorial legal instrumentation encompassing both energy legislation, information security, their financial banking regulation and environmental regulation. To raise the legal certainty and popular trust, Wongthongtham et al. (2021) recommend the establishment of blockchain energy authorities or interagency task forces that will guide the implementation process, audit it, and arise to resolve any of its disputes.

Altogether, literature comes to the same conception of a strategic regulatory framework of blockchain energy trading, which has to be adaptive, inclusive, and climate-aligned. The new law needs to focus on environmental responsibility, digital and energy justice, and interoperability. Just decentralized energy systems and scalable systems would never flourish under systems of law like they exist currently that remain stuck in the traditional paradigm or paradigms and fail to respond to the needs of decentralized automated and participatory blockchain technology.

However, the existing legal frameworks are challenged by the question of whether or not such tokens should enforce the status of commodities, securities, utility tokens or a new category of asset. This kind of classification is not just a mere technicum of taxonomy exercise but it essentially defines disclosure responsibilities, licensing requirements, tax treatment, and investor protection requirements. In the European Union, regulation could be through the Markets in Crypto Assets Regulation (MiCA); and in the United States, securities may regulate tokenized energy assets through the set of tests to check investment agreements. This regulatory opaqueness creates overlapping compliance requirements either side of the energy law, financial regulation, and therefore increases the entry barriers of decentralised platforms. Therefore, stricter statutory guidelines are needed to distinguish between functional energy tokens, which are used as a tool in functional trading, and speculative financial instrument, which will reduce the ambiguity of law and achieve a relative response in regulation.

MAIN FINDINGS

The research concludes that blockchain-based decentralized energy trading is already developing at a greater pace than the legal framework aimed to oversee them, thus introducing notable gaps concerning regulation and institutional issues. Regulatory frameworks, which are mostly constructed to suit centralized markets, are not well prepared to regulate the decentralized peer-to-peer systems operated by prosumers and transacted by tokens. Among its key findings, the tokenisation of energy and the method of security in enforcing smart contract are not clearly clarified in most jurisdictions and therefore pose issues of uncertainty of compliance by players in the market. Additionally, the prosumers do not have a legal status in many jurisdictions and therefore cannot enjoy fair opportunities in energy markets, and this limitation of access affects the inclusive power of blockchain platforms.

The disproportionate global response by the regulators also features in the research. European Union has placing some effort shifting towards energy communities and consumer involvement in the form of legislation such as Clean Energy Package, yet implementation by a national level is variable. Its disaggregated regulatory environment registers a high level of innovation state-by-state in the United States, but there is no cohesive federal policy on decentralized energy. Singapore and Australia are among the first countries in the Asia-Pacific region that have frontrunner status in terms of experimentation in sandboxes and integration of blockchain at the exploration stage in the Brunei energy governance, but encounter classification and scaling challenges.

The most important finding is that regulatory sandboxes can provide beneficial transitional means, but long-term governance would need more lasting, evolving, adopting structures. These should facilitate the legal interoperability, cross border coordination, and compatibility with digital and climate policies. Inclusion, climate, and digital equality turn out to be the cornerstones of the future regulatory design. Lastly, the research concludes that effective regulatory change relies on institutional capacity-building, interagency coordination, and the principle-directed approach to law that favors the aspects of innovation and consumer protection and environment-friendly. The energy blockchain seems to have huge potential in democratizing energy systems, decreasing emissions, and increasing decentralized resilience, but without such reforms, this potential will simply not be achieved.

DISCUSSIONS

As the literature indicates, the blockchain technology is changing the energy markets because it makes possible the decentralized, peer-to-peer energy trading models that disrupt the traditional regulatory frameworks. Though revolutionary in nature, current laws are ill in most aspects in governing such decentralized systems. They were prepared to accommodate centralized industries with vertically integrated operations and not in a decentralized system characterized with a disjointed ecosystem of sellers and buyers that sell and buy power across smart contracts and tokenized energy goods. The main regulatory issue inheres in the legal uncertainty about the classification of energy tokens (considering them as commodities or securities or utility tokens) and taking smart contracts as enforceable documents, particularly, cross-jurisdictionally. Also, lack of definite legal status and rights by prosumers limit their exemptions in energy markets especially in areas where licensing regimes and rules of joining the grid are advantageous to incumbent utilities.

Although other nations are trying out this approach using regulatory sandboxes and digital innovation hubs, their is a tendency to have these programs be smaller and temporary. Other jurisdictions such as the EU, US and Asia-Pacific are variously prepared. EU has come up with enabling directives in the package of clean energy however implementation is unequal. US is characterized by state-based experimentation which is fragmented, and there is no comprehensive national framework, whereas Asia-Pacific nations, especially Singapore and Australia, are showing leadership via regulatory pilot sandboxes and market environments that are welcoming to blockchain technology.

To accommodate the scalability and climate-alignment of the energy trading using Blockchains, the literature suggests the flexibility and principle-based regulatory frameworks. They should stimulate interoperability at legal, technical, or market levels, as well as anchoring on energy justice and environmental sustainability. Institutions need to reconcile energy, financial, data and environmental law into an overall system of governance. Finally, the choice of balancing innovation and inclusivity, privacy, consumer protection, and societal shared objective of a low-carbon energy future is the key to regulatory change success.

SUGGESTIONS

- Authorities must use principle-based and technology-neutral pieces of legislation, which can ensure flexibility and adaptation in the fast-moving energy markets, including emerging technologies, such as blockchain.
- It is also necessary that these governments will create legal certainty over the classification of tokenised energy assets so that they are made to stand out as distinct from other financial tools or commodities and can ensure significant regulatory clarity and simplicity on the part of the developers and prosumers.
- The enforcement of smart contracts should be targeted by the revision of the existing contract law, which accepts digitally automated agreements and states the prerequisites that make them competent in relation to legal enforcement, dispute settlement, and consumer defense.
- Prosumers should be formally recognized and have legal protection in terms of generating, trading, and consuming energy through the same process across legal jurisdiction to simplify the procedures to get a license and enter the grid and simplify the access process to the grid through small-scale participation.
- Regulatory sandboxes are to be transformed into less transitory models offering a perpetual possibility to innovate within the established framework of legislation, with institutional control, transparency, and feedback mechanisms.
- This is to be done by creation of interoperability standards that will streamline legal, technical and operations matters across blockchain platforms, making them completely integrated and scalable into regional and national energy systems.
- There is the need to combine energy law, data protection, financial regulation, environmental policies in institutional structures to build structures that have

coalescing governance in decentralized energy systems where interdisciplinary approaches are applied.

- The public-private partnership needs to be engaged so that infrastructure and pilot programs as well as blockchain related energy application research can be funded especially in underserved areas towards encouraging digital and energy inclusion.
- Regulatory reforms should be coupled with consumer education programs and digital literacy initiatives so that all interested parties have the opportunity to be active participants in the decentralized energy market including the low-income populations.
- Blockchain energy regulations are to include environmental protection, such as low-energy consensus mechanism requirements and connecting platforms with renewable energy tracking and carbon accounting systems.

FUTURE IMPLICATION

Future implications of this research indicate that the decentralized energy trading using blockchain model is going to take center stage in transforming the energy sector globally. With the changes in the regulations and laws, they will have to facilitate more than the innovation in technology, but also in pursuit of wider aspirations of climate resilience, energy equity and digital equity. The paper examines the possibility of blockchain to make energy more democratic, cut carbon emissions, and improve grid performance when legal frameworks are sensitive and inclusive. Relevant policymaking in the future will overshadow interoperability, data sovereignty, and prosumer empowerment, changing the ways to generate, exchange and regulate energy. Also, due to the combination of blockchain with AI and IoT technologies, regulators could be encouraged to introduce a more flexible, cross-sectoral approach to governance. With countries aiming to fulfill their climate targets and energy transition aspirations the results of the given work serve as a reminder of the importance of adding a legal framework that may accommodate decentralized, low-carbon technologies. In addition, the study will open up the possibility of trying to make use of this type of inquiry in further academic research on the legal side of digital energy markets and provide a source of comparative regulatory analysis, construction of institutional designs, and international collaboration on decentralized energy regulation. Finally, this work is part of the landscape of a clearer, greener, and more equal energy future that will be provided by technological and legal creativity.

CONCLUSION

In this work, the author has found out that with the introduction of blockchain-powered decentralized energy trading, a complete rethink of the current laws and regulations regarding this sphere of activity is required. Long held energy law constructs based on centralized energy production, linear energy market structures are too limited to regulate the predicaments presented by decentralized energy production, decentralized energy market participation and decentralized energy asset transactions. The results demonstrate why the issue of smart contract enforceability needs to be clarified with regard to legal governance, the nature of energy tokens, and fair entry to the new marketplace participants. Regulatory sandboxes are a useful testing box but different solutions need permanent and adaptive legal frameworks which would be principle-based, interoperable and technologically neutral. Global comparison of the jurisdiction releases a diverse picture: in some areas, more innovative models are being developed, and in others, the system moves slowly, under the influence of institutional inertia or regulatory uncertainty. Notably, the research highlights the fact that the legal architectures of the future should not just be versatile to the innovations but also should support such things as exclusiveness, climate-proof, and consumer-safe. The institutions need to cooperate inter-sectorally, in energy, finance, environment, and data to form cohesive, flexible, and resilient to climate regulatory ecosystems. Finally, it is clear that blockchain in energy markets potentially promises transformational capacities, which, however, can be achieved by only progressive legal changes that can strike the right balance between innovation, social responsibility, and environmental concerns. This research provides a guideline to how this can influence such reforms so as to benefit the future of the decentralized, sustainable energy systems.

REFERENCES

Dong, J., Ning, P., Zhao, H., & Song, C. (2025). Decentralized Peer-to-Peer Energy Trading: A Blockchain-Enabled Pricing Paradigm. *Journal of King Saud University. Computer and Information Sciences*, *8*(1), 25. DOI: 10.1007/s44443-025-00025-2

Energies. (2022). *Towards Blockchain-Based Energy Trading: A Smart Contract Implementation of Energy Double Auction and Spinning Reserve Trading*. Energies, 15(11), 4084. DOI: 10.3390/en15114084

Hua, W., Luo, F., Du, L., Chen, S., Kim, T., Morstyn, T., Robu, V., & Zhou, Y. (2022). Blockchain technologies empowering peer-to-peer trading in multi-energy systems: From advanced technologies towards applications. *IET Smart Grid*, *5*(4), 219–222. DOI: 10.1049/stg2.12081

Hua, W., Luo, F., Du, L., & Qadrdan, M. (2022). Blockchain & AI for Prosumers in Smart Grids: A Review. *IET Smart Grid*, *5*(4), 219–222. DOI: 10.1049/stg2.12081

Iglesias, B., & Guerrero, J. (2022). Blockchain-Based Peer-to-Peer Energy Trading Between Wind Producers and Prosumers. *Frontiers in Energy Research*, *10*, 923292. DOI: 10.3389/fenrg.2022.923292

Islam, S. N. (2024). A Review of Peer-to-Peer Energy Trading Markets: Enabling Models and Technologies. *Energies*, *17*(7), 1702. DOI: 10.3390/en17071702

Javadi, M.. (2022). Blockchain-Based Peer-to-Peer Transactive Energy Management Scheme for Smart Grid System. *Sensors (Basel)*, *22*(13), 4826. DOI: 10.3390/s22134826 PMID: 35808325

Khorasany, M., Dorri, A., Razzaghi, R., & Jurdak, R. (2020). *Lightweight Blockchain Framework for Location-aware Peer-to-Peer Energy Trading. arXiv preprint.* /arXiv.2005.14520DOI: 10.48550

Rahimi, A., Khorasany, M., Dorri, A., & Jurdak, R. (2022). Lightweight Blockchain Framework for Location-Aware P2P Energy Trading. *Scientific Reports*, *12*, 18603. DOI: 10.1038/s41598-022-18603-z

Rahman, M., Chowdhury, S., Shorfuzzaman, M., Hossain, M. K., & Hammoudeh, M. (2023). Peer-to-Peer Power Energy Trading in Blockchain Using Efficient Machine Learning Model. *Sustainability (Basel)*, *15*(18), 13640. DOI: 10.3390/su151813640

Romano, L., & Marino, A. (2022). Blockchain, Data Protection & P2P Energy Trading: GDPR Implications. *Sustainability*, *14*(23), 16305. DOI: 10.3390/su142316305

Tushar, W., Yuen, C., Saha, T. K., Morstyn, T., Chapman, A. C., Alam, M. J. E., Hanif, S., & Poor, H. V. (2021). Peer-to-peer energy systems for connected communities: A review of recent advances and emerging challenges. *Applied Energy*, *282*, 116131. DOI: 10.1016/j.apenergy.2020.116131

Wang, N., Zhou, X., Lu, X., Guan, Z., Wu, L., Du, X., & Guizani, M. (2019). *When Energy Trading meets Blockchain in Electrical Power System: The State of the Art. arXiv preprint*. /arXiv.1902.07233DOI: 10.48550

Wang, S., Taher, A. F., Wang, J., Kvaternik, K., & Hahn, A. (2019). Energy Crowdsourcing and Peer-to-Peer Energy Trading in Blockchain-Enabled Smart Grids. *IEEE Transactions on Smart Grid*, *10*, 5501–5511. DOI: 10.1109/TSG.2019.2895172

Wongthongtham, P., Marrable, D., Abu-Salih, B., Liu, X., & Morrison, G. (2021). Blockchain-enabled Peer-to-Peer energy trading. *Computers & Electrical Engineering*, *94*, 107299. DOI: 10.1016/j.compeleceng.2021.107299

Xue, F., Chang, K., Li, W., Wang, Q., Zhao, H., Zhang, H., Ni, Y., & Xia, W. (2022). Blockchain Smart Contract-Enabled Secure Energy Trading for Electric Vehicles. *Energies*, *15*(18), 6733. DOI: 10.3390/en15186733

Zhu, Y., & Salam, R.. (2024). Transaction-Loss-Aware Energy Trading in Microgrids. *Frontiers in Energy Research*, *12*, 1397975. DOI: 10.3389/fenrg.2024.1397975

Chapter 5
Blockchain in Smart Governments:
Legal Architectures for Decentralised Energy Trading

Francisco R. Trejo-Macotela
http://orcid.org/0000-0003-2133-3456
Universidad Politécnica de Pachuca, Mexico

ABSTRACT

This chapter examines the role of blockchain as a transformative instrument in the digital energy transition, highlighting its capacity to support decentralised trading, automate compliance, and enhance institutional transparency. It analyses how blockchain aligns with smart-governance principles by embedding regulatory logic into digital infrastructures while requiring adaptive legal frameworks to address liability, data protection, and market oversight. Through comparative case studies—including Brooklyn Microgrid, Powerledger, and WePower—the chapter illustrates the interplay between technological innovation, regulatory experimentation, and policy integration. Ethical dimensions such as fairness, environmental sustainability, and algorithmic accountability are addressed as essential conditions for legitimate decentralisation. The findings demonstrate that blockchain's effectiveness depends not on technological capability alone but on the coherence of governance, legal interoperability, and societal inclusion.

1. INTRODUCTION

The accelerating digital transformation of the energy sector is reshaping conventional models of generation, distribution, and consumption. As systems transition

DOI: 10.4018/979-8-3373-5742-3.ch005

from centralised infrastructures towards distributed networks, governance structures must evolve accordingly. Blockchain has emerged as a secure, decentralised mechanism for recording transactions and validating ownership, enabling new forms of energy trading and coordination aligned with smart-governance principles (Andoni et al., 2019). Global priorities—carbon neutrality, prosumer participation, and transparency—reinforce the need to redesign legal and institutional frameworks originally crafted for centralised utilities (International Energy Agency, 2024; European Commission, 2023). Distributed generation, peer-to-peer (P2P) exchanges, and smart-contract-based governance challenge existing regulatory regimes, requiring coherent legal structures to prevent market fragmentation and safeguard trust (Bassey et al., 2024).

Blockchain's immutability, transparency, and automated verification support equitable access to reliable data and correspond with smart-governance objectives of accountability and interoperability (Andoni et al., 2019; Kouhizadeh et al., 2019). However, decentralised trading raises complex legal questions. Traditional markets operate under strict licensing, consumer-protection, and taxation rules, while smart contracts introduce novel uncertainties around liability. Tokenised energy assets further disrupt established conceptions of property and value (Diestelmeier, 2019). A stable legal architecture integrating standards, norms, and institutional practices is therefore required for lawful implementation.

Within smart-government frameworks, legal design becomes central. Smart governments integrate policy, technology, and law to support adaptive and transparent governance (Gil-Garcia et al., 2020). Blockchain-based energy systems demand regulatory approaches balancing innovation with privacy, market integrity, and consumer rights (Paudler et al., 2022). Regulatory sandboxes—such as OFGEM pilots including Piclo Flex and Electron—enable controlled experimentation (Office of Gas and Electricity Markets, 2023). EU initiatives, including the Clean Energy Package and the Digitalising the Energy System Action Plan, similarly advance distributed ledger applications for Guarantees of Origin and cross-border trade (European Commission, 2023).

The absence of global harmonisation remains a major constraint. Divergent legal traditions produce inconsistent regulatory treatment of smart contracts and tokenised assets, jeopardising interoperability and legal certainty. A framework capable of reconciling national laws with global interoperability requirements is therefore essential.

Blockchain integration also intersects with expanding data-governance challenges. Increased digitalisation heightens concerns regarding cybersecurity, privacy, and algorithmic accountability. Compliance with regimes such as the GDPR is indispensable, particularly where blockchain's immutability conflicts with data-erasure rights (European Data Protection Board, 2025). Smart governments must therefore balance transparency with data protection. At the policy level, blockchain can strengthen participatory and equitable energy markets by enabling traceability, automated settlement, renewable-credit tracking, and community-based energy initiatives, functioning simultaneously as a technological enabler and governance instrument.

This chapter advances three propositions: (1) blockchain's value is inseparable from broader governance reform; (2) its effective deployment requires clear legal definitions of rights and responsibilities; and (3) these frameworks must align with smart-governance principles to ensure sustainable and socially equitable decentralised energy trading. The chapter proceeds with analyses of smart-governance foundations, blockchain attributes, legal architectures, regulatory frameworks, case studies, and policy recommendations.

1.1. Methodological and Theoretical Positioning

This chapter adopts a doctrinal and comparative legal methodology, complemented by normative governance analysis. Its primary objective is not to provide technical engineering modelling or econometric simulation, but rather to interrogate the evolving legal architecture that underpins blockchain-enabled decentralised energy trading. The analysis proceeds through three interconnected lenses. First, a doctrinal legal approach is employed to examine statutory instruments, regulatory frameworks, and emerging policy documents governing distributed ledger technologies within energy markets. This includes analysis of digital-asset regulation, smart-contract enforceability, data protection regimes, and energy market design reforms. By systematically interpreting these sources, the chapter identifies areas of coherence, fragmentation, and legal uncertainty. Secondly, a comparative regulatory assessment is undertaken across selected jurisdictions, including the European Union, the United Kingdom, the United States, Australia, and selected emerging economies. The comparative dimension is not intended to provide an exhaustive jurisdictional survey; rather, it seeks to highlight structural differences in regulatory philosophy—integrated governance, experimental sandboxes, sectoral fragmentation, and risk-based proportionality—and to evaluate their implications for decentralised energy markets. Thirdly, the chapter advances a normative governance framework grounded in smart-governance theory and algorithmic regulation scholarship. It engages with debates surrounding code-based regulation (Lessig, 2006), algorithmic accountability (Yeung, 2018), and responsibility gaps in automated systems (Matthias, 2004). Rather than adopting a techno-deterministic stance, the analysis critically evaluates whether blockchain can legitimately function as a regulatory instrument without displacing democratic oversight and constitutional safeguards. The chapter therefore positions itself at the intersection of energy law, digital governance, and regulatory theory. Its contribution lies in articulating a structured legal architecture for decentralised energy trading that integrates enforceability, interoperability, accountability, and sustainability. While empirical pilot projects are examined illustratively, the central ambition is conceptual clarification and governance design. In this respect, the chapter contributes to ongoing scholarly debates concerning the transformation of public regulation in digitally mediated markets and the constitutional implications of embedding normative rules within technical infrastructures (Yeung, 2018; Lessig, 2006).

2. SMART GOVERNANCE AND THE DIGITAL ENERGY TRANSITION

Smart governance has become a defining paradigm in contemporary public administration, reflecting a move from hierarchical decision-making to data-driven, adaptive regulatory models. Within the energy transition, it denotes the institutional capacity to integrate emerging technologies—blockchain, artificial intelligence (AI), and the Internet of Things (IoT)—into coherent structures of regulation, coordination, and accountability (Gil-Garcia et al., 2020; Meijer & Bolívar, 2016). This evolution represents not only technological modernisation but also a transformation of governance logics, where digital infrastructures and regulatory mechanisms jointly support transparency, participation, and sustainability.

2.1. Conceptual Foundations of Smart Governance

Smart governance employs digital technologies to improve efficiency, transparency, and inclusiveness in governmental processes. Operating at the intersection of public management, data science, and civic participation, it enables evidence-based policymaking and adaptive regulation. Unlike earlier e-government initiatives focused on service digitalisation, smart governance seeks cross-sectoral integration through interoperable platforms, thereby aligning technological progress with broader sustainability objectives (Ansell & Gash, 2018). A significant element of this model is its capacity to coordinate complex, distributed systems. Governance becomes polycentric, with public, private, and hybrid actors contributing to regulatory functions—an arrangement well suited to decentralised energy systems (Meijer & Bolívar, 2016). The emergence of distributed energy resources (DERs) and prosumer-led markets requires regulatory flexibility and technological interoperability, attributes supported by blockchain's transparency, auditability, and consensus-based verification (Kouhizadeh et al., 2019).

2.2. Smart Governments as Enablers of the Digital Energy Transition

The digitalisation of the energy sector is reshaping state–market–citizen relations. Smart governments act as orchestrators, promoting innovation while safeguarding market integrity, consumer rights, and social equity. Digitalisation is recognised as a key decarbonisation enabler (International Energy Agency, 2024), yet its benefits depend on sound governance frameworks addressing data sharing, cybersecurity, and interoperability. National strategies illustrate this alignment. The EU's Digitalising the Energy System Action Plan (European Commission, 2023) and the United Kingdom's Energy Digitalisation Strategy integrate energy and digital policies to establish standards and coordination mechanisms. Regulatory sandboxes—such as OFGEM's Piclo Flex and Electron—provide controlled environments for testing blockchain-based models (Office of Gas and Electricity Markets, 2023).

Blockchain's decentralised structure contributes to distributed accountability, reducing information asymmetries and enabling verifiable transaction records (Andoni et al., 2019). For public authorities, these features support enhanced monitoring and compliance while shifting regulatory roles from direct data control to oversight of secure, partially self-regulating ecosystems (Ansell & Gash, 2018).

2.3. Data Governance and Interoperability

Data governance is foundational to smart governance. The energy sector relies on continuous data flows—from smart meters, IoT devices, and digital twins—to balance system operations. Blockchain can function as a trust infrastructure for such data, ensuring integrity and authenticity while supporting compliance with privacy frameworks such as the GDPR (European Data Protection Board, 2025).

Interoperability remains a central challenge. Efficient digital energy systems require seamless data exchange among utilities, regulators, grid operators, and consumers. While blockchain offers standardised data structures and cryptographic verification, these benefits depend on alignment with existing legal and technical standards. Trusted data spaces have therefore gained prominence as federated environments enabling secure data sharing under defined governance rules (European Commission, 2023). Such mechanisms underpin advanced trading models—peer-to-peer (P2P) and community markets—where smart contracts automate exchanges according to predetermined legal and technical conditions.

2.4. Smart Governance, Sustainability, and Citizen Participation

Smart governance also embodies normative ambitions concerning sustainability and inclusion. In energy systems, this translates into enabling citizens to participate as prosumers in decentralised renewable generation. Blockchain supports these models by providing transparent records of production and consumption and enabling tools such as tokenised incentives, microgrid coordination, and renewable energy certificates (RECs) (Diestelmeier, 2019).

As digitalisation expands, issues of equity, accessibility, and digital literacy become increasingly relevant. Smart governments must ensure that technological innovations do not exacerbate inequalities but instead promote fair access to data, transparent algorithmic processes, and equitable distribution of benefits (Paudler et al., 2022). Blockchain's decentralisation can support participatory governance approaches where decision-making is shared through common digital rules rather than hierarchical institutions.

Overall, smart governance provides the institutional foundation for blockchain-based energy systems. By combining regulatory foresight, technological adaptability, and civic inclusion, it enables the development of secure, sustainable, and democratically governed decentralised energy markets.

3. BLOCKCHAIN FOUNDATIONS IN ENERGY SYSTEMS

The global energy landscape is undergoing a profound transformation, driven by the convergence of digitalisation, decarbonisation, and decentralisation. Within this paradigm, blockchain technology has emerged as a fundamental enabler of new energy trading architectures that are transparent, secure, and democratically governed. Its capacity to support peer-to-peer (P2P) energy exchanges, automate settlements, and ensure the traceability of renewable generation introduces a level of integrity and efficiency that conventional centralised systems cannot easily achieve (Andoni et al., 2019; Bassey et al., 2024).

Blockchain provides a distributed and immutable ledger that records every transaction across multiple nodes in a network, ensuring data integrity and eliminating the need for a single central authority. In the energy domain, this capability facilitates the creation of decentralised energy markets in which consumers, prosumers, and aggregators can transact electricity directly based on predefined contractual conditions encoded as smart contracts. Such architectures are closely aligned with the principles of smart governance, where transparency, accountability, and interoperability are prerequisites for effective regulation and policy coherence.

3.1. Core Technological Principles

At its core, blockchain operates as a distributed ledger technology (DLT) composed of a network of nodes that collectively validate, record, and store transactions. Each transaction is encrypted, timestamped, and grouped into blocks that are linked together using cryptographic hashes, forming an immutable chronological chain. The consensus mechanisms—such as Proof of Work (PoW), Proof of Stake (PoS), or Proof of Authority (PoA)—ensure that all participants in the network agree on the validity of each block without relying on a central intermediary (Fatima & Arshad, 2025).

In decentralised energy markets, these features provide several advantages:

- **Transparency and Trust:** Every participant can verify energy transactions on the ledger, reducing the risk of data manipulation.
- **Automation:** Smart contracts execute predefined conditions automatically, enabling instantaneous energy settlements between peers.
- **Traceability:** Energy certificates and carbon credits can be tokenised and tracked from generation to consumption, ensuring compliance with sustainability standards.
- **Security:** Cryptographic verification and decentralisation minimise vulnerabilities to cyberattacks.

Blockchain's structure, when combined with IoT sensors and smart meters, ensures that energy data are both accurate and verifiable. This is particularly critical for the certification of renewable energy, where guarantees of origin (GOs) or renewable energy certificates (RECs) depend on trustworthy data regarding production and consumption (Zheng et al., 2023). See Figure 1 for an integrated depiction of the principal digital tools that support smart energy governance.

Figure 1. Digital tools enabling smart energy governance

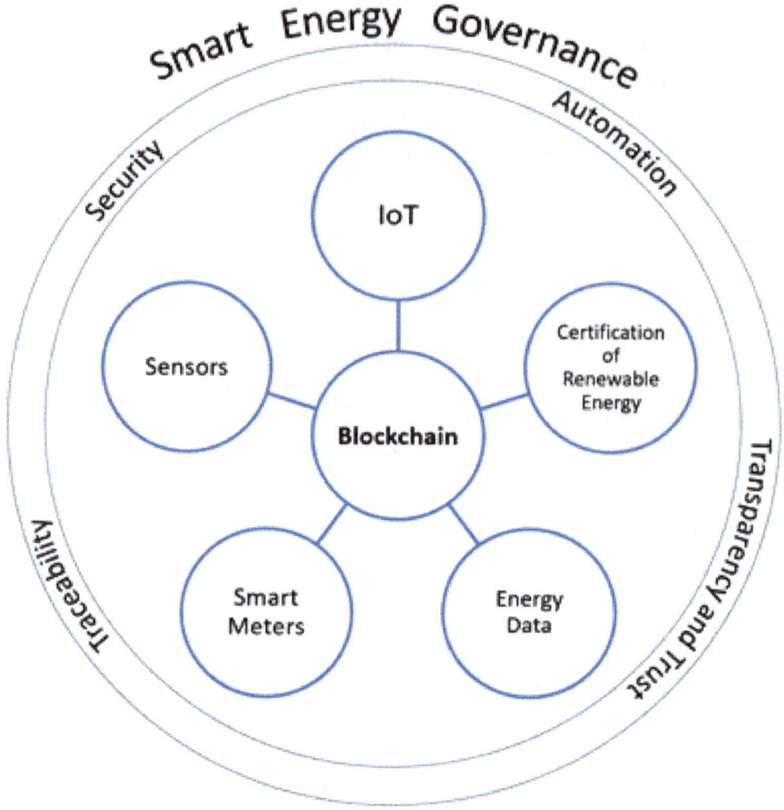

3.2. Integration with Smart Grids and Distributed Energy Resources (DERs)

Smart grids are dynamic energy networks that integrate advanced digital communication, automation, and control systems to optimise electricity flows. Blockchain complements these infrastructures by offering a decentralised layer for secure transaction management and data exchange. Through oracles—interfaces that link off-chain data (e.g., from sensors or metering devices) with on-chain smart contracts—blockchain systems can validate real-time energy production and consumption, thus enabling automatic market settlements.

In the context of distributed energy resources, blockchain facilitates the coordination of small-scale renewable producers, storage systems, and consumers within local energy communities. The participatory process is visualised in Figure 2, which maps the iterative cycle of consultation, co-design and evaluation enabled by digital platforms.

Figure 2. Citizen participation cycle in energy policy-making

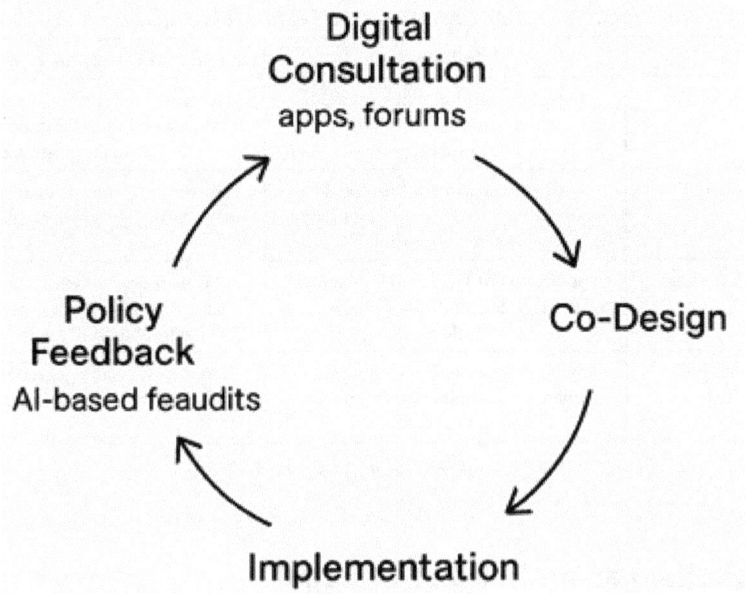

By embedding economic incentives through tokenisation, blockchain enables participants to monetise surplus generation and trade it transparently within microgrids. The integration of blockchain into such systems promotes transactive energy—a paradigm in which market operations occur at the distribution level, fostering flexibility and grid stability (Mengelkamp et al., 2018).

This integration also aligns with regulatory objectives. For instance, the European Union's Clean Energy Package encourages digital innovations that enhance market participation and transparency in energy exchanges (European Commission, 2023). Blockchain can operationalise these goals by providing verifiable records of compliance, environmental impact, and market transactions, thereby reducing regulatory overhead while maintaining accountability.

3.3. Market Models Enabled by Blockchain

Blockchain supports several innovative market configurations that differ fundamentally from traditional wholesale or retail electricity markets (Table 1). The diversity of these market models reflects blockchain's flexibility as both a technical and institutional innovation. However, their successful implementation depends on the establishment of comprehensive legal architectures that define the rights, obligations, and liabilities of all participating actors.

Table 1. Emerging blockchain-based energy market models and their legal implications

Market Model	Main Features	Legal and Governance Implications
Peer-to-Peer (P2P) Trading	Direct energy exchange between prosumers using smart contracts and real-time pricing.	Requires redefinition of licensing frameworks and liability provisions for automated contracts.
Community Energy Markets	Energy sharing within local cooperatives or municipalities using tokenised credits.	Involves cooperative legal models and data-sharing agreements under GDPR.
Wholesale Blockchain Platforms	Integration of DLT into balancing, settlement, and auditing processes in wholesale markets.	Raises cross-border regulatory concerns and requires harmonised interoperability standards.
Renewable Energy Certificates (REC) Tokenisation	Digital representation of renewable generation units or emissions reductions.	Requires formal legal recognition of tokens as tradable or financial assets.

(Adapted from Andoni et al., 2019; Diestelmeier, 2019; Bassey et al., 2024)

3.4. Blockchain as an Enabler of Legal Transparency and Compliance

Blockchain's immutability and auditability introduce enhanced mechanisms for regulatory verification. Conventional oversight depends on centralised reporting and manual audits, processes vulnerable to delay, asymmetry, and error. Distributed ledgers, by contrast, provide regulators with continuous and tamper-resistant visibility of market activity, enabling full traceability whereby each kilowatt-hour traded or certified is recorded through a verifiable, shared register (Kouhizadeh et al., 2019).

Smart contracts further advance compliance by embedding regulatory conditions directly within transactional logic. Restrictions on trading volumes, emissions, or licensing parameters can be automated, transforming compliance from reactive enforcement to proactive, rule-based governance. This aligns with smart-governance ambitions in which digital infrastructures function as instruments of transparency and public accountability.

Figure 3 illustrates the layered integration of blockchain, smart governance, and energy infrastructure. The upper layer defines legal and regulatory principles; the blockchain layer translates these into cryptographic and programmable functions; and the infrastructural layer operationalises them within physical energy systems.

Together, these components produce a transparent, audit-ready, and legally aligned decentralised marketplace consistent with sustainability mandates.

Figure 3. Conceptual model of blockchain integration in smart energy governance

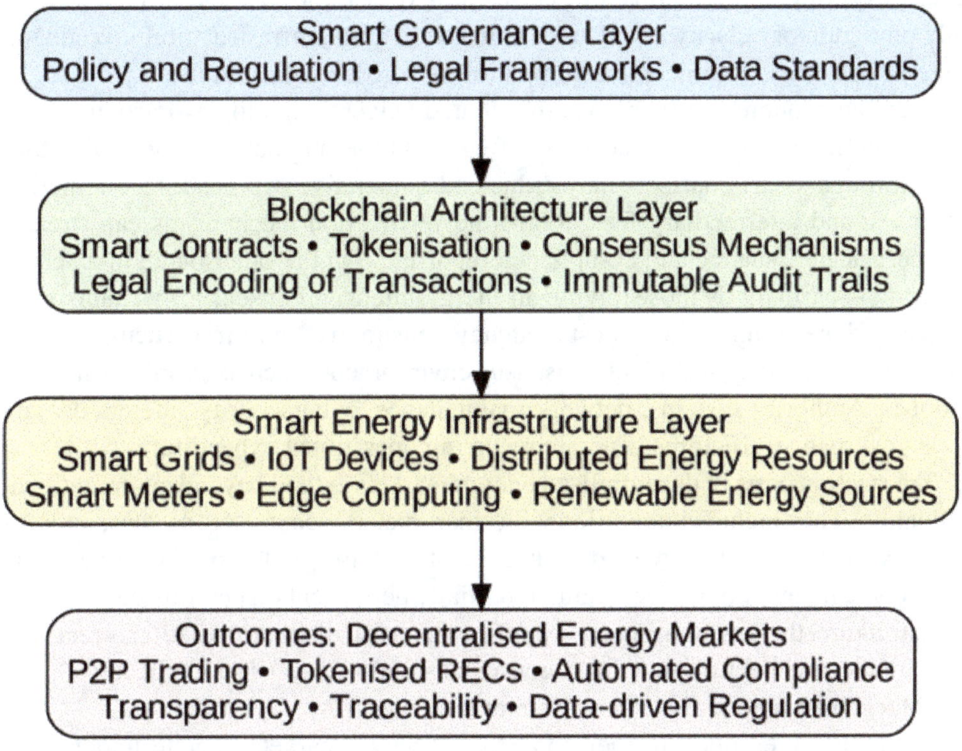

(Author's adaptation based on Andoni et al., 2019; European Commission, 2023)

Blockchain therefore contributes not merely to system optimisation but to a reconfiguration of market governance, embedding legal norms within digital architectures and supporting broader aims of energy democracy. However, its deployment necessitates comprehensive legal safeguards addressing data protection, consumer rights, and cross-border interoperability.

The subsequent section develops this further by outlining a Legal Architecture for Decentralised Energy Trading, including taxonomies, liability structures, and compliance mechanisms required to align blockchain innovation with prevailing regulatory systems.

3.5. Economic and Operational Feasibility Constraints

While blockchain is frequently presented as an efficiency-enhancing infrastructure for decentralised energy markets, its economic and operational feasibility warrants careful scrutiny. The viability of peer-to-peer (P2P) trading models depends not only on regulatory clarity but also on transaction costs, infrastructural investment, system scalability, and incentive compatibility across market actors.

From an economic perspective, distributed ledger technologies introduce both cost-reduction and cost-displacement effects. On the one hand, automated settlement through smart contracts may reduce administrative overhead, reconciliation expenses, and intermediary fees. Real-time verification mechanisms can streamline certificate issuance and compliance reporting. On the other hand, blockchain networks—particularly those relying on energy-intensive consensus mechanisms—may introduce computational costs, latency constraints, and infrastructure redundancy. Even in energy-efficient consensus environments such as Proof-of-Stake or Proof-of-Authority systems, network maintenance, cybersecurity safeguards, and node governance structures generate recurring operational expenditures.

Transaction costs in decentralised energy markets extend beyond computational validation. They include integration with smart meters, secure data oracles, interoperability layers with legacy grid-management systems, and compliance auditing. In practice, many pilot projects remain partially dependent on centralised back-end infrastructure, thereby limiting the extent to which full disintermediation is realised. This hybrid architecture may be economically rational, yet it complicates claims that blockchain inherently lowers system-wide costs.

Scalability presents a further constraint. Energy markets operate under strict temporal parameters, particularly in balancing and ancillary service contexts where settlement intervals are measured in seconds or minutes. Distributed consensus protocols must therefore reconcile decentralisation with throughput capacity. Latency issues may undermine real-time grid responsiveness if not carefully managed. Moreover, as network participation increases, governance complexity expands, potentially introducing coordination frictions that offset theoretical efficiency gains.

Incentive compatibility is equally central to feasibility. Prosumers must perceive tangible economic benefits to participate in P2P trading schemes. Where regulatory frameworks impose grid charges, balancing responsibilities, or taxation without proportional compensation mechanisms, decentralised trading may lose attractiveness relative to conventional retail supply arrangements. Aggregators and distribution system operators must also retain sufficient economic incentives to maintain infrastructure reliability. Absent clear revenue models, blockchain-based marketplaces risk remaining experimental rather than systemic.

Empirical evidence from early pilot initiatives suggests that while blockchain can enhance transparency and automate micro-settlement processes, large-scale commercial deployment requires alignment between digital architecture and tariff structures (Mengelkamp et al., 2018; Andoni et al., 2019; Bassey et al., 2024). Without integration into formal market design—particularly wholesale balancing markets and capacity mechanisms—decentralised exchanges may remain peripheral.

Finally, economic feasibility intersects with governance legitimacy. If blockchain-based markets disproportionately benefit technologically sophisticated or asset-owning participants, distributive imbalances may emerge. Equity considerations therefore influence not only ethical assessment but also long-term market stability.

In sum, the economic viability of blockchain-enabled decentralised energy trading cannot be assumed ex ante. It depends on carefully calibrated regulatory frameworks, scalable technical design, sustainable incentive structures, and integration with existing grid operations. Legal architecture alone is insufficient; operational realism must inform governance innovation.

3.6. Political Economy, Institutional Resistance and Governance Friction

Technological innovation in energy markets does not occur in a politically neutral environment. Blockchain-enabled decentralised trading interacts with entrenched institutional arrangements, legacy infrastructure investments, and established regulatory hierarchies. Accordingly, the feasibility of decentralised energy architectures must be analysed not only through technical and economic lenses, but also through political economy dynamics.

Incumbent utilities, transmission operators, and vertically integrated energy companies possess significant structural influence within regulatory processes. Historically centralised market design has conferred upon these actors defined roles in balancing, settlement, and risk management. The introduction of peer-to-peer (P2P) trading platforms potentially reconfigures these functions by redistributing coordination and verification responsibilities to digital infrastructures. Such redistribution may generate institutional resistance where perceived revenue streams, asset utilisation, or regulatory privileges are threatened. Empirical analyses of energy transitions demonstrate that incumbent actors often adapt strategically to preserve influence within evolving governance frameworks (Sovacool et al., 2019).

Regulatory capture risks further complicate decentralised innovation. Where blockchain platforms rely upon regulatory exemptions, sandbox participation, or bespoke licensing arrangements, the asymmetry of technical expertise between regulators and private developers may create dependency relationships. Yeung (2018) warns that algorithmic regulation can obscure normative choices within technical design, potentially shifting decision-making authority from public institutions to code architects. In energy contexts, this dynamic may manifest in tariff algorithms, prioritisation rules, or automated compliance mechanisms embedded within smart contracts. Without transparent oversight, such embedded rules may entrench private governance preferences under the appearance of neutral automation.

Governance friction also emerges at municipal and distribution-network levels. Local grid operators must ensure voltage stability, congestion management, and safety compliance. Decentralised energy trading platforms that operate without real-time coordination with distribution system operators (DSOs) may inadvertently introduce balancing pressures or settlement inconsistencies. As Mengelkamp et al. (2018) demonstrate in microgrid case studies, market design must integrate physical grid constraints to avoid destabilisation. Consequently, decentralisation does not eliminate coordination; rather, it redistributes it across digital and physical layers.

Decentralised Autonomous Organisations (DAOs) have been proposed as potential governance vehicles for community energy systems. In theory, DAOs enable collective decision-making through token-weighted voting and programmable treasury management (Reijers et al., 2016). Applied to energy markets, such structures could facilitate participatory tariff setting, surplus allocation, or renewable asset financing. However, DAO governance raises unresolved legal questions concerning liability, fiduciary duties, and jurisdictional recognition. The diffusion of responsibility within token-based governance models may amplify the "responsibility gap" identified in automated systems (Matthias, 2004). Until formal legal personality and accountability mechanisms are clarified, DAO-based energy governance is likely to remain experimental.

Political economy considerations are particularly salient in emerging economies. In jurisdictions where regulatory institutions face capacity constraints or where grid infrastructure remains unevenly developed, decentralised digital trading may encounter implementation barriers unrelated to technological design. Limited cybersecurity resources, inconsistent enforcement capacity, and reliance on state-owned utilities may restrict the operational autonomy of blockchain platforms. Moreover, energy justice concerns arise where decentralised trading disproportionately benefits asset-owning urban households while excluding rural or low-income populations (Sovacool et al., 2019).

These dynamics underscore a central insight: decentralised energy governance is co-produced by technology, law, and institutional power structures. Blockchain does not inherently democratise energy markets; rather, its governance outcomes depend on how regulatory authority, infrastructural control, and economic incentives are distributed. A legally coherent architecture must therefore account for institutional adaptation, regulatory transparency, and participatory safeguards if decentralised trading is to avoid reproducing existing asymmetries under digital form.

4. LEGAL ARCHITECTURE FOR DECENTRALISED ENERGY TRADING

Legal architecture denotes the principles, rules, and institutional arrangements governing blockchain-enabled decentralised energy markets. It integrates statutory provisions, regulatory standards, and technical norms that define the status of actors, the validity of transactions, and the treatment of digital energy assets (Diestelmeier, 2019). Within such markets, the architecture must provide legal certainty while maintaining sufficient flexibility to accommodate technological innovation. Smart contracts, distributed ledgers, and tokenised assets often operate across jurisdictions, necessitating clarity on what constitutes a legally valid transaction, who bears liability, and how oversight can be preserved without negating decentralisation.

4.1. Legal Taxonomy of Actors and Transactions

Decentralised trading alters traditional market hierarchies, distributing responsibilities among prosumers, aggregators, DSOs, regulators, and technology providers. Table 2 outlines these actors and their respective legal obligations. Such plurality requires hybrid governance models combining contractual autonomy with regulatory supervision. Because blockchain diffuses operational responsibility, legal frameworks must integrate contract, tort, and administrative principles to allocate accountability effectively.

Table 2. Taxonomy of actors and their legal roles in decentralised blockchain-based energy markets

Actor	Functional Role in Blockchain Market	Legal Status / Obligations	Examples of Regulatory Relevance
Prosumers	Generate and trade surplus renewable energy via smart contracts.	Must comply with grid codes, taxation, and licensing norms.	Licensing exemptions for small producers (EU Renewable Energy Directive 2018/2001).
Aggregators	Coordinate multiple DERs or consumers for trading or balancing purposes.	Act as intermediaries; may need authorisation as market participants.	EU Regulation 2019/943 (Electricity Market Design).
Distribution System Operators (DSOs)	Manage network access, ensure reliability, facilitate data exchange.	Subject to unbundling rules and data protection obligations.	Energy Efficiency Directive 2023/1791.

continued on following page

Table 2. Continued

Actor	Functional Role in Blockchain Market	Legal Status / Obligations	Examples of Regulatory Relevance
Regulators / Public Authorities	Oversee compliance, ensure consumer protection and market transparency.	Mandated to provide legal certainty and interoperability frameworks.	OFGEM's Regulatory Sandbox (UK, 2022).
Technology Providers	Develop blockchain platforms and manage smart contracts.	Potential joint liability for failures or data breaches.	GDPR and cybersecurity obligations under NIS2 Directive.

(Adapted from Diestelmeier, 2019; Office of Gas and Electricity Markets, 2023)

4.2. Smart Contracts and Legal Enforceability

Smart contracts automate pricing, settlement, and verification in decentralised markets (Werbach & Cornell, 2017). Their legal status, however, remains uneven. While instruments such as the UK Electronic Transactions Act (2019) recognise digital signatures, the enforceability of code-based agreements is still debated (Clack et al., 2016). Liability for coding errors or defective execution is equally uncertain, requiring frameworks capable of attributing responsibility within distributed systems. A dual-layer model—pairing on-chain automation with off-chain legal documentation—is increasingly accepted, supported by the UK Law Commission (2021). Given globally distributed nodes, regulators are also developing standard clauses on applicable law and dispute resolution (European Commission, 2024).

4.3. Dispute Resolution Mechanisms in Blockchain-Based Energy Trading

The automation of transactions through smart contracts does not eliminate the possibility of dispute; rather, it alters its form, timing, and locus. In decentralised energy markets, disputes may arise from coding errors, oracle malfunction, inaccurate metering data, tariff misapplication, token misclassification, cybersecurity breaches, or conflicts between on-chain execution and off-chain contractual intent. Accordingly, the legal architecture of blockchain-enabled energy trading must incorporate structured dispute-resolution mechanisms compatible with both automated execution and public regulatory oversight.

Smart contracts are frequently described as self-executing and immutable (Werbach & Cornell, 2017). Yet their operational finality can generate rigidity where factual circumstances deviate from coded assumptions. As Mik (2017) observes, code-based agreements often lack interpretive flexibility, and errors may propagate automatically once triggered. In energy markets—where settlement depends on real-time data inputs—faulty oracle transmissions or meter inaccuracies can produce cascading contractual consequences. Without corrective mechanisms, automated settlement may prioritise procedural certainty over substantive fairness.

Several dispute-resolution models have emerged in response to these challenges:

(1) Dual-Layer Contractual Architecture. A widely endorsed approach combines on-chain automation with off-chain legal agreements. Under this model, the smart contract governs operational settlement, while a traditional written contract specifies governing law, jurisdiction, liability allocation, and remedies in case of malfunction. The UK Law Commission (2021) recognises that English contract law can accommodate such hybrid structures, provided intention, certainty, and consideration are present. This layered approach preserves enforceability within established judicial systems while retaining the efficiency benefits of automated execution.

(2) Embedded Arbitration Clauses and Digital Escrow. Smart contracts may incorporate pre-programmed arbitration triggers. Upon detection of anomalous conditions—such as data inconsistency or contested delivery—execution may pause and funds be placed in escrow pending adjudication. Online dispute resolution (ODR) mechanisms, including digital arbitration panels, have been proposed as scalable solutions for high-frequency micro-transactions (Werbach & Cornell, 2017; Mik, 2017). However, questions remain regarding due process safeguards, enforceability of arbitral awards across jurisdictions, and compatibility with mandatory consumer-protection rules. In regulated energy markets, purely private arbitration may be insufficient. Consumer energy law often mandates access to public redress mechanisms and ombudsman services. Therefore, blockchain platforms operating in retail contexts must ensure compatibility with statutory complaint procedures and sector-specific dispute bodies.

(3) Regulatory Supervisory Intervention. In sandbox environments and licensed digital marketplaces, regulators may retain supervisory override authority. This model treats blockchain platforms not as fully autonomous ecosystems but as regulated infrastructures subject to administrative correction. While such intervention enhances consumer protection and systemic stability, it may partially reintroduce centralised oversight into ostensibly decentralised systems. The legitimacy of supervisory override depends on clearly defined thresholds and procedural transparency.

(4) Algorithmic Accountability and Reversibility Mechanisms. The principle of algorithmic accountability requires that automated systems remain contestable and reviewable (Yeung, 2018). In energy markets, this implies the existence of audit trails, explainable code structures, and controlled reversibility functions.

Although full immutability is often portrayed as a core blockchain virtue, limited reversibility—through governance voting mechanisms or multi-signature administrative controls—may be necessary to prevent irreversible harm. Such mechanisms must be carefully circumscribed to avoid arbitrary interference while ensuring legal proportionality.

4.3.1. Consumer Protection and Equity Considerations

Retail energy customers are typically protected by mandatory disclosure obligations, cooling-off rights, and regulated complaint channels. Embedding energy trading into smart contracts cannot circumvent these protections. Where decentralised platforms intermediate between prosumers and consumers, operators may assume joint liability for unfair terms or misleading pricing algorithms. Failure to integrate consumer safeguards risks regulatory sanction and reputational harm.

Moreover, unequal access to legal recourse may exacerbate distributive imbalances. Digitally sophisticated participants may be better positioned to challenge algorithmic outcomes, while vulnerable users may lack the technical literacy required to identify coding anomalies. Effective dispute mechanisms must therefore combine technical auditability with accessible procedural remedies.

4.3.2. Cross-Border Complexity

Blockchain networks often operate across jurisdictions, raising questions of applicable law and forum selection. Energy trading, however, remains territorially regulated. Conflicts may arise where a smart contract executed across distributed nodes interfaces with national grid obligations or tax regimes. Clear governing-law clauses and jurisdictional stipulations are essential to mitigate fragmentation. Absent such clarity, enforcement risks may deter institutional participation.

4.3.3. Concluding Observations

Dispute resolution in blockchain-based energy markets requires reconciling automation with legal contestability. Smart contracts do not displace adjudication; they necessitate its redesign. A robust legal architecture must therefore integrate hybrid contractual models, accessible redress mechanisms, regulatory supervision, and algorithmic transparency. Without such safeguards, the efficiency gains of decentralised trading may be undermined by legitimacy deficits and enforceability uncertainty.

4.4. Tokenisation of Energy Assets

Tokenisation converts physical or intangible energy assets into tradable digital tokens. These may represent energy units, infrastructure ownership, or compliance instruments such as Guarantees of Origin. Regulatory treatment depends on classification under financial, energy, or property law, as summarised in Table 3. While the MiCA Regulation (EU 2023/1114) provides partial guidance, many energy-specific tokens fall outside its scope (European Parliament & Council of the European Union,2023). Tokenisation enhances transparency and market liquidity, yet raises questions of digital possession, ownership transfer, and taxation (Werbach, 2021).

Table 3. Typology and legal implications of energy tokenisation

Token Type	Underlying Asset	Legal Classification	Regulatory Implications
Utility Token	Energy unit (kWh) or service access	May be exempt from financial regulation if non-speculative	Requires energy trading licence or consumer protection compliance
Asset Token	Share in generation infrastructure	May qualify as security under MiFID II	Subject to prospectus and disclosure rules
Compliance Token	Renewable energy or emission certificate	Regulated under energy or environmental law	Must align with national GOs and verification systems

4.5. Liability Models and Risk Allocation

Blockchain alters conventional liability chains by dispersing operational responsibility among network participants (Vyas et al., 2022). Three principal models have emerged:

- **Platform liability**, where operators of private or consortium blockchains assume responsibility for system failures (Finck, 2019);
- **Shared network responsibility**, distributing liability among participants through predefined contractual rules (Werbach & Cornell, 2017);
- **Regulatory oversight**, where authorities act as supervisory nodes, particularly within sandboxes (Office of Gas and Electricity Markets, 2023).

All models must align with wider frameworks for consumer protection, data security, and cybersecurity, including NIS2 obligations for critical energy infrastructures.

4.6. Data Governance, Privacy, and Compliance

Data governance is central to legality in decentralised systems. Blockchain's immutability presents tensions with GDPR requirements such as the right to erasure (European Data Protection Board, 2025; Finck, 2019). Hybrid solutions—off-chain storage, pseudonymisation, encryption, and privacy-by-design—are increasingly deployed to reconcile these demands. While blockchain enhances regulatory auditability, automated oversight must remain subject to human review, consistent with principles set out in the EU Artificial Intelligence Act (2024).

4.7. Sandboxing and Adaptive Regulation

Regulatory sandboxes allow blockchain applications to operate under controlled conditions, enabling experimental learning and iterative rule-making (OECD, 2025). Notable examples include OFGEM's Energy Regulation Sandbox, Singapore's EMA sandbox for blockchain-based RECs, and EU pilots under the European Blockchain Services Infrastructure (European Commission, 2023). These initiatives exemplify adaptive, collaborative governance approaches suited to rapidly evolving digital energy systems.

4.8. Towards a Coherent Regulatory Ecosystem

The interaction of digital, financial, and energy law demands regulatory coherence to avoid fragmentation. Table 4 summarises emerging frameworks across leading jurisdictions. Cross-border interoperability—harmonised data, cybersecurity, taxation, and certification rules—is increasingly seen as essential. Efforts by the European Blockchain Partnership (European Commission, 2024) reflect moves toward shared legal templates and standardisation.

Table 4. Comparative overview of regulatory ecosystems for blockchain-based energy trading

Jurisdiction / Framework	Key Legal Instruments	Regulatory Approach	Relevance to Decentralised Energy Trading
European Union	Clean Energy Package; MiCA Regulation 2023/1114; GDPR; NIS2	Integrated energy–digital governance	Recognises energy tokens; harmonises cybersecurity and data standards.
United Kingdom	Energy Act 2023; Law Commission Report on Smart Contracts; OFGEM Sandbox	Experimental and adaptive	Provides legal recognition of smart contracts and controlled testing environments.
Australia	National Electricity Rules; ASIC FinTech Sandbox; Powerledger Regulations	Innovation-led	Enables blockchain energy trading pilots under flexible regulatory schemes.
Singapore	EMA Sandbox; Payment Services Act (2019)	Risk-based and technology-neutral	Allows limited peer-to-peer trading and tokenised RECs under compliance protocols.
United States	FERC Regulations; SEC/FinCEN guidance; State-level sandboxes	Fragmented and sectoral	Encourages innovation but lacks federal uniformity for energy tokenisation.

(compiled from European Commission, 2023; Office of Gas and Electricity Markets, 2023; OECD, 2025)

4.9. Discussion: From Legal Certainty to Algorithmic Accountability

Blockchain shifts compliance from ex post enforcement to ex ante, code-embedded regulation, enhancing efficiency but transferring normative authority to software design (Lessig, 2006). Smart governments must therefore maintain human-centred oversight, ensuring that algorithmic processes remain anchored in democratic and constitutional principles. The objective is not to replace law with code but to ensure legal and technical interoperability in mutually reinforcing ways.

4.10. Summary of Key Legal Principles

The emergent legal architecture rests on five principles:

1. **Legal certainty** in the recognition and enforceability of blockchain transactions;
2. **Transparency and accountability**, supported by auditable ledgers;
3. **Data sovereignty and privacy**, achieved through privacy-by-design compliance;
4. **Technological neutrality**, ensuring innovation is not constrained;
5. **Interoperability**, enabling cross-border energy and data flows.

Collectively, these principles support a regulatory environment in which blockchain can operate within smart-governance frameworks. A legally coherent, privacy-respecting, and interoperable architecture will be critical for scaling decentralised energy markets while safeguarding consumer protection, market integrity, and democratic accountability.

5. REGULATORY FRAMEWORKS AND GAPS IN BLOCKCHAIN ENERGY MARKETS

The rapid emergence of blockchain-based energy trading systems has challenged existing regulatory paradigms across jurisdictions. Energy law, traditionally designed for centralised markets, must now accommodate distributed generation, peer-to-peer (P2P) exchanges, and digital asset management. The resulting tension between innovation and regulation highlights the need for adaptive, risk-based, and interoperable frameworks that ensure legal certainty while fostering technological progress.

This section analyses current regulatory approaches adopted in major jurisdictions, identifies critical legal gaps, and proposes pathways for harmonising blockchain governance within decentralised energy markets.

5.1. European Union: Integrated Energy–Digital Governance

The European Union (EU) has taken a leading role in integrating energy and digital policies under the European Green Deal and the Digital Strategy for Europe. The EU's Clean Energy for All Europeans Package (2019) introduced fundamental reforms in market design, consumer empowerment, and decentralised energy systems. Within this framework, blockchain is explicitly recognised as a potential enabler of transparency and traceability in energy data and certificates (European Commission, 2023).

The Digitalising the Energy System Action Plan (2023) sets out measures to develop secure and interoperable data spaces for energy. It envisions distributed ledger technologies as instruments for automating the certification of renewable energy and for facilitating cross-border trading of Guarantees of Origin (GOs). Simultaneously, the Markets in Crypto-Assets Regulation (MiCA) (EU 2023/1114) establishes a harmonised regime for digital tokens, addressing custody, disclosure, and consumer protection (European Parliament & Council of the European Union, 2023). While MiCA focuses primarily on financial and payment tokens, its definitional flexibility allows future extension to energy-related tokens such as RECs or GOs.

However, regulatory fragmentation persists among Member States regarding the treatment of blockchain-based energy platforms. Some countries, such as Germany and the Netherlands, have adopted technology-neutral approaches, allowing energy communities to experiment with DLT under existing market rules (BMWE/BMWK, 2023). Others lack clear legal status for smart contracts or tokenised assets. The absence of uniform data governance standards further complicates cross-border energy transactions, particularly concerning privacy and cybersecurity (European Data Protection Board, 2025).

5.2. United Kingdom: Experimental and Adaptive Regulation

The United Kingdom (UK) has established itself as a pioneer in experimental energy governance. Following its departure from the EU, the UK adopted an innovation-led regulatory approach, characterised by flexibility and stakeholder engagement. The Energy Act 2023 and OFGEM's Regulatory Sandbox Framework enable innovative energy products and services to operate under temporary exemptions or modified rules.

Within this framework, blockchain projects—such as Electron and Piclo Flex—have explored decentralised market models that integrate DLT for grid flexibility services and peer-to-peer energy exchanges (Office of Gas and Electricity Markets, 2023). These experiments inform permanent regulatory reforms by providing empirical evidence on performance, risk, and consumer outcomes.

In parallel, the UK Law Commission Report on Smart Contracts (2021) confirmed that English contract law is sufficiently robust to accommodate blockchain-based agreements. This recognition provides legal certainty for automated transactions in energy markets. However, data privacy and cybersecurity regulation remain governed by the UK GDPR and the Data Protection Act 2018, which may conflict with blockchain's immutability. To mitigate this, UK regulators encourage "privacy-by-design" principles and off-chain data management.

Despite progress, challenges persist in harmonising the regulatory treatment of energy tokens. The Financial Conduct Authority (FCA) distinguishes between utility and security tokens, but no formal classification exists for energy tokens that represent renewable generation or environmental value. The lack of clarity limits market scalability and investor confidence.

5.3. Australia: Innovation-Led and Market-Driven Frameworks

Australia's regulatory approach exemplifies a market-driven model that encourages innovation while maintaining oversight. The Australian Energy Regulator (AER) and the Australian Securities and Investments Commission (ASIC) jointly operate regulatory sandboxes for energy and fintech innovations. The Powerledger project, one of the earliest blockchain energy trading platforms, was launched under these schemes, enabling households to trade renewable energy within microgrids (Mengelkamp et al., 2018).

Legally, the National Electricity Rules provide flexibility for distributed energy resources (DERs) and pilot projects, while consumer protection remains under the National Energy Retail Law. However, energy tokens are not yet recognised as financial products unless they have speculative features (ASIC, n.d.). This creates ambiguity regarding taxation, asset classification, and consumer rights.

Data governance in Australia follows a risk-based model under the Privacy Act 1988 and the Consumer Data Right (Energy) Rules (2022), which promote interoperability and user control of energy data. Blockchain platforms that integrate these standards can achieve compliance by providing transparent consent mechanisms and secure identity management. Yet, cross-jurisdictional recognition of energy certificates remains limited.

5.4. Singapore: Technology-Neutral and Risk-Based Regulation

Singapore's Energy Market Authority (EMA) adopts a technology-neutral stance towards blockchain. The Energy Regulatory Sandbox allows startups and utilities to trial novel business models without full compliance burdens. Notably, the EMA's Renewable Energy Certificate (REC) Marketplace incorporates blockchain to verify authenticity and prevent double-counting.

The Payment Services Act (2019) and Monetary Authority of Singapore (MAS) FinTech Regulatory Sandbox regulate tokenised assets that may have payment or security characteristics. Singapore's approach emphasises proportionality: regulation is scaled to the risk level of the innovation. The result is a supportive environment for blockchain experimentation while maintaining legal certainty in consumer protection, anti-money laundering (AML), and cybersecurity.

Nevertheless, certain gaps persist. The absence of explicit legislative definitions for smart contracts and digital energy assets may create interpretive challenges. Furthermore, cross-sectoral coordination between energy and financial regulators is essential to avoid regulatory overlap or gaps in oversight.

5.5. United States: Fragmented and Sectoral Regulation

In the United States, regulatory competence for blockchain-based energy trading is divided between federal and state authorities. The Federal Energy Regulatory Commission (FERC) governs interstate energy transactions, while individual states regulate local distribution and retail activities. This division creates jurisdictional complexity for decentralised energy markets.

Blockchain pilots—such as Brooklyn Microgrid in New York—operate primarily under state-level experimental licences. Some states, including California and Texas, have initiated blockchain task forces to explore applications in energy and emissions tracking. However, there is no federal framework recognising energy tokens or smart contracts (Guerra & Prybutok, 2025).

Financial oversight by the Securities and Exchange Commission (SEC) and Financial Crimes Enforcement Network (FinCEN) applies to tokenised energy assets that qualify as securities or payment instruments. This patchwork approach introduces uncertainty for investors and operators, limiting large-scale deployment. Moreover, data protection remains fragmented, governed by state-level laws such as the California Consumer Privacy Act (CCPA), creating inconsistency in compliance requirements across jurisdictions.

5.6. Latin America and Emerging Economies

In Latin America, blockchain adoption in the energy sector is at a formative stage, but several countries are experimenting with regulatory sandboxes and pilot projects.

- **Chile** has launched Energy Lab Chile, an open innovation platform that includes blockchain pilots for energy traceability under the supervision of the Superintendencia de Electricidad y Combustibles (SEC).
- **Colombia** introduced a Regulatory Sandbox for Renewable Energy through Decree 2106 (2021), enabling distributed energy and blockchain trials.
- **Brazil's National Electric Energy Agency (ANEEL)** has recognised blockchain's potential for decentralised generation, particularly in peer-to-peer solar trading (ANEEL, 2023).

These initiatives show an emerging commitment to digital energy innovation. However, most Latin American jurisdictions lack comprehensive data protection and cybersecurity frameworks comparable to the EU's GDPR. This regulatory gap may hinder international interoperability and the scalability of blockchain-based systems.

5.7. Identified Legal and Regulatory Gaps

Despite global experimentation, several structural gaps remain evident across jurisdictions:

- **Lack of Legal Definition for Energy Tokens:** Most jurisdictions have yet to define the legal nature of tokenised energy units. This creates uncertainty regarding ownership, taxation, and transfer rights.
- **Ambiguity in Smart Contract Enforcement:** While common law systems (e.g., UK, US) tend to recognise digital contracts, civil law jurisdictions lack explicit provisions ensuring their enforceability, especially for automated transactions.
- **Cross-Border Interoperability:** Inconsistent regulatory standards impede cross-jurisdictional trading and recognition of renewable energy certificates. Harmonisation is required through mutual recognition agreements or regional directives (European Commission, 2024).
- **Data Protection Conflicts:** The immutability of blockchain records conflicts with data erasure and rectification rights under privacy laws such as GDPR and CCPA. Technical–legal alignment through hybrid architectures is essential (Finck, 2019).
- **Liability Allocation:** Decentralised governance obscures accountability in case of fraud, technical malfunction, or cybersecurity breaches. Legal frameworks need clear delineation of responsibilities between users, developers, and platform operators (Mik, 2017).
- **Regulatory Fragmentation:** The coexistence of sectoral, financial, and energy regulators creates overlap and ambiguity. Integrated governance mechanisms are needed to streamline decision-making and ensure coherent oversight (OECD, 2025).

5.8. Pathways for Regulatory Convergence

To address these gaps, smart governments should pursue regulatory convergence based on three complementary strategies:

- **Adaptive Regulation and Sandboxing:** Expand the use of regulatory sandboxes as dynamic testing environments where blockchain energy projects can operate under temporary, risk-adjusted exemptions. Lessons from these pilots should inform permanent reforms (Office of Gas and Electricity Markets, 2023).
- **Legal Interoperability Frameworks:** Develop international standards for token classification, smart contract enforceability, and data exchange. The European Blockchain Services Infrastructure (EBSI) and International Electrotechnical Commission (IEC) initiatives provide precedents for standardisation.

- **Hybrid Governance Models:** Combine state oversight with decentralised self-regulation through codes of conduct and auditable compliance protocols. Blockchain's transparency can enable regulators to adopt continuous monitoring rather than periodic audits (Yeung, 2018).
- **Ethical and Human-Centred Regulation:** Incorporate principles of fairness, non-discrimination, and human oversight in algorithmic regulation. The EU Artificial Intelligence Act (2024) provides a framework for integrating ethics into technical governance of automated systems (Yeung, 2018; Matthias, 2004).

Table 5 distils cross-jurisdictional regulatory tendencies, revealing persistent disparities in smart-contract recognition, token classification, and data-governance alignment. These contrasts underscore the need for coherent, interoperable legal frameworks to support scalable decentralised energy markets.

Table 5. Comparative summary of regulatory orientations, recognition levels, and legal gaps across selected jurisdictions

Jurisdiction	Regulatory Orientation	Legal Recognition of Smart Contracts / Tokens	Energy Sandbox / Pilot Projects	Key Legal Gap
European Union	Integrated digital–energy governance	Partial (MiCA, Clean Energy Package)	EBSI, GO Blockchain Pilots	Fragmented national implementation
United Kingdom	Experimental, adaptive	Recognised (Law Commission 2021)	OFGEM Sandbox, Electron	Unclear token classification
Australia	Market-driven innovation	Conditional (ASIC FinTech Rules)	Powerledger, AER Sandbox	Ambiguous asset classification
Singapore	Risk-based, tech-neutral	Implicit via Payment Services Act	EMA REC Marketplace	Lack of explicit energy token regulation
United States	Fragmented, sectoral	State-level only	Brooklyn Microgrid	No federal recognition
Latin America	Emerging, exploratory	Minimal recognition	Chile, Colombia, Brazil pilots	Weak data protection and interoperability

(compiled from European Commission, 2023; Office of Gas and Electricity Markets, 2023; OECD, 2025)

The global regulatory landscape for blockchain-based decentralised energy trading is characterised by experimentation, divergence, and gradual convergence. While the European Union and the United Kingdom lead in policy integration and adaptive regulation, most jurisdictions are still navigating the boundaries between innovation and compliance. Achieving coherence requires regulatory interoperability—the capacity of legal systems to communicate through shared principles, standards, and digital infrastructures. Figure 4 illustrates the layered governance architecture underpinning smart-grid implementations, highlighting how regulatory, technological, and operational domains interact. The diagram emphasises that effective decentralised energy systems depend on coherent coordination across these interdependent governance layers.

Figure 4. Governance layers for smart grid implementation

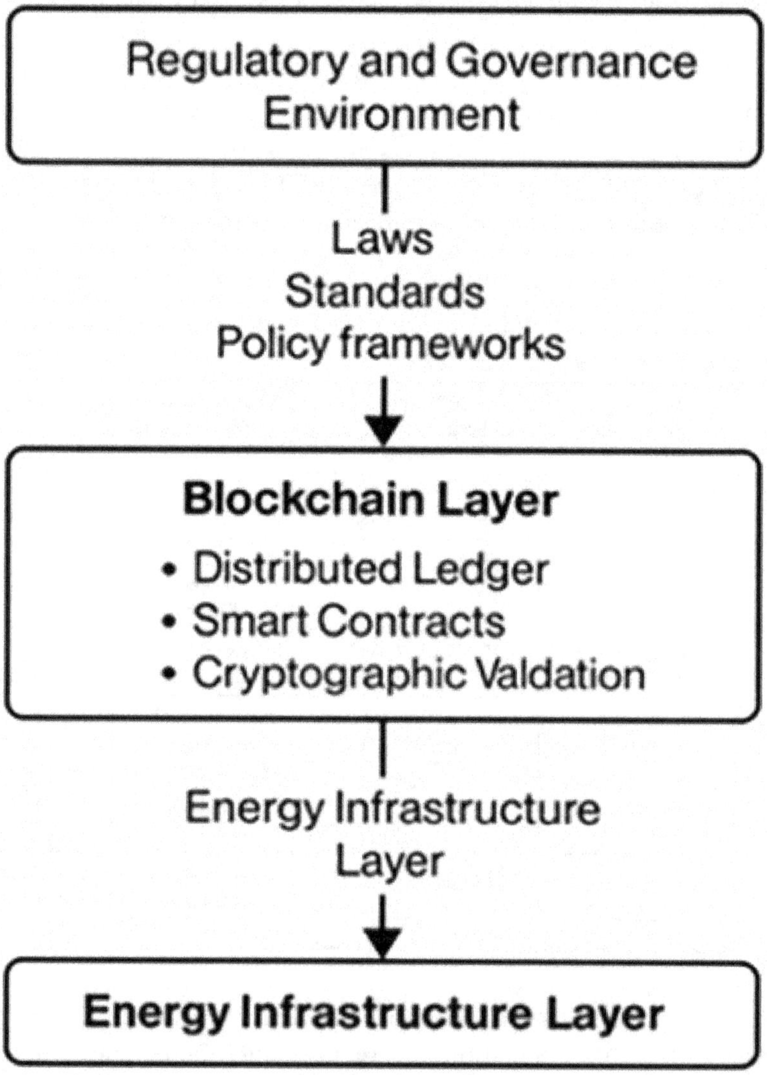

Smart governments play a critical role as orchestrators of this transition. By combining evidence-based policymaking, regulatory foresight, and multi-stakeholder collaboration, they can establish governance ecosystems that are both resilient and innovation-friendly. The next section illustrates these principles in practice through case studies of blockchain implementation in decentralised energy trading.

6. CASE STUDIES AND PRACTICAL IMPLEMENTATIONS

Practical deployments of blockchain in decentralised energy systems reveal how technical design, market structure, and legal regulation interact in real operational settings. Pilot projects demonstrate both the transformative potential of distributed ledgers and the constraints imposed by existing regulatory architectures. This section synthesises three influential initiatives—Brooklyn Microgrid (United States), Powerledger (Australia), and WePower (European Union)—to illustrate the implementation of peer-to-peer (P2P) trading, tokenisation, and automated compliance across diverse jurisdictions.

6.1. Brooklyn Microgrid (United States)

The Brooklyn Microgrid (BMG), launched in 2016 by LO3 Energy, is a pioneering attempt to facilitate local solar energy exchange through a blockchain-based P2P platform (Mengelkamp et al., 2018). Participating households use smart contracts to transact surplus photovoltaic generation with neighbours.

Technological Architecture

BMG employs a permissioned blockchain integrated with smart meters functioning as oracles that verify generation and consumption in real time. Smart contracts execute trades automatically, with settlement governed by local tariffs and user preferences (Tushar et al., 2020).

Regulatory Context

Operating under the New York State Public Service Commission and protected by the Reforming the Energy Vision (REV) demonstration framework, BMG was permitted to experiment without contravening retail-licensing rules. Nonetheless, U.S. federal and state fragmentation restricted full monetisation: transactions could be simulated via energy tokens but not implemented as direct retail sales (Guerra & Prybutok, 2025).

Legal Implications

BMG highlighted the need for regulatory categories capturing energy communities and digital marketplaces, as well as for sandboxing mechanisms balancing innovation and consumer protection.

Key Lessons

- P2P trading requires regulatory flexibility.
- Blockchain can complement regulated infrastructures when liability and data governance are clear.
- Federal fragmentation inhibits scale.

6.2. Powerledger (Australia)

Powerledger, established in Perth in 2016, offers a comprehensive blockchain ecosystem integrating P2P trading, renewable energy certificate (REC) issuance, and carbon-credit tokenisation.

Technological Design

The platform uses a dual-layer architecture: a permissioned chain for high-volume transactions and Ethereum for tokenisation. Its POWR token enables access to platform services, while smart contracts automate energy accounting and REC issuance (Fatima & Arshad, 2025). IoT-enabled metering provides real-time, oracle-verified data.

Regulatory Environment

Australia's regulatory stance is strongly innovation-oriented. Powerledger has operated within the ASIC FinTech Sandbox and the AER Energy Innovation Sandbox, ensuring compliance with both financial and energy law. Consumer data governance is defined by the Consumer Data Right (Energy) regime (ASIC, n.d.).

Legal and Economic Implications

POWR tokens are classified as utility assets—permitted so long as use remains non-speculative (ASIC, n.d.). Powerledger's model demonstrates how regulatory collaboration and sandboxing facilitate legally compliant tokenisation.

Key Lessons

- Sandboxing mitigates regulatory risk.
- Tokenised markets can enhance environmental compliance.
- Co-governance between regulators and innovators supports scalability.

6.3. WePower (European Union)

Founded in 2017, WePower represents one of the most advanced integrations of blockchain with EU renewable-energy governance. It enables corporates to procure renewable power directly from generators via smart-contract-based tokenisation.

Technological Framework

WePower tokenises each MWh of renewable generation into a WePower Energy Token (WPR), containing metadata on origin and carbon intensity. Tokens interface with national registries for Guarantees of Origin (GOs), creating a unified system for verification and compliance (European Commission, 2023).

Regulatory Integration

Operating under the Clean Energy Package and the Renewable Energy Directive (2018/2001), WePower benefits from harmonised EU digitalisation strategies, including the European Blockchain Services Infrastructure. However, under MiCA, tokens representing physical commodities remain outside conventional financial classifications, creating interpretative challenges.

Governance Implications

Smart contracts embed regulatory requirements directly into the transaction layer, supporting real-time compliance and reducing administrative burdens (Kouhizadeh et al., 2019).

Key Lessons

- Interoperability with national certification systems is essential.
- Tokenisation strengthens sustainability reporting.
- EU policy coherence accelerates adoption.

6.5. Concluding Remarks

Several themes recur across the three initiatives:

- **Sandboxing as a transitional regulatory tool.** Controlled experimentation allowed each project to be tested without full market exposure, providing regulators with empirical insights (OECD, 2025).
- **Legal interoperability and standardisation.** Tokenisation and smart-contract design require harmonised standards to avoid jurisdictional and technical isolation.
- **Hybrid governance models.** Blockchain supports polycentric arrangements where regulators, utilities, and consumers share oversight (Ansell & Gash, 2018).
- **Consumer empowerment.** Enhanced control over energy data operationalises energy citizenship, though equitable participation demands safeguards (Paudler et al., 2022).
- **Scalability and legal certainty.** Definitions of energy tokens, cross-border rules, and licensing frameworks remain essential for systemic expansion.

Table 6 contrasts leading blockchain energy pilots, showing how differing regulatory models, token strategies, and governance arrangements shape each project's scalability and legal robustness across jurisdictions.

Table 6. Comparative summary of blockchain energy trading case studies

Case Study	Jurisdiction	Regulatory Approach	Blockchain Functionality	Legal Innovation	Main Challenge
Brooklyn Microgrid	United States	State-level sandbox (REV Programme)	Peer-to-peer energy exchange	Demonstration licensing	Fragmented federal regulation
Powerledger	Australia	Multi-agency sandbox (AER & ASIC)	Tokenised P2P trading and RECs	Utility token classification	Ambiguity in taxation and financial scope
WePower	European Union	Integrated policy framework (Clean Energy Package, MiCA)	Tokenisation of renewable contracts and GOs	Interoperability with national registries	Limited token recognition under financial law

(Compiled from Mengelkamp et al., 2018; Guerra & Prybutok, 2025)

The case studies collectively illustrate that blockchain can operate simultaneously as a technological enabler and an instrument of regulatory transformation. Their divergent trajectories show that success depends less on the ledger itself than on the coherence between legal frameworks, governance structures, and system design. Ultimately, decentralised energy markets evolve through a co-adaptive process in which law, technology, and policy continually reshape one another, setting the stage for the ethical and legal analyses developed in the subsequent section.

7. CHALLENGES AND ETHICAL CONSIDERATIONS

The adoption of blockchain within decentralised energy systems constitutes a significant socio-technical shift, reshaping legal, economic, and ethical norms. Although gains in transparency, efficiency, and participatory governance are well recognised, the ethical implications remain comparatively underexamined. The principal challenges arise in relation to trust and transparency, accountability, data governance, social equity, and environmental sustainability.

7.1. Trust and Transparency in Decentralised Markets

Blockchain is frequently presented as a technological substitute for institutional trust, offering immutable and auditable transaction histories in peer-to-peer markets (Werbach, 2021). Yet transparency does not invariably equate to intelligibility. Technical visibility may obscure understanding for non-experts, creating a transparency–interpretability paradox.

In decentralised energy markets, where smart contracts and tokenised assets mediate decisions, transparency must therefore extend to explainability. Smart governments bear a responsibility to ensure that users comprehend how algorithmic processes shape market outcomes. Trust should be treated as a public asset, requiring mechanisms for independent certification, auditing, and redress (European Commission, 2023).

7.2. Accountability and Liability in Algorithmic Systems

Decentralisation disrupts established hierarchies of responsibility. Automated execution by smart contracts introduces attribution challenges when errors occur (Werbach & Cornell, 2017). Proposals to distribute liability among network participants or recognise limited algorithmic agency (Latif et al., 2021) have gained little regulatory traction. The European Parliament continues to insist on traceable human oversight for autonomous systems.

Ethically, dispersed agency risks creating “responsibility gaps” (Matthias, 2004). Clear ex ante duties, audit trails, and contestability procedures are essential to preserve human accountability within algorithmic governance.

7.3. Data Governance and Privacy

Decentralised energy systems generate sensitive personal and operational data (Kouhizadeh et al., 2021). Blockchain immutability, while beneficial for verification, conflicts with GDPR rights to rectification and erasure (Finck, 2019). Hybrid architectures—off-chain storage combined with hashed on-chain references—mitigate but do not eliminate these tensions. The European Data Protection Board (2025) emphasises that blockchain operators remain joint controllers under GDPR.

Ethically, data ownership and value distribution are central concerns. Without safeguards, private actors can exploit granular consumption datasets, reinforcing informational asymmetries (Zheng et al., 2023). Data-sovereignty models and co-operative governance structures can help ensure equitable control.

7.4. Social Equity and Inclusion

Decentralised energy trading aspires to support energy democratisation, yet risks entrenching inequalities if access to digital infrastructure and literacy is uneven (Paudler et al., 2022; Sovacool et al., 2019). Low-income and rural communities may struggle to participate in technologically demanding markets. Regulatory frameworks should therefore embed social-inclusion criteria, encompassing financial accessibility, simplified interfaces, and targeted education.

Algorithmic trading systems also raise distributive-justice concerns: efficiency-optimised mechanisms may disadvantage small-scale participants. Safeguards against speculation and market manipulation remain essential.

7.5. Environmental Sustainability and Ethical Technology Design

Blockchain's environmental impact is contingent on its consensus mechanism. Energy-intensive Proof-of-Work systems are incompatible with sustainability objectives (Stoll et al., 2019). More efficient protocols—PoS, PoA, DPoS—significantly reduce energy consumption (Goodkind et al., 2020) and are increasingly adopted in energy platforms. Ethical deployment requires sustainability-by-design principles, including renewable energy use in data-centre operations (European Commission, 2023), as well as consideration of hardware lifecycles and e-waste (Kouhizadeh et al., 2019).

7.6. Governance Ethics and Algorithmic Accountability

Blockchain embeds governance directly into code, challenging traditional notions of deliberation and legitimacy (Lessig, 2006). Algorithmic systems that influence pricing or allocation fall within the scope of high-risk AI under the EU Artificial Intelligence Act (2024), requiring human oversight, transparency of design, and independent auditability. Ethical legitimacy also necessitates stakeholder participation in governance processes (Yeung, 2018). The OECD Principles on AI (ASIC, n.d.) provide a broader normative basis for aligning blockchain governance with public values.

7.7. Concluding Remarks

Table 7 presents the main ethical and regulatory challenges in blockchain-enabled decentralised energy markets. Blockchain-enabled decentralised energy trading presents significant ethical and regulatory challenges that extend beyond technical optimisation. The illusion of technological neutrality obscures the reality that design choices embed normative commitments. Sustainable and equitable deployment therefore requires ethics as a foundational design principle, embedded from protocol architecture to public policy. Achieving transparent, accountable, and socially just energy systems will depend on aligning digital innovation with democratic values and long-term sustainability.

Table 7. Ethical and regulatory challenges in blockchain-enabled decentralised energy markets

Ethical Domain	Core Challenge	Implication for Governance	Policy Mechanism
Transparency and Trust	Technical opacity of blockchain	Ensuring epistemic accessibility and interpretability	Public certification of blockchain protocols
Accountability and Liability	Responsibility gaps in smart contracts	Maintaining human oversight and auditability	Legal frameworks defining shared responsibility
Data Governance and Privacy	Conflict between immutability and GDPR rights	Aligning blockchain design with data protection	Hybrid on-/off-chain storage and joint control mechanisms
Social Equity and Inclusion	Digital divide and market exclusion	Equitable access and fair distribution	Inclusive regulation and education initiatives
Environmental Sustainability	Energy-intensive consensus mechanisms	Carbon neutrality and lifecycle ethics	Mandatory sustainability-by-design requirements

(compiled from Finck, 2019; European Commission, 2023; OECD, 2025; Kouhizadeh et al., 2021)

8. POLICY RECOMMENDATIONS AND FUTURE DIRECTIONS

Blockchain's emergence as a core infrastructure for decentralised energy markets requires a substantial reconfiguration of regulatory and governance models. As control shifts from centralised institutions to distributed socio-technical networks, smart governments must provide coherence, legal certainty, and sustainability. This section outlines policy directions across five strategic domains—legal harmonisation, data governance, technological interoperability, social inclusion, and international cooperation—supported by comparative insights summarised in Figure 5.

Figure 5. Comparative matrix of smart energy strategies by region

8.1. Legal Harmonisation and Adaptive Regulation

Proactive regulation is essential for rapidly evolving blockchain ecosystems, replacing traditional reactive approaches (OECD, 2025). Institutions such as the EU Regulatory Foresight Framework (2023) and the UK Regulatory Horizon Council demonstrate how anticipatory oversight can identify impacts before deployment. Adaptive legal frameworks—revised through delegated regulation—enable iterative adjustment as evidence emerges from controlled experimentation. Regulatory sandboxes in the UK and Singapore illustrate how structured pilots (Office of Gas and Electricity Markets, 2023) facilitate learning and reduce compliance risks.

Legal clarity on smart contracts and energy tokenisation remains crucial. Policymakers should formalise the enforceability of smart contracts (UK Law Commission, 2021), categorise token types, and adopt model legislative clauses aligned with frameworks such as MiCA. These measures reduce uncertainty, stabilise markets, and support responsible innovation.

8.2. Data Governance and Privacy Protection

Decentralised markets rely on granular energy data, necessitating ethical governance mechanisms. Energy data trusts—neutral stewards managing access, consent, and value distribution—offer an effective institutional model (Hardjono & Pentland, 2021). Privacy-by-design must underpin blockchain infrastructures through off-chain data storage, zero-knowledge proofs, and dynamic consent arrangements, ensuring compliance with GDPR and comparable regimes. Strengthened cybersecurity is required for systems classified as critical infrastructure under NIS2 (2023), including enforced encryption, authentication, and coordinated incident reporting.

8.3. Technological Standardisation and Interoperability

Scalability depends on standardised protocols for smart contracts, energy data structures, and tokenisation (ISO/TC 307; IEC 61850; I-REC). International alignment through ISO, IEC, and CENELEC would advance technical and regulatory convergence. Secure digital identities and interoperable registries are essential for cross-platform participation, following models such as the EU Digital Identity Wallet. Public procurement favouring open-source development, in line with the EU Open Data Directive, would promote transparency and prevent vendor lock-in.

8.4. Social Inclusion and Capacity-Building

To avoid reinforcing inequalities, governments must guarantee equitable access to blockchain-enabled markets. Measures include subsidised digital infrastructure, accessible user interfaces, and public digital-literacy programmes. EU energy community initiatives exemplify inclusive governance models. Workforce capability is equally crucial: competence centres and reskilling initiatives (International Energy Agency, 2024) should cultivate interdisciplinary expertise in technology, law, and cybersecurity. Embedding ethics-by-design within innovation frameworks strengthens responsible experimentation (Birhane, 2021).

8.5. International Collaboration and Global Governance

Cross-border energy and data flows demand harmonised legal regimes. Mutual recognition agreements for smart contracts and tokenised certificates, together with model laws (e.g., OECD), would reduce uncertainty and facilitate interoperability. Global standards for ethical and sustainable blockchain—coordinated by UN/CEFACT and IRENA—should integrate principles aligned with SDG 7 and SDG 13. Enhanced North–South cooperation, supported by institutions such as the World Bank and IDB, is necessary to prevent technological dependency and to strengthen regulatory capacity (Sovacool et al., 2019).

8.6. Vision for the Future: Towards Algorithmic Governance in Energy Systems

Blockchain's evolution will accelerate the shift towards algorithmic regulation, where compliance mechanisms are embedded within code (Yeung, 2018). Hybrid governance—combining explainable algorithms with human oversight—will be essential to maintain constitutional integrity and public legitimacy. Decentralised Autonomous Organisations (DAOs) offer possibilities for community-level energy governance, but their viability requires explicit legal recognition and liability structures (Reijers et al., 2016). Beyond trading, blockchain could support integrated sustainability monitoring, linking energy production to environmental metrics within Blockchain-for-SDG frameworks (Thanasi-Boçe & Hoxha, 2025).

8.7. Concluding Remarks

Table 8 summarizes the strategic policy recommendations for blockchain-based decentralised energy systems. The Blockchain-enabled decentralisation constitutes both a technological shift and a profound transformation of governance. Smart governments must construct adaptive regulatory ecosystems that evolve with innovation while safeguarding accountability, equity, and sustainability. The future of energy governance will be shaped by co-regulatory models in which legal norms, technological design, and societal values co-evolve. A harmonised, ethically grounded, and human-centred blockchain ecosystem will enable efficient markets while reinforcing the democratic and environmental commitments fundamental to the digital energy transition.

Table 8. Summary of strategic policy recommendations for blockchain-based decentralised energy systems

Policy Domain	Key Recommendation	Expected Outcome
Legal Harmonisation	Adaptive frameworks recognising smart contracts and energy tokens	Legal certainty and innovation
Data Governance	Energy data trusts and privacy-by-design certification	Consumer trust and GDPR compliance
Standardisation	Interoperable blockchain/tokenisation standards (ISO/IEC)	Cross-border integration
Social Inclusion	Digital literacy and access programmes	Reduced digital divide
International Collaboration	MRAs and global ethics standards	Harmonised governance
Algorithmic Governance	Explainable AI with hybrid oversight	Transparent automated compliance

(compiled from OECD, 2025; European Commission, 2023; Yeung, 2018)

Table 9. Integrated legal–operational risk matrix for blockchain-enabled decentralised energy trading

Risk Category	Specific Risk	Legal Implication	Operational Impact	Governance Response
Regulatory Fragmentation	Divergent national approaches to blockchain and digital assets	Jurisdictional uncertainty; enforceability conflicts	Reduced cross-border scalability	Harmonised regulatory standards; mutual recognition frameworks
Smart Contract Rigidity	Coding errors or inflexible execution logic	Contractual disputes; liability allocation ambiguity	Automatic mis-settlement; consumer harm	Hybrid contracts; embedded arbitration clauses; supervisory override
Data Protection Conflict	Immutable ledgers vs. GDPR rights (erasure, rectification)	Compliance risk; administrative sanctions	System redesign costs; reputational harm	Off-chain storage; data minimisation; privacy-by-design architectures

continued on following page

Table 9. Continued

Risk Category	Specific Risk	Legal Implication	Operational Impact	Governance Response
Scalability Constraints	Limited transaction throughput; latency in consensus	Failure to meet settlement obligations	Grid imbalance; reduced market confidence	Layer-2 solutions; permissioned networks; throughput optimisation
Cybersecurity Vulnerability	Oracle manipulation; node compromise	Liability exposure; regulatory investigation	Financial losses; system instability	Multi-signature validation; real-time audit trails; cybersecurity standards
Institutional Resistance	Utility and DSO opposition to decentralised trading	Policy inertia; delayed licensing	Restricted pilot expansion	Inclusive regulatory dialogue; sandbox frameworks
Economic Non-Viability	Insufficient incentive alignment for prosumers	Platform insolvency; market exit	Low participation rates	Tariff reform; integration into wholesale markets
DAO Governance Ambiguity	Lack of recognised legal personality	Unclear fiduciary duties; responsibility gaps	Governance paralysis; investor uncertainty	Legal recognition frameworks; defined accountability rules
Consumer Protection Gaps	Algorithmic pricing opacity	Breach of consumer law obligations	Loss of trust; litigation	Mandatory disclosure standards; accessible redress mechanisms
Energy Justice Concerns	Exclusion of low-income or rural actors	Inequitable distributional outcomes	Social legitimacy deficit	Targeted inclusion policies; community ownership structures
Cross-Border Settlement Issues	Conflicting tax and energy regulations	Double taxation or non-compliance	Operational fragmentation	Clear governing-law clauses; regulatory cooperation mechanisms

Table 9 illustrates that blockchain-enabled decentralised energy trading entails multi-layered risk interactions across legal, technical, economic, and institutional domains. Crucially, no identified risk is purely technological; each carries regulatory and governance consequences. Conversely, most legal uncertainties generate operational reverberations. This interdependence underscores the necessity of an integrated regulatory architecture rather than sector-specific reform in isolation. Effective governance therefore requires simultaneous attention to enforceability, scalability, cybersecurity resilience, institutional adaptation, and distributive equity. The objective is not risk elimination—an unattainable goal—but structured risk allocation consistent with public law principles and energy market stability.

9. CONCLUSIONS

Blockchain-enabled decentralised energy trading represents neither a purely technological disruption nor an inevitable pathway towards democratised energy governance. Rather, it constitutes a structural reconfiguration of how regulatory authority, market coordination, and contractual enforcement are distributed across digital and institutional domains. The central argument advanced in this chapter is that such reconfiguration cannot be understood through technological optimism alone; it requires the deliberate construction of a coherent legal architecture capable of integrating automation with public oversight.

The analysis has demonstrated that decentralised trading systems operate at the intersection of energy regulation, digital governance, and contractual doctrine. Smart contracts may enhance transactional efficiency, yet they also introduce rigidity and accountability challenges. Distributed ledgers may increase transparency, yet they generate data protection tensions and jurisdictional complexity. Peer-to-peer market models may expand participation, yet their economic sustainability depends upon tariff alignment, grid integration, and incentive compatibility. In each instance, technological design and legal structure are mutually constitutive rather than sequential.

A core insight emerging from this study is that decentralisation does not eliminate governance; it redistributes it. Decision-making authority may shift from central utilities to code-based protocols, from administrative agencies to platform governance mechanisms, or from public adjudication to embedded algorithmic rules. Without careful institutional design, such shifts risk obscuring responsibility, entrenching asymmetries, or generating regulatory fragmentation. The legitimacy of blockchain-enabled markets therefore depends upon maintaining contestability, accountability, and proportionality within automated systems.

The integrated risk framework developed in this chapter underscores the multidimensional nature of decentralised energy governance. Legal enforceability, operational scalability, cybersecurity resilience, institutional adaptation, and distributive equity are interdependent variables. Addressing one dimension in isolation is insufficient. A fragmented regulatory response may inadvertently undermine market stability or public trust.

From a policy perspective, the findings suggest that effective governance should not seek either to prohibit decentralised experimentation or to relinquish oversight in the name of innovation. Instead, regulators should pursue calibrated integration: hybrid contractual models, interoperable digital standards, sandbox experimentation with clear exit pathways, and explicit accountability mechanisms embedded within platform design. Such an approach recognises the transformative potential of distributed ledger technologies while preserving the foundational principles of energy law and administrative accountability.

Ultimately, blockchain does not automatically democratise energy systems, nor does it inherently destabilise them. Its governance implications are contingent upon institutional choices. By articulating a structured legal architecture for decentralised energy trading, this chapter contributes to the broader debate concerning how digitally mediated infrastructures can be aligned with sustainability objectives, regulatory coherence, and constitutional safeguards. The challenge ahead is not merely technological refinement, but principled integration.

REFERENCES

Agência Nacional de Energia Elétrica (ANEEL). (2023). 2022 management report: Integrated report. Brasília, DF: ANEEL. Retrieved October 2, 2025, from https://www2.aneel.gov.br/cedoc/aprt20236810_2.pdf

Andoni, M., Robu, V., Flynn, D., Abram, S., Geach, D., Jenkins, D., McCallum, P., & Peacock, A. (2019). Blockchain technology in the energy sector: A systematic review of challenges and opportunities. *Renewable & Sustainable Energy Reviews*, *100*, 143–174. DOI: 10.1016/j.rser.2018.10.014

Ansell, C., & Gash, A. (2018). Collaborative platforms as a governance strategy. *Journal of Public Administration: Research and Theory*, *28*(1), 16–32. DOI: 10.1093/jopart/mux030

Australian Securities and Investments Commission (ASIC). (n.d.). Enhanced regulatory sandbox (ERS). Retrieved October 3, 2025, from https://www.asic.gov.au/for-business-and-companies/innovation-hub/enhanced-regulatory-sandbox-ers/

Bassey, K. E., Rajput, S. A., & Oyewale, K. (2024). *Peer-to-peer energy trading: Innovations, regulatory challenges, and the future of decentralized energy systems*. World Journal Of Advanced Research and Reviews., DOI: 10.30574/wjarr.2024.24.2.3324

Birhane, A. (2021). Algorithmic injustice: A relational ethics approach. *Patterns (New York, N.Y.)*, *2*(2), 100205. DOI: 10.1016/j.patter.2021.100205 PMID: 33659914

Clack, C. D., Bakshi, V. A., & Braine, L. (2016). Smart contract templates: Foundations, design landscape and research directions. arXiv Preprint. https://doi.org//arXiv.1608.00771DOI: 10.48550

Diestelmeier, L. (2019). Changing power: Shifting the role of electricity consumers with blockchain technology – Policy implications for EU electricity law. *Energy Policy*, *128*, 189–196. DOI: 10.1016/j.enpol.2018.12.065

European Commission. (2023). Digitalisation of the energy system. Retrieved October 14, 2025, from https://energy.ec.europa.eu/topics/eus-energy-system/digitalisation-energy-system_en

European Commission. (2024). European Blockchain Partnership. Retrieved October 5, 2025, from https://digital-strategy.ec.europa.eu/en/policies/blockchain-partnership

European Data Protection Board. (2025). Guidelines 02/2025 on processing of personal data through blockchain technologies (Public consultation). Retrieved October 15, 2025, from https://www.edpb.europa.eu/our-work-tools/documents/public-consultations/2025/guidelines-022025-processing-personal-data_en

Fatima, S., & Arshad, M. J. (2025). A comprehensive review of blockchain and machine learning integration for peer-to-peer energy trading in smart grids. *IEEE Access : Practical Innovations, Open Solutions*, *13*, 92756–92782. DOI: 10.1109/ACCESS.2025.3572174

Federal Ministry for Economic Affairs and Climate Action (BMWE/BMWK). (2023). The climate-neutral electricity system platform – in dialogue for a new market design. https://www.bundeswirtschaftsministerium.de/Redaktion/EN/Dossier/the-climate-neutral-electricity-system-platform.html

Finck, M. (2019). Blockchain and the general data protection regulation: Can distributed ledgers be squared with European data protection law? Max Planck Institute for Innovation & Competition Research Paper No. 18-01. DOI: 10.2139/ssrn.3080322

Gil-Garcia, J. R., Gasco-Hernandez, M., & Pardo, T. A. (2020). Beyond transparency, participation, and collaboration? A reflection on the dimensions of open government. *Public Performance & Management Review*, *43*(3), 483–502. DOI: 10.1080/15309576.2020.1734726

Goodkind, A. L., Jones, B. A., & Berrens, R. P. (2020). Cryptodamages: Monetary value estimates of the air pollution and human health impacts of cryptocurrency mining. *Energy Research & Social Science*, *59*, 101281. DOI: 10.1016/j.erss.2019.101281

Guerra, K., & Prybutok, V. (2025). Blockchain technology: An analysis of economic, technical, and legal implications. *International Review of Law Computers & Technology*, •••, 1–21. DOI: 10.1080/13600869.2025.2497632

Hardjono, T., & Pentland, A. (2021). Data cooperatives: Towards a foundation for decentralized personal data management. Computers and Society. https://doi.org//arXiv.1905.08819DOI: 10.48550

International Energy Agency. (2024). World Energy Outlook 2024 [Report]. https://www.iea.org/reports/world-energy-outlook-2024

Kouhizadeh, M., Sarkis, J., & Zhu, Q. (2021). Blockchain and the circular economy: Critical reflections and future research directions. *Production Planning and Control*, *32*(13), 1041–1054. DOI: 10.1080/09537287.2020.1732906

Kouhizadeh, M., Zhu, Q., & Sarkis, J. (2019). Blockchain and the circular economy: Potential tensions and critical reflections from practice. *Production Planning and Control, 31*(11–12), 950–966. DOI: 10.1080/09537287.2019.1695925

Latif, S., Idrees, Z., Huma, Z., & Ahmad, J. (2021). Blockchain technology for the industrial Internet of Things: A comprehensive survey on security challenges, architectures, applications, and future research directions. *Transactions on Emerging Telecommunications Technologies, 32*(11), e4337. DOI: 10.1002/ett.4337

Lessig, L. E. (2006). *Code and other laws of cyberspace, version 2.0*. Basic Books.

Matthias, A. (2004). The responsibility gap: Ascribing responsibility for the actions of learning automata. *Ethics and Information Technology, 6*(3), 175–183. DOI: 10.1007/s10676-004-3422-1

Meijer, A., & Bolívar, M. P. R. (2016). Governing the smart city: A review of the literature on smart urban governance. *International Review of Administrative Sciences, 82*(2), 392–408. DOI: 10.1177/0020852314564308

Mengelkamp, E., Gärttner, J., Rock, K., Kessler, S., Orsini, L., & Weinhardt, C. (2018). Designing microgrid energy markets: A case study: The Brooklyn Microgrid. *Applied Energy, 210*, 870–880. DOI: 10.1016/j.apenergy.2017.06.054

Mik, E. (2017). Smart contracts: Terminology, technical limitations and real-world complexity. *Law, Innovation and Technology, 9*(2), 269–300. DOI: 10.1080/17579961.2017.1378468

OECD. (2025). *Regulating for the future: OECD Regulatory Policy Outlook 2025*. OECD Publishing., DOI: 10.1787/56b60e39-

Office of Gas and Electricity Markets. (2023). Consultation on updates to Data Best Practice Guidance and Digitalisation Strategy and Action Plan Guidance (Consultation OFG1163) [PDF]. https://www.OFGEM.gov.uk/sites/default/files/2023-02/Consultation%20on%20updates%20to%20Data%20Best%20Practice%20Guidance%20and%20Digitalisation%20Strategy%20and%20Action%20Plan%20Guidance.pdf

Paudler, H. J. A., Schwanitz, V. J., & Wierling, A. (2022). Realizing a human-centered digitalization of the energy sector. *Open Research Europe, 2*, 140. DOI: 10.12688/openreseurope.15129.1 PMID: 37645332

Reijers, W., O'Brolcháin, F., & Haynes, P. (2016). Governance in blockchain technologies & social contract theories. Ledger, 1, 134 151. DOI: 10.5195/ledger.2016.62

Sovacool, B. K., Sovacool, B. K., Martiskainen, M., Hook, A., & Baker, L. (2019). Decarbonization and its discontents: A critical energy justice perspective on four low-carbon transitions. *Climatic Change, 155*(4), 581–619. DOI: 10.1007/s10584-019-02521-7

Stoll, C., Klaaßen, L., & Gallersdörfer, U. (2019). The carbon footprint of bitcoin. *Joule, 3*(7), 1647–1661. DOI: 10.1016/j.joule.2019.05.012

Thanasi-Boçe, M., & Hoxha, J. (2025). Blockchain for Sustainable Development: A Systematic Review. *Sustainability (Basel), 17*(11), 4848. DOI: 10.3390/su17114848

Tushar, W., Saha, T. K., Yuen, C., Smith, D., Poor, H. V., & Morstyn, T. (2020). Peer-to-peer energy systems for connected communities: A review of recent advances and emerging challenges. *Applied Energy, 282*, 116131. DOI: 10.1016/j.apenergy.2020.116131

Vyas, N., Beije, A., & Krishnamachari, B. (2022). *Blockchain and the supply chain: Concepts, strategies and practical applications* (2nd ed.). Kogan Page.

Werbach, K. (2021). *The blockchain and the new architecture of trust*. MIT Press.

Werbach, K., & Cornell, N. (2017). Contracts Ex Machina. *Duke Law Journal, 67*(2), 313–382. https://scholarship.law.duke.edu/dlj/vol67/iss2/2/

Yeung, K. (2018). Algorithmic regulation: A critical interrogation. *Regulation & Governance, 12*(4), 505–523. DOI: 10.1111/rego.12158

Zheng, Y., Zhang, R., Xu, J., Gong, Y., You, Y., & Zhang, Y. (2023). *Research on the data governance in power industry based on data owner. In 2023 Panda Forum on Power and Energy (PandaFPE)*. IEEE., DOI: 10.1109/PandaFPE57779.2023.10140515

KEY TERMS AND DEFINITIONS

Algorithmic Accountability: A governance principle requiring that automated systems—including smart contracts and blockchain-based trading algorithms—remain transparent, auditable, and subject to human oversight. In decentralised energy markets, algorithmic accountability ensures that embedded regulatory logic does not undermine legal rights, fairness, or democratic oversight.

Blockchain (Distributed Ledger Technology – DLT): A decentralised digital ledger system that records transactions across multiple nodes in a network using cryptographic validation and consensus mechanisms. In decentralised energy markets, blockchain ensures immutability, transparency, traceability, and tamper-resistant recording of energy transactions, certificates, and smart contract executions without reliance on a central intermediary.

Decentralised Energy Trading (Peer-to-Peer Energy Markets): A market structure in which electricity producers (prosumers) and consumers transact directly with one another through digital platforms, often supported by blockchain infrastructure. This model reduces reliance on central utilities while requiring adapted legal frameworks for licensing, liability, and consumer protection.

Guarantees of Origin (GOs) / Renewable Energy Certificates (RECs): Certifiable instruments that verify the renewable source of electricity generation. When tokenised on blockchain platforms, these certificates can be digitally tracked, preventing double counting and enhancing transparency in sustainability reporting and cross-border renewable energy trade.

Legal Architecture for Decentralised Energy Markets: The integrated framework of statutory provisions, regulatory standards, contractual norms, and technical protocols that collectively define rights, obligations, liability structures, and compliance mechanisms in blockchain-enabled energy systems. A coherent legal architecture reconciles innovation with consumer protection, data governance, cross-border interoperability, and constitutional safeguards.

Prosumer: An energy market participant who simultaneously produces and consumes electricity, typically through distributed renewable energy installations such as rooftop solar photovoltaic systems. Prosumers may trade surplus generation through blockchain-enabled platforms under regulated grid and market conditions.

Regulatory Sandbox: A controlled legal environment in which innovative technologies or business models are tested under temporary regulatory exemptions or modified compliance requirements. In blockchain-based energy systems, sandboxes enable experimentation with peer-to-peer trading, smart contracts, and tokenisation while safeguarding consumer protection and market integrity.

Smart Contract: A self-executing digital agreement encoded in software that automatically performs predefined actions when specified conditions are met. Within decentralised energy trading, smart contracts automate pricing, settlement, compliance verification, and energy token transfers, operating as programmable instruments of regulatory and contractual governance.

Smart Governance: A governance paradigm that integrates digital technologies, regulatory foresight, data-driven policymaking, and participatory mechanisms to enhance transparency, accountability, and sustainability. In the energy sector, smart governance aligns legal frameworks with digital infrastructures to support decentralised, interoperable, and citizen-centred energy systems.

Tokenisation of Energy Assets: The process of digitally representing physical or intangible energy-related assets—such as kilowatt-hours (kWh), renewable energy certificates (RECs), or shares in generation infrastructure—through blockchain-based tokens. Tokenisation enhances liquidity, traceability, and fractional ownership while raising legal classification and regulatory compliance questions.

Chapter 6
Regulatory Sandboxes for Legal Testing of Blockchain-Based Decentralized Energy Trading Models

Wildan Kamil
http://orcid.org/0009-0007-4113-2252
Universitas Esa Unggul, Indonesia

Gusti Muhamad Sardana
http://orcid.org/0009-0001-4281-2432
Universitas Esa Unggul, Indonesia

Binastya Anggara Sekti
http://orcid.org/0000-0001-5489-4888
Universitas Esa Unggul, Indonesia

ABSTRACT

The rapid development of blockchain-based decentralized energy trading has introduced peer-to-peer (P2P) market models that challenge traditional centralized electricity regulation. Existing energy regulatory frameworks, designed for vertically integrated utility systems, often lack the flexibility to accommodate prosumers, smart contracts, and automated transactions. This book examines regulatory sandboxes as adaptive legal frameworks for testing blockchain-driven decentralized energy trading models under controlled regulatory conditions. It analyzes how sandboxes enable experimentation with innovative energy business models while safeguarding

DOI: 10.4018/979-8-3373-5742-3.ch006

consumer protection, grid reliability, data privacy, and market integrity. Through legal and policy-oriented analysis and comparative international case studies, the book explores sandbox design principles and key regulatory challenges across energy law, data protection, competition law, and liability regimes. The findings show that regulatory sandboxes reduce legal uncertainty and support evidence-based regulatory development for markets.

1. INTRODUCTION

The global energy sector is undergoing a fundamental transformation driven by the convergence of digitalization, decarbonization, and decentralization. At the forefront of this revolution is blockchain technology, which promises to reshape traditional energy trading models by enabling peer-to-peer (P2P) transactions, enhancing transparency, and facilitating the integration of distributed energy resources (DERs) such as rooftop solar panels, battery storage systems, and electric vehicles (Burger & Weinmann, 2022; Evens et al., 2023). Blockchain-driven decentralized energy trading platforms allow prosumers consumers who also produce energy to directly buy and sell electricity without intermediaries, creating more efficient markets and empowering individuals to participate actively in the energy transition (Seven et al., 2020). This technological innovation has gained significant momentum globally, with pilot projects and commercial implementations emerging across developed and developing nations alike. However, the disruptive nature of these blockchain-based energy models challenges existing regulatory frameworks that were designed for centralized, utility-dominated electricity systems, creating an urgent need for adaptive regulatory approaches that can foster innovation while maintaining grid stability, consumer protection, and public interest safeguards.

The regulatory challenges posed by blockchain-driven decentralized energy trading are multifaceted and complex, spanning across energy law, data protection, consumer rights, competition policy, and financial regulation (Pradhan et al., 2022). Traditional energy regulations typically assume a hierarchical structure where licensed utilities generate, transmit, and distribute electricity to passive consumers under strict regulatory oversight. In contrast, blockchain-enabled P2P trading platforms create dynamic, multi-directional energy flows among numerous participants, blurring the lines between producers, consumers, and traders (Heng et al., 2022). Key regulatory concerns include: determining the legal status of prosumers who engage in energy trading; establishing appropriate licensing requirements for platform operators; ensuring fair access to grid infrastructure; protecting consumer data and privacy in transparent blockchain systems; maintaining electricity quality and reliability standards; preventing market manipulation and ensuring fair competition; and addressing liability issues in automated, smart contract-driven transactions. Moreover, the cross-border nature of blockchain networks raises jurisdictional questions, particularly in regions with interconnected grids and varying national regulations. These regulatory uncertainties create significant barriers to entry for innovative startups and deter investment in potentially transformative energy technologies, while simultaneously exposing early adopters to legal risks and potential penalties for non-compliance with ambiguous or outdated regulations (Baig et al., 2023).

In response to these challenges, regulatory sandboxes have emerged as an innovative policy instrument that allows controlled testing of novel business models and technologies in a live market environment with regulatory flexibility and oversight (Thanasi-Boçe & Hoxha, 2025). Originating in the financial technology (fintech) sector, regulatory sandboxes provide a "safe space" where innovators can experiment with new products and services under relaxed regulatory requirements, while regulators observe, learn, and gather evidence to inform future policy decisions. In the energy sector, regulatory sandboxes enable blockchain-based trading platforms to operate on a limited scale geographically, temporally, or in terms of participant numbers without requiring full compliance with all existing regulations, provided that adequate consumer safeguards are in place (Fatima & Arshad, 2025). This approach offers multiple benefits: it reduces regulatory barriers for innovators while managing risks through controlled experimentation; it provides regulators with practical insights into emerging technologies and business models, enabling evidence-based policy development; it protects consumers through enhanced monitoring and safeguards during the testing phase; and it accelerates the pace of innovation by providing clearer pathways from concept to market. Several jurisdictions, including the United Kingdom, Australia, Singapore, and various European Union member states, have implemented energy-focused regulatory sandboxes with varying designs and outcomes, offering valuable lessons for policymakers worldwide seeking to balance innovation with consumer protection and market integrity (Jogunola et al., 2024).

This chapter aims to provide a comprehensive analysis of regulatory sandboxes as legal frameworks for testing innovative energy models in blockchain-driven decentralized energy trading, examining their design principles, implementation experiences, challenges, and effectiveness in bridging the gap between technological innovation and regulatory adaptation (Papapostolou et al., 2025). The chapter pursues several specific objectives: first, to establish a theoretical foundation by reviewing the concepts of regulatory sandboxes, blockchain technology in energy systems, and the regulatory gaps that necessitate innovative policy approaches; second, to analyze the design elements and best practices of energy-focused regulatory sandboxes through comparative examination of international implementations; third, to identify and examine the key legal and regulatory challenges that arise in sandbox testing of blockchain-based energy trading platforms, including issues related to energy law, data protection, consumer rights, and cross-border transactions; fourth, to evaluate the effectiveness of regulatory sandboxes in achieving their stated goals of fostering innovation, informing regulatory development, and protecting public interest; and fifth, to develop evidence-based policy recommendations for regulators, industry participants, and other stakeholders involved in the governance of emerging energy technologies (Grigoryan, 2024). The scope of this chapter encompasses both technical and legal dimensions of blockchain-enabled energy trading, while maintaining a primary focus on regulatory frameworks and policy implications rather than detailed technical specifications of blockchain protocols or energy systems engineering (Yahaya et al., 2021).

The remainder of this chapter is structured as follows: Section 2 presents the theoretical framework and reviews relevant literature on regulatory sandboxes, blockchain technology, and decentralized energy trading, establishing the conceptual foundation for subsequent analysis. Section 3 examines the specific characteristics of blockchain-based energy models that necessitate regulatory innovation and the limitations of traditional regulatory approaches in accommodating these innovations. Section 4 articulates the key design principles for effective energy-focused regulatory sandboxes, including eligibility criteria, scope definition, safeguard mechanisms, and exit strategies. Section 5 provides in-depth case studies of regulatory sandbox implementations across multiple jurisdictions, analyzing their approaches, outcomes, and lessons learned. Section 6 explores the major legal and regulatory challenges that arise in sandbox testing, examining issues across energy law, data protection, consumer rights, competition policy, and jurisdictional matters. Section 7 investigates the technical-legal interface, focusing on how legal frameworks must adapt to accommodate automated smart contract execution, grid integration requirements, and settlement mechanisms. Section 8 analyzes stakeholder perspectives and the importance of multi-stakeholder governance in sandbox design and operation. Section 9 synthesizes success factors and best practices derived from international experiences, while Section 10 discusses future directions for regulatory evolution and presents actionable policy recommendations. Finally, Section 11 concludes by summarizing key findings and identifying priorities for future research and policy development in this rapidly evolving field.

2. THEORETICAL FRAMEWORK AND LITERATURE REVIEW

This chapter adopts a conceptual framework that positions regulatory sandboxes as a mediating governance mechanism between blockchain-enabled peer-to-peer energy trading and adaptive regulatory development. As illustrated in Figure 1, the framework integrates technological innovation, market participation, and legal experimentation within a controlled regulatory environment designed to contain systemic risks, protect consumers, and generate regulatory learning. This framework guides the subsequent theoretical discussion and comparative analysis by linking blockchain technology, decentralized energy markets, and regulatory adaptation within a unified governance perspective.

Figure 1. Conceptual framework of blockchain-based decentralized energy trading and regulatory sandbox governance

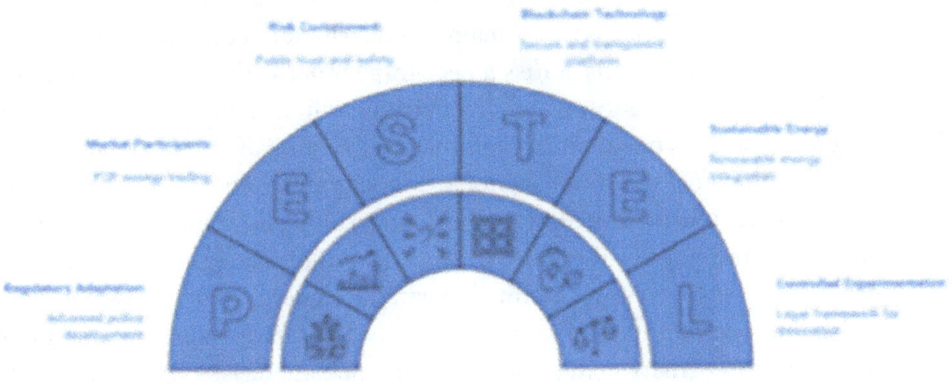

Regulatory sandboxes represent an adaptive and experimental regulatory approach that departs from traditional rigid, ex-ante rule-making (Burger & Weinmann, 2022). Introduced by the UK Financial Conduct Authority (FCA) in 2016, a regulatory sandbox is defined as a framework that allows innovators to test new products and services in a controlled environment under close regulatory supervision. Core principles of regulatory sandboxes include regulatory flexibility, controlled experimentation, continuous regulatory oversight, consumer protection, and regulatory learning to inform future policy development (Papapostolou et al., 2025). The primary objectives of this approach are to accelerate innovation, enable regulators to better understand emerging technologies, and safeguard public interests. The sandbox model is theoretically grounded in the concepts of responsive and adaptive regulation, which emphasize flexibility and iterative learning in response to technological change (Grigoryan, 2024).

Peer-to-peer (P2P) energy trading enables direct electricity transactions between producers and consumers without traditional utility intermediaries and has emerged alongside the diffusion of distributed energy resources such as rooftop solar and energy storage systems (Seven et al., 2020). P2P energy trading is commonly defined as a decentralized market arrangement in which prosumers can buy and sell electricity according to their individual preferences and needs. This model offers potential benefits including higher utilization of renewable energy, reduced transmission losses, increased consumer empowerment, and lower transaction costs (Wang et al., 2019). Existing studies identify several P2P market architectures, ranging from fully decentralized models to community-based and hybrid configurations. The shift from passive consumers to active prosumers fundamentally challenges conventional utility business models and existing approaches to grid management (Heng et al., 2022).

Blockchain technology is widely regarded as a key enabler of P2P energy trading due to its transparency, immutability, and ability to support disintermediated transactions (Zhou et al., 2022). Blockchain is a distributed ledger technology that records and verifies transactions across a peer-to-peer network without reliance on a central authority. In the energy sector, blockchain facilitates transparent recording of energy generation and consumption, automated execution of transactions through smart contracts, low-cost microtransactions, and energy tokenization (Baig et al., 2023). Smart contracts are particularly important, as they allow automated pricing, settlement, and coordination of energy trades. Nevertheless, significant technical challenges remain, including scalability constraints, energy-intensive consensus mechanisms, interoperability with legacy energy systems, and secure integration of off-chain data such as smart-meter readings. These characteristics position blockchain not merely as a technical tool, but as a governance-enabling infrastructure that reshapes market interactions and regulatory assumptions in decentralized energy systems (Grigoryan, 2024).

Within this conceptual framework, the rapid development of blockchain-based P2P energy systems has exposed a regulatory gap, as existing energy regulation was designed for centralized systems characterized by vertically integrated utilities and passive consumers. Current regulatory frameworks often fail to clearly define the legal status of prosumers, platform operators, and peer-to-peer transactions, while also raising concerns related to grid access, data protection, and consumer rights. This regulatory uncertainty creates barriers to scaling innovative energy solutions beyond pilot projects and into commercial deployment.

The academic literature suggests that regulatory sandboxes offer a promising middle ground between prohibiting innovation and prematurely imposing rigid regulatory structures. Studies in both the financial and energy sectors indicate that sandboxes can support experimentation with novel business models, including blockchain-based energy trading, while generating regulatory learning. However, scholars also highlight limitations, such as restricted scalability, potential competitive distortions, and uncertain long-term impacts. Overall, the literature reveals a need for further research on the effectiveness of regulatory sandboxes in addressing the complex regulatory challenges posed by blockchain-enabled P2P energy systems.

This gap motivates the analytical focus of this chapter on regulatory sandboxes as adaptive governance mechanisms for blockchain-enabled energy markets.

3. THE NEED FOR REGULATORY INNOVATION IN ENERGY SECTOR

Conventional energy regulatory frameworks face structural limitations in accommodating blockchain-based decentralized energy trading because they are grounded in assumptions that no longer fully reflect contemporary technological and market realities (Kumar & Chittora, 2024). These frameworks were largely developed in the twentieth century to address electricity systems characterized by centralized generation, vertically integrated utilities, and natural monopolies in transmission and distribution networks (Syahputra, 2022). Their primary objectives were to ensure system reliability, universal service provision, affordable pricing, and protection of end-users through centralized control and extensive regulatory oversight. As a result, traditional energy regulation is built around centralized grid operation by transmission and distribution system operators, strict functional separation between monopoly network activities and competitive market segments, comprehensive licensing regimes, tariff structures designed to recover infrastructure costs through volumetric consumption charges, and consumer protection mechanisms based on a bilateral utility–consumer relationship (Schneiders & Shipworth, 2021).

These regulatory architectures struggle to accommodate the defining features of blockchain-enabled peer-to-peer (P2P) energy trading (Yusuf et al., 2024). One fundamental tension arises from the assumption that commercial transactions directly correspond to physical power flows an assumption often described as “electrons follow contracts”. In decentralized P2P systems, however, electricity flows are determined by physical network conditions rather than by individual peer-to-peer contracts, creating a decoupling between contractual arrangements and actual power flows (Kumar & Chittora, 2024). Furthermore, traditional regulation often treats distributed energy resources (DERs) primarily as sources of disruption rather than as providers of valuable grid services, such as demand flexibility, voltage control, and frequency regulation, which can contribute positively to system stability. The rigid categorization of actors into generators, suppliers, network operators, and consumers also becomes increasingly untenable when prosumers simultaneously generate, consume, store, and trade electricity, and when blockchain platforms enable dynamic and fluid market participation (Schneiders & Shipworth, 2021). In addition, existing regulatory frameworks were not designed to address data governance challenges related to blockchain transparency or liability concerns arising from automated smart contract execution.

Blockchain-based energy trading systems also possess a set of technological and operational characteristics that further differentiate them from conventional electricity markets and require specialized regulatory consideration (Boumaiza, 2024). These systems are fundamentally disintermediated, reducing or eliminating the role of traditional trusted intermediaries such as licensed suppliers or centralized market operators, thereby challenging regulatory models that allocate responsibilities and obligations primarily through intermediaries. They rely heavily on smart contracts that automate trading, settlement, and in some cases control of physical assets in near real time, raising complex questions regarding accountability and liability when automated processes malfunction or generate unintended outcomes. At the same time, the transparency of blockchain ledgers creates opportunities for improved traceability and regulatory oversight but also poses risks to privacy and the protection of commercially sensitive information.

Moreover, blockchain-based energy trading often involves a high volume of microtransactions executed at unprecedented speed and frequency, which existing market monitoring, reporting, and compliance systems may not be capable of processing efficiently (Srinivas et al., 2025). The tokenization of energy further blurs the boundary between energy markets and financial markets, potentially triggering financial services regulation alongside energy-specific rules. Many blockchain platforms also facilitate localized energy trading within communities or microgrids, raising questions about how such local markets should interact with wholesale and retail markets and how system-wide reliability and cost allocation can be maintained. In addition, the potential for blockchain platforms to operate across national borders, particularly in interconnected electricity systems, creates regulatory challenges that nationally oriented energy frameworks are ill-equipped to address. Collectively, these features indicate that blockchain-based energy trading cannot be effectively governed by simply extending existing regulatory categories.

The resulting legal and regulatory uncertainty creates significant risks for innovators and early adopters. Firms developing blockchain-based energy trading platforms often face unclear or overlapping regulatory requirements, with limited guidance on licensing obligations, grid access rights, or compliance pathways. Uncertainty persists regarding whether prosumers should be regulated as electricity suppliers, how platform operators should be classified, whether energy tokens constitute financial instruments, how liability should be allocated for automated transactions, and whether existing data protection frameworks are compatible with blockchain's immutability. These uncertainties generate multiple forms of risk, including compliance risk, investment risk, reputational risk, and strategic risk, which can discourage innovation and delay commercialization. Smaller firms and start-ups are particularly vulnerable, as they typically lack the legal resources and risk tolerance of incumbent utilities, potentially reinforcing existing market power structures.

Beyond uncertainty, existing regulatory structures also create substantial market entry barriers. Licensing regimes often impose administrative and financial requirements that are disproportionate to the scale of blockchain-based P2P trading initiatives (Junlakarn et al., 2022). Grid access and connection procedures are frequently designed for large centralized generators and may impose technical standards or fees that undermine the economic viability of small-scale trading. Tariff designs based on volumetric charging further complicate P2P trading by raising concerns about cost recovery and cross-subsidization, while market rules and technical standards tailored to centralized systems may exclude small prosumers from meaningful participation. These barriers, combined with the structural advantages of incumbent utilities, hinder the transition of blockchain energy innovations from pilot projects to large-scale deployment.

Given the complexity and uncertainty surrounding blockchain-enabled energy systems, there is a strong rationale for regulatory innovation based on a *learning-by-doing* approach. Traditional ex-ante regulatory design assumes that risks, impacts, and optimal rules can be identified in advance, an assumption that is increasingly unrealistic for rapidly evolving technologies. Regulators often face knowledge gaps, timing challenges, and uncertainty regarding long-term system effects, which can result either in premature restrictions or in regulatory vacuums. A learning-based approach emphasizes iterative rule development, empirical evidence generation, regulatory humility, and adaptive capacity.

Regulatory sandboxes operationalize this approach by providing a controlled environment in which blockchain-based energy trading models can be tested under regulatory supervision. Through sandbox experimentation, regulators and innovators can jointly assess technical performance, market behavior, consumer impacts, and systemic risks, while generating evidence to inform future regulatory frameworks. This process supports more flexible, informed, and responsive regulation, enabling energy governance structures to evolve alongside technological innovation rather than lag behind it.

4. DESIGN PRINCIPLES OF ENERGY-FOCUSED REGULATORY SANDBOXES

This section translates the conceptual foundations discussed earlier into a set of practical design principles for energy-focused regulatory sandboxes. Drawing on established regulatory sandbox practices and energy sector-specific considerations, the proposed design framework consists of six interrelated components: eligibility criteria, scope and boundaries, consumer protection safeguards, monitoring and reporting mechanisms, exit and transition strategies, and stakeholder engagement.

These components, illustrated in Figure 2, collectively aim to balance innovation enablement, risk containment, and regulatory learning in blockchain-enabled peer-to-peer energy markets.

Figure 2. Key design components of an energy-focused regulatory sandbox for blockchain-enabled peer-to-peer energy trading

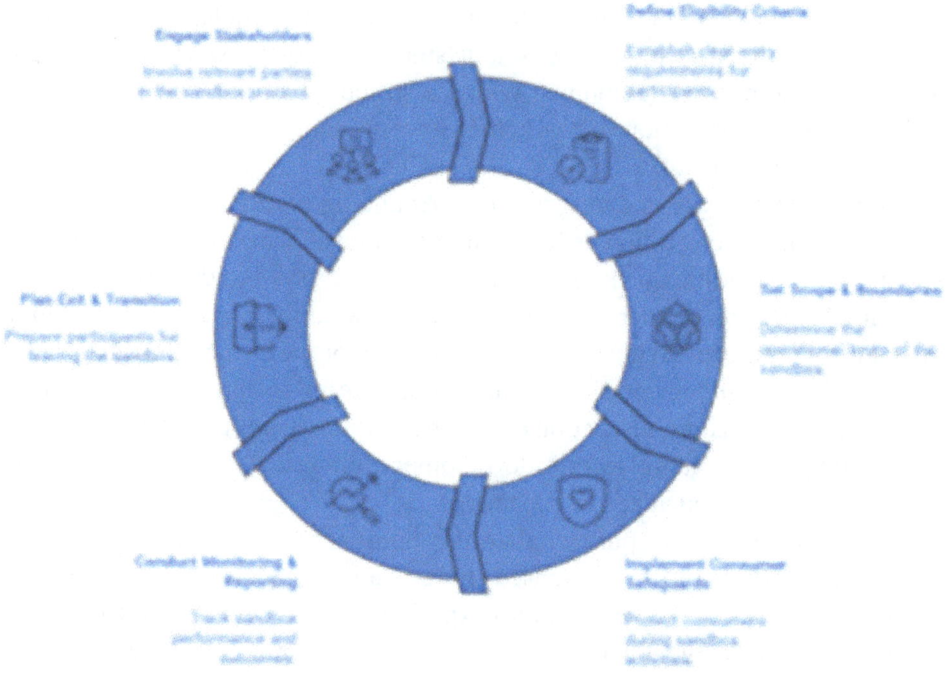

4.1 Eligibility Criteria: Ensuring Meaningful and Responsible Innovation

Clear eligibility criteria are a foundational element of energy-focused regulatory sandboxes, as they determine who may participate and under what conditions. Properly designed criteria ensure that sandboxes support *genuine innovation* rather than regulatory avoidance, while safeguarding public interests and regulatory credibility. Eligibility requirements perform several critical functions: they filter applicants whose innovations cannot reasonably be deployed under existing regulations, prioritize projects with meaningful learning potential for regulators, allocate limited supervisory resources efficiently, and reduce consumer risk by excluding participants lacking technical, financial, or organizational capacity.

Common eligibility requirements in energy sandboxes include proof of regulatory novelty, meaning the innovation cannot proceed without regulatory relief, expected consumer or system benefits, such as lower costs, increased flexibility, or environmental gains, and business and technical readiness, demonstrated through prototypes, proof-of-concept testing, or pilot results. Applicants are also typically required to demonstrate adequate financial capacity to complete testing and compensate consumers in case of losses, as well as a clear commitment to consumer protection. For blockchain-based energy trading, additional eligibility considerations often include cybersecurity capability, interoperability with grid infrastructure, technical expertise in both blockchain and energy systems, and credible scalability potential beyond the sandbox phase. Importantly, eligibility frameworks must balance rigor with accessibility so that innovative start-ups are not excluded by excessive administrative burdens.

4.2 Scope and Boundaries: Managing Risk While Enabling Learning

Defining clear sandbox boundaries is essential to balancing innovation, risk containment, and regulatory learning. Boundaries typically apply across four dimensions: geography, time, scale, and technology. Geographical limits restrict sandbox activities to specific regions, distribution network areas, or microgrids, allowing regulators and system operators to monitor grid impacts while limiting systemic risk. Temporal limits define how long sandbox participation may last, usually between six months and two years, preventing indefinite operation under relaxed regulatory conditions and ensuring periodic regulatory review.

Scale boundaries constrain the number of participants, energy volumes, financial exposure, or installed capacity involved in testing. These limits are particularly important in energy sandboxes, as overly large-scale experiments could affect grid stability or market fairness, while overly small-scale tests may fail to generate meaningful insights. Technological boundaries may also apply, for example restricting the type of blockchain architecture used, though excessive prescription risks constraining innovation. To maintain legitimacy and effectiveness, sandbox boundaries should be clearly defined ex ante, transparently communicated, and flexibly adjustable based on observed outcomes.

4.3 Consumer Protection Safeguards: Preserving Trust in Regulatory Experimentation

Consumer protection is the ethical core of regulatory sandboxes. While sandboxes relax certain regulatory requirements, they must maintain or strengthen consumer safeguards through alternative mechanisms. Key safeguards include informed consent, ensuring that consumers clearly understand the experimental nature of participation and associated risks; transparency obligations regarding pricing, data use, and platform functionality; and financial protections such as insurance, compensation funds, or guarantees against losses arising from platform failures or incorrect smart contract execution.

For blockchain-based energy platforms, additional safeguards are often necessary. These include cybersecurity requirements to protect against attacks that could disrupt grid operations or compromise personal data; technical reliability standards to ensure continuity of electricity supply; transparency regarding algorithmic pricing and smart contract logic; and protections against market manipulation or abuse by dominant participants. Exit rights are especially important, allowing consumers to withdraw from sandbox participation at any time without penalty and return to conventional supply arrangements. Safeguards must be proportionate: insufficient protection risks consumer harm and loss of public trust, while excessive requirements may undermine the feasibility of innovation testing.

4.4 Monitoring and Reporting: Generating Evidence for Regulatory Learning

Effective monitoring and reporting systems enable regulators to oversee sandbox activities, detect emerging risks, and generate empirical evidence for future regulatory development. Monitoring typically covers operational performance (transaction volumes, platform reliability, grid impacts), consumer outcomes (costs, satisfaction, complaints), and risk indicators such as cybersecurity incidents or financial instability. In blockchain-based energy sandboxes, monitoring may leverage the transparency of distributed ledgers to gain detailed insights into trading behavior and system performance.

Reporting requirements operationalize monitoring by specifying the content, frequency, and format of information provided to regulators. These usually include periodic reports, incident notifications, and final evaluation reports summarizing outcomes and lessons learned. While blockchain offers unprecedented data transparency and integrity, it also raises challenges related to data volume, analytical capacity, and privacy protection. Monitoring frameworks should therefore be proportionate, technology-enabled where possible, and designed to produce insights that can inform broader regulatory reform without imposing excessive burdens on participants.

4.5 Exit Strategies and Transition Pathways

Clear exit strategies are essential to ensure orderly completion of sandbox experiments and to enable successful innovations to scale beyond the sandbox. Exit scenarios include successful transition to full authorization, early termination due to failure or risk, regulatory intervention, and expiration of the testing period. For successful blockchain-based energy platforms, transition pathways may involve regulatory reform, individual licensing under modified conditions, or hybrid approaches combining both.

These pathways should be defined at the outset, including success criteria, evidence requirements, evaluation processes, and timelines. For unsuccessful experiments, exit strategies must ensure consumer protection through orderly wind-down procedures, continuity of electricity supply, settlement of outstanding transactions, and preservation of data and lessons learned. Well-designed exit mechanisms enhance regulatory credibility and investor confidence while preserving flexibility to adapt regulatory approaches as evidence accumulates.

4.6 Stakeholder Engagement: Enhancing Legitimacy and Learning

Meaningful stakeholder engagement throughout the sandbox lifecycle strengthens legitimacy, improves design quality, and broadens the value of sandbox learning. Relevant stakeholders include regulators, system operators, utilities, technology providers, consumers, data protection authorities, financial regulators, researchers, and policymakers. Engagement supports transparency, identifies potential risks early, and ensures that diverse perspectives are reflected in regulatory outcomes.

Engagement mechanisms should vary by sandbox phase, including consultation during design, targeted input during application assessment, ongoing dialogue during operation, and broad consultation during evaluation and regulatory reform. For blockchain energy sandboxes, engagement is particularly important for addressing technical integration, market design, data governance, and consumer protection issues. Best practices include early and continuous engagement, transparent communication, inclusive participation mechanisms, and public reporting on how stakeholder input influences regulatory decisions.

5. CASE STUDIES: GLOBAL IMPLEMENTATION OF ENERGY REGULATORY SANDBOXES

This section examines the implementation of energy regulatory sandboxes across selected jurisdictions, with particular attention to initiatives accommodating blockchain-enabled peer-to-peer (P2P) energy trading. The case studies highlight how variations in institutional capacity, market structure, governance arrangements, and national policy priorities influence sandbox design, implementation processes, and regulatory outcomes. Collectively, these cases demonstrate that regulatory sandboxes are not uniform policy instruments but context-sensitive regulatory tools whose effectiveness depends strongly on local regulatory, technical, and market conditions.

To facilitate cross-jurisdictional comparison, Table 1 summarizes the objectives, scope, outcomes, and policy lessons of regulatory sandbox implementations in selected developed and emerging energy markets.

Table 1. Comparative case study of energy regulatory sandboxes

Jurisdiction	Regulatory Authority	Sandbox Focus & Scope	Key Outcomes	Policy Lessons
United Kingdom	Ofgem	Time-bound sandbox for energy innovation, including P2P trading and flexibility markets	Enhanced regulatory learning and controlled experimentation	Clear exit strategies are essential for scalability
Australia	Australian Energy Regulator (AER)	Pilot-based sandbox with geographic and capacity limits for DER and local energy markets	Improved DER integration and local market testing	Coordination with network operators is critical
Singapore	Energy Market Authority (EMA)	Highly controlled sandbox supporting digital energy and blockchain-based settlement	Rapid deployment and operational efficiency	Strong central coordination enables fast experimentation
European Union	National energy regulators	Fragmented national sandboxes under EU internal energy market framework	Diverse experimentation across member states	Greater regulatory harmonization is needed
Emerging Markets	National regulators and development agencies	Small-scale, donor-supported pilots for local energy access and P2P trading	Improved access and localized innovation	Sandbox design must be context-sensitive

As shown in Table 1, while regulatory sandboxes share common objectives across jurisdictions, their design, scope, and outcomes vary significantly depending on regulatory capacity, market structure, and policy priorities.

5.1 United Kingdom

The United Kingdom is widely regarded as a global pioneer in regulatory innovation in the energy sector. The Office of Gas and Electricity Markets has developed a comprehensive innovation ecosystem that includes the Innovation Link (2017), the Regulatory Sandbox (2019), and multiple innovation funding mechanisms. The Innovation Link functions as a low-threshold entry point for innovators, offering tailored regulatory guidance and early-stage clarification of licensing and compliance requirements. This mechanism has been particularly valuable for blockchain-based energy innovators whose business models often fall outside traditional regulatory categories.

The Regulatory Sandbox represents a more formalized experimentation framework, allowing firms to test innovative products and services with temporary regulatory modifications for periods typically ranging from 12 to 24 months. Crucially, sandbox trials are conducted with real consumers under close regulatory supervision, enabling Ofgem to gather empirical evidence on risks, benefits, and systemic implications.

Several blockchain-related energy initiatives have contributed to regulatory learning in the UK. Projects such as Electron explored blockchain-based infrastructure for system registration, switching, and settlement, while community-level initiatives most notably the Repowering London/Verv project tested blockchain-enabled local energy trading within social housing contexts. These trials demonstrated tangible benefits, including improved utilization of local renewable generation and modest cost reductions for participants, while also exposing significant challenges related to system integration, consumer comprehension, and cost allocation for network usage.

Regulatory learning from these initiatives has led to incremental but meaningful adaptations. Ofgem acknowledged that the traditional binary distinction between licensed suppliers and consumers was poorly suited to P2P trading, prompting the issuance of interpretive guidance clarifying when limited trading activities trigger licensing obligations. Sandbox trials also informed reforms to data access frameworks, improvements in switching processes, and ongoing network charging reform aimed at aligning tariffs with actual system costs.

Nevertheless, the UK experience also highlights the limits of incremental reform. While sandboxes facilitated learning and dialogue, they could not fully resolve structural tensions between centralized market design and decentralized trading models. As such, the UK case illustrates both the strengths of regulatory sandboxes as learning instruments and their constraints when confronting potentially transformative technological change.

5.2 Australia

Australia represents one of the most ambitious global contexts for regulatory experimentation in decentralized energy systems, driven by exceptionally high rooftop solar penetration, strong consumer engagement, and growing pressure on legacy regulatory frameworks. The Australian Energy Market Commission (AEMC) introduced a formal energy regulatory sandbox in 2019, enabling trial projects to operate under modified rules for up to two years, with a strong emphasis on consumer benefit and system reliability.

Australia has become a focal point for blockchain-based energy innovation, most notably through Power Ledger, which has conducted multiple P2P trading trials across urban and regional settings. These projects demonstrated that blockchain platforms can process high transaction volumes reliably and automate settlement effectively. However, they also exposed integration challenges with legacy metering infrastructure, retailer billing systems, and the National Electricity Market's settlement architecture.

From a regulatory perspective, sandbox experiences revealed fundamental policy dilemmas. Network tariffs based on volumetric consumption incentivize local trading but risk undermining cost recovery for shared infrastructure. Consumer protection regimes designed for bilateral retailer-consumer relationships proved ill-suited to multi-party trading platforms. Moreover, uncertainty persisted regarding the legal classification of P2P platforms as retailers, intermediaries, or a novel regulatory category.

Australia's response has combined sandbox-based experimentation with systemic reform. At the national level, reforms have included the introduction of the Consumer Data Right for energy, enhanced regulation of embedded networks, and development of frameworks for coordinating distributed energy resources with system operators. State-level variation has further enabled policy experimentation, though it has also introduced regulatory fragmentation.

Overall, the Australian case demonstrates the importance of embedding sandboxes within broader reform agendas. Sandboxes alone cannot resolve structural issues related to tariff design, data governance, and system coordination, but they can provide critical empirical foundations for reform in highly decentralized energy systems.

5.3 Singapore

Singapore's approach to energy regulatory sandboxes is distinguished by strong central coordination and integration with the national Smart Nation strategy. The Energy Market Authority (EMA) launched its energy sandbox in 2018 to support experimentation while maintaining Singapore's high standards of system reliability and consumer protection. Unlike larger jurisdictions, Singapore benefits from compact geography, advanced digital infrastructure, and strong inter-agency coordination, enabling close monitoring and rapid regulatory feedback.

Blockchain energy projects in Singapore have focused on peer-to-peer solar trading, renewable energy certificate tracking, and regional energy transactions. Initiatives led by SP Group and Electrify demonstrated the technical feasibility of blockchain-based settlement and attribute tracking, particularly within planned smart districts such as Punggol.

However, sandbox outcomes also reflected Singapore's structural constraints. Limited rooftop solar potential, an already competitive retail market, and high grid efficiency reduced the economic attractiveness of P2P trading relative to other jurisdictions. Consumer engagement remained concentrated among early adopters, with limited mainstream uptake.

Regulatory insights from sandbox trials led to clarifications on licensing thresholds, data governance requirements, and cybersecurity standards. Importantly, close coordination between the EMA and the Monetary Authority of Singapore helped resolve ambiguities surrounding energy tokens and financial regulation, reducing uncertainty for innovators.

Singapore's experience illustrates the value of strategically integrated sandboxes aligned with national innovation objectives. At the same time, it underscores how geographic and market conditions constrain domestic scalability, positioning Singapore primarily as a technology development and export hub rather than a large-scale deployment environment.

5.4 European Union

The European Union employs a multi-level regulatory approach that combines EU-wide legal principles with national implementation. The Clean Energy for All Europeans Package (2018–2019) established the legal foundation for P2P energy trading through concepts such as active customers and citizen energy communities, while explicitly encouraging regulatory experimentation.

Implementation varies significantly across member states. Germany has favored informal experimentation through interpretive guidance and selective enforcement tolerance, enabling numerous blockchain energy pilots but generating legal uncertainty. Projects such as local P2P trading initiatives and the Enerchain wholesale platform demonstrated technical potential while highlighting tensions with unbundling rules, GDPR compliance, and renewable support schemes.

By contrast, the Netherlands has adopted a formalized sandbox framework embedded in legislation, granting time-limited exemptions with clear conditions.

Dutch projects emphasize community energy and cooperative models, supported by explicit provisions on licensing, network charging, and consumer protection.

At the EU level, regulatory diversity has enabled comparative learning but also created fragmentation that constrains cross-border scalability. The EU experience highlights the challenge of balancing harmonization with flexibility, a tension that remains central to the future of blockchain-enabled energy markets in multi-level governance systems.

5.5 Emerging Markets

Emerging markets present fundamentally different conditions for blockchain energy sandboxes, shaped by infrastructure deficits, regulatory capacity constraints, and development priorities. In many contexts, decentralized energy systems are not marginal innovations but potential solutions to persistent energy access challenges.

Pilot initiatives in countries such as Thailand, the Philippines, South Africa, and Chile demonstrate growing interest in sandbox-based experimentation, particularly in off-grid and mini-grid contexts. These settings reduce regulatory barriers while offering opportunities for leapfrogging centralized infrastructure models.

However, emerging markets face significant constraints, including limited regulatory expertise, weak digital infrastructure, low consumer purchasing power, and resistance from incumbent utilities. As a result, sandbox implementation has often been informal and donor-supported rather than embedded in permanent regulatory frameworks.

Despite these challenges, emerging markets offer unique opportunities for high-impact innovation. Blockchain-enabled community energy systems can simultaneously address access, affordability, and resilience objectives, provided sandboxes are aligned with development goals and supported by capacity-building initiatives.

6. LEGAL AND REGULATORY CHALLENGES

Blockchain-based peer-to-peer (P2P) energy trading fundamentally challenges the foundations of existing legal and regulatory frameworks, which were primarily designed for centralized electricity systems characterized by clearly defined market roles, linear power flows, and identifiable regulatory subjects. The decentralized, automated, and cross-sectoral nature of blockchain-enabled energy platforms exposes structural misalignments across multiple regulatory domains, including energy law, data protection regimes, consumer protection frameworks, competition law, and jurisdictional governance arrangements. These challenges underscore the limitations of existing regulatory approaches and highlight the need for adaptive mechanisms capable of addressing technological innovation while safeguarding public interests.

To synthesize these challenges and examine how regulatory sandboxes respond to them in practice, Table 2 maps key regulatory problem areas to the mitigation mechanisms commonly employed within sandbox environments, while also highlighting their remaining limitations.

Table 2. Regulatory challenges and sandbox-based mitigation mechanisms in blockchain-enabled energy markets

Regulatory Domain	Key Challenge	Sandbox Mechanism	Remaining Limitations
Energy law	Prosumer licensing and role ambiguity	Temporary licensing exemptions and regulatory guidance	Lack of permanent licensing frameworks
Data protection	Tension between blockchain immutability and GDPR requirements	Use of off-chain data storage and controlled data access	Persistent legal uncertainty across jurisdictions
Consumer protection	Dispute resolution and liability allocation	Platform-level liability rules and enhanced disclosure	Limited scalability beyond sandbox environments
Competition law	Risk of market concentration and platform dominance	Participation caps and transaction limits	Potential network effects after sandbox exit

6.1 Energy Law Perspectives

- **Licensing and authorization**

Traditional energy law is built on the assumption that electricity activities can be clearly separated into generation, transmission, distribution, and supply, each subject to licensing and regulatory oversight. Blockchain-based P2P trading disrupts this model by enabling prosumers to act simultaneously as generators, suppliers, and consumers. Applying generation or supply licensing requirements to individual prosumers is legally problematic because such regimes were never intended for large numbers of small-scale actors and would impose disproportionate administrative and economic burdens.

Similarly, platform operators occupy an ambiguous legal position. While they often claim to function merely as intermediaries facilitating transactions, regulators may view their control over pricing mechanisms, market rules, and settlement processes as constituting supply activity. This ambiguity creates legal uncertainty that discourages investment and limits scalability, as market participants cannot reliably predict regulatory obligations or compliance risks.

- **Grid access and connection rights**

Although P2P trading is commercially decentralized, it remains physically dependent on centralized grid infrastructure. Existing grid access frameworks were designed for predictable, one-directional flows and a limited number of regulated users. In contrast, P2P trading introduces highly dynamic, bidirectional flows driven by localized market signals.

This raises unresolved legal questions regarding export rights, curtailment priority, cost allocation, and responsibility for grid impacts. The absence of clear legal rules governing these issues creates uncertainty for both network operators and P2P participants, complicating grid planning and potentially undermining system stability.

- **Quality, reliability, and liability**

Electricity law traditionally assigns responsibility for quality and reliability to licensed suppliers who exercise technical control over supply portfolios. P2P trading fragments this responsibility across multiple actors individual prosumers, platform operators, and network operators without a clear allocation of legal liability.

When quality deviations or supply interruptions occur, existing legal frameworks provide no clear answer as to who bears responsibility. Smart contracts further complicate matters by automating decisions that may prioritize economic optimization over system reliability, raising concerns about resilience and accountability.

- **Unbundling and market structure**

Unbundling rules seek to prevent conflicts of interest by separating monopoly network activities from competitive generation and supply. Blockchain platforms blur these boundaries by enabling integrated coordination of generation, trading, storage, and demand response.

This challenges the effectiveness of existing unbundling regimes: while blockchain transparency may reduce some risks of discrimination, control over physical infrastructure and platform governance can still be leveraged anti-competitively. Regulators therefore face unresolved questions about whether to extend, adapt, or fundamentally rethink unbundling rules for decentralized energy platforms.

6.2 Data Protection and Privacy

Blockchain's core features immutability, transparency, and decentralization directly conflict with data protection principles, particularly under GDPR. Energy data, especially smart meter data, constitutes personal data because it reveals detailed behavioral patterns. Recording such data on immutable, distributed ledgers challenges rights to erasure, rectification, and purpose limitation.

Decentralized architectures also obscure the identification of data controllers and processors, undermining accountability. Even pseudonymized blockchain data may be re-identified through correlation with external datasets, amplifying privacy risks.

Technical solutions such as permissioned blockchains, off-chain storage, and privacy-enhancing cryptography can mitigate some risks but involve trade-offs between transparency, decentralization, and legal compliance. From a regulatory perspective, the persistence of legal uncertainty suggests that data protection remains a critical constraint on large-scale deployment of blockchain energy trading.

6.3 Consumer Protection

- **Disclosure and transparency**

P2P energy trading introduces unfamiliar technical and commercial risks that traditional disclosure frameworks are ill-equipped to address. Consumers must understand not only prices and contract terms, but also platform mechanics, pricing algorithms, reliability arrangements, and data usage practices.

The regulatory challenge lies in providing meaningful disclosure without overwhelming consumers with technical complexity. Sandbox experiments demonstrate the need for simplified, layered disclosures that balance accessibility with informational completeness.

- **Dispute resolution**

Existing consumer redress mechanisms assume a bilateral relationship between supplier and consumer. Blockchain platforms distribute responsibility across multiple actors and rely on automated smart contracts that are difficult to reverse.

This fragmentation complicates accountability and remedies, necessitating redesigned dispute resolution models that assign clear responsibility often to platform operators while preserving regulatory oversight and procedural fairness.

- **Vulnerable consumers**

P2P trading risks exacerbating inequality by favoring digitally literate homeowners with capital to invest in generation assets. Vulnerable consumers may face exclusion, higher risks, or increased bill volatility.

Effective regulation must therefore incorporate safeguards such as bill caps, guaranteed backup supply, participation restrictions, and monitoring of distributional impacts to ensure that innovation does not undermine social policy objectives.

- **Price transparency**

Algorithmic and dynamic pricing mechanisms undermine the predictability and comparability of energy costs. Without regulatory intervention, consumers may struggle to understand how prices are formed or assess whether outcomes are fair.

Transparency obligations, algorithm audits, and consumer protection limits are necessary to prevent opacity, manipulation, and excessive volatility.

6.4 Competition Law

Despite their decentralized rhetoric, blockchain energy platforms can generate new forms of market power through network effects, control over platform governance, data dominance, and access restrictions. Large prosumers, aggregators, and vertically integrated utilities may also exploit strategic advantages.

Competition law must therefore address risks of exclusionary practices, algorithmic collusion, and leveraging of monopoly positions, while avoiding premature intervention that could stifle innovation. Ensuring a level playing field requires proportionate regulation, interoperability, and vigilant oversight of both incumbents and new entrants.

6.5 Cross-Border Issues

Blockchain's borderless architecture challenges jurisdiction-based regulatory systems. Determining applicable law for contracts, energy regulation, data protection, competition enforcement, and taxation becomes complex when transactions, nodes, and parties span multiple jurisdictions.

Cross-border P2P trading also raises strategic concerns related to energy security, regulatory arbitrage, and enforcement capacity. Without coordination, platforms may exploit regulatory fragmentation, undermining consumer protection and market integrity.

Addressing these challenges requires international cooperation, harmonization of core standards, and potentially the development of transnational regulatory frameworks tailored to decentralized energy systems.

7. TECHNICAL-LEGAL INTERFACE IN SANDBOX TESTING

Regulatory sandboxes function as experimental environments where technological design choices and legal norms interact directly. In blockchain-based peer-to-peer (P2P) energy trading, this interaction is particularly intense because technical architectures smart contracts, automated control systems, and decentralized data infrastructures do not merely support legal arrangements but actively perform legally relevant functions. As a result, technical decisions made during sandbox testing have immediate consequences for contract validity, liability allocation, grid safety, consumer protection, and regulatory accountability.

7.1 Smart Contract Legality

- **Legal recognition and contract formation**

Smart contracts challenge core assumptions of contract law by replacing ongoing human decision-making with automated code execution. Traditional doctrines of offer, acceptance, and intention to create legal relations presuppose human agency at the moment contractual obligations arise. In contrast, smart contracts in energy trading may execute transactions automatically based on predefined parameters and external data inputs, without contemporaneous human awareness or consent.

This raises a fundamental legal question: whether contractual intent can be validly expressed ex ante through code, and whether subsequent automated execution should be treated as legally binding performance. While some jurisdictions have explicitly recognized smart contracts and blockchain-based signatures, many rely on technology-neutral contract law principles that were not designed for autonomous systems. In sandbox contexts, this legal ambiguity creates uncertainty for participants and regulators alike, as the enforceability of transactions may only be clarified ex post through litigation.

- **Enforceability and doctrinal friction**

Smart contracts intensify tensions between legal flexibility and technical rigidity. Contract law doctrines such as mistake, frustration, and unconscionability exist precisely to allow courts to adjust or invalidate agreements when strict enforcement would be unjust or contrary to public policy. Blockchain immutability, however, prevents easy reversal or modification of executed transactions, even where legal grounds for non-enforcement exist.

In energy trading, this problem is amplified by volatility and system constraints. Automated execution at extreme prices, execution during grid emergencies, or settlement based on faulty data may be technically correct but legally or commercially unreasonable. Sandbox testing therefore reveals that enforceability is not merely a legal question but a system design issue: platforms must incorporate ex ante safeguards such as emergency stop functions, dispute resolution hooks, and compensation mechanisms to preserve space for legal intervention.

- **Liability in autonomous environments**

Liability allocation is one of the most complex issues at the technical–legal interface. Smart contracts distribute causation across developers, platform operators, data oracles, users, and physical infrastructure. Traditional fault-based liability models struggle to identify a single responsible actor when harm results from emergent system behavior rather than discrete human decisions.

Sandbox experiments highlight competing liability models, ranging from strict platform liability to insurance-based risk pooling. From a regulatory perspective, the key insight is that liability rules shape incentives: unclear or diffused liability risks underinvestment in security and system robustness, while overly burdensome liability may deter innovation. Sandboxes thus play a critical role in testing whether proposed liability allocations strike an appropriate balance between accountability and innovation.

7.2 Grid Integration Issues

- **Technical standards as legal obligations**

Grid codes and technical standards are not merely engineering guidelines; they are legally enforceable instruments designed to protect system integrity and public safety. Blockchain-based P2P trading challenges these standards by introducing decentralized, algorithm-driven dispatch decisions that may not align with traditional assumptions about controllability and predictability.

Sandbox testing demonstrates that economic optimization through automated trading can conflict with technical constraints such as voltage stability, frequency control, and protection coordination. This reveals a critical legal implication: compliance with grid codes cannot be assumed simply because individual devices meet technical certification requirements. Instead, regulators must assess whether *system-level behavior* induced by trading algorithms remains compliant.

- **Interoperability and regulatory risk**

Interoperability failures create not only technical inefficiencies but also legal and competition risks. Proprietary platforms that lock participants into closed ecosystems may undermine consumer choice, hinder regulatory oversight, and concentrate market power. At the same time, premature standardization may entrench suboptimal technologies.

Sandboxes therefore function as testing grounds for proportional interoperability obligations, allowing regulators to observe whether open interfaces and data standards are necessary to prevent lock-in without stifling innovation.

- **Security and systemic stability**

When blockchain platforms coordinate large numbers of distributed resources, cybersecurity and system stability become matters of public interest rather than private risk. Automated, synchronized responses to price or grid signals can amplify disturbances rather than mitigate them, especially in the absence of human oversight.

Sandbox regimes respond by imposing scale limits, emergency override requirements, and enhanced security obligations. These measures reflect an emerging regulatory principle: autonomy in energy systems must be bounded by mechanisms that preserve ultimate human and institutional control over critical infrastructure.

7.3 Metering and Settlement

- **Measurement as a legal foundation**

Accurate metering underpins not only billing but also legal trust in market outcomes. P2P trading requires granular, bidirectional, and near-real-time measurement that exceeds the design assumptions of traditional metering systems. Sandbox testing reveals that measurement errors, even if small, can undermine confidence and trigger disputes when multiplied across numerous microtransactions.

From a legal perspective, metering accuracy is inseparable from consumer protection and market integrity. Regulators therefore increasingly treat metering standards, certification, and auditability as preconditions for lawful operation of P2P platforms.

- **Verification and the oracle problem**

The oracle problem illustrates a core vulnerability of blockchain systems: they cannot independently verify physical reality. In energy trading, this creates risks of fraud, misallocation, and false claims of renewable supply. Sandbox experiments show that purely technical solutions are insufficient; institutional trust mechanisms—

such as certified meters, third-party validation, and regulatory oversight—remain indispensable.

This underscores a broader insight: blockchain does not eliminate the need for trusted institutions but redistributes trust across technical and legal layers.

- **Settlement and system reconciliation**

Automated settlement through smart contracts challenges established practices of billing, taxation, balancing responsibility, and network charging. While real-time settlement improves efficiency, it complicates consumer comprehension, dispute resolution, and integration with existing market structures.

Sandbox experience consistently demonstrates that scalable deployment requires hybrid models in which blockchain coordinates local transactions while interfacing with traditional systems for balancing, taxation, and final settlement. Legally, this reinforces the view that blockchain platforms should complement rather than replace existing regulatory institutions.

8. STAKEHOLDER PERSPECTIVES AND MULTI-STAKEHOLDER GOVERNANCE

Blockchain-based peer-to-peer (P2P) energy trading and the use of regulatory sandboxes to test such innovations give rise to a complex constellation of stakeholder interests, institutional constraints, and governance challenges. Because these systems sit at the intersection of energy regulation, digital technology, financial systems, and data governance, no single actor possesses sufficient authority or expertise to govern them unilaterally. Consequently, understanding stakeholder perspectives is essential for designing regulatory sandboxes and post-sandbox governance frameworks that are both effective and legitimate.

8.1 Regulatory Perspectives

From a regulatory standpoint, blockchain-based P2P energy trading presents a fundamental balancing challenge between enabling innovation and safeguarding core public interest objectives. Energy regulators are typically mandated to ensure security and reliability of electricity supply, protect consumers, promote efficient and competitive markets, support long-term infrastructure investment, and advance policy goals such as decarbonization and affordability. Blockchain-enabled P2P trading simultaneously promises progress toward several of these objectives while also introducing novel risks that existing regulatory frameworks were not designed to address.

On the positive side, regulators increasingly recognize that P2P trading may enhance consumer engagement, improve utilization of distributed renewable energy resources, stimulate innovation in retail energy markets, and support climate policy objectives by facilitating local renewable energy consumption. These perceived benefits motivate regulatory openness toward experimentation, particularly through regulatory sandboxes that allow controlled testing under supervisory oversight.

At the same time, regulators remain cautious due to a range of concerns. These include potential consumer protection risks arising from complex pricing mechanisms and automated transactions; threats to system reliability if decentralized trading undermines coordinated grid operation; equity concerns if P2P participants avoid paying fair network costs; cybersecurity vulnerabilities associated with digital platforms; and regulatory capacity constraints given limited in-house expertise on blockchain technologies. Regulators are also subject to political pressures, industry lobbying, and international policy diffusion, all of which shape their risk tolerance and institutional posture.

As a result, the dominant regulatory approach has been one of cautious experimentation: allowing blockchain-based energy trading within narrowly defined sandbox environments, imposing strict participation limits, requiring extensive reporting, and prioritizing consumer safeguards. Large-scale deployment is generally deferred until empirical evidence demonstrates that risks can be managed without undermining core regulatory objectives.

8.2 Industry Perspectives

Industry perspectives on blockchain-based energy trading are highly heterogeneous and strongly shaped by actors' positions within the energy value chain. Incumbent utilities often occupy an ambivalent or conflicted position. Traditional utility business models based on centralized generation, volumetric sales, and regulated monopolies may be disrupted by P2P trading, which enables customers to bypass conventional supply arrangements and reduces reliance on centralized intermediaries.

Some utilities therefore adopt defensive strategies, lobbying against sandbox initiatives, emphasizing system risks, and advocating network charges that reduce the economic attractiveness of P2P trading. In contrast, more adaptive utilities pursue strategic engagement, participating directly in sandboxes to explore new roles as platform operators, flexibility service providers, or coordinators of distributed energy resources. Distribution network operators, whose revenues are less dependent on energy sales volumes, are often more receptive to P2P trading than generation or retail entities.

Start-ups developing blockchain energy platforms are typically strong proponents of regulatory sandboxes, viewing them as essential pathways to market entry in highly regulated environments. However, these firms frequently express frustration with lengthy approval processes, restrictive sandbox conditions, uncertain exit pathways, and perceived advantages enjoyed by incumbents with established regulatory relationships. Technology firms from outside the energy sector such as blockchain infrastructure providers and IoT companies generally support regulatory experimentation but may disengage if energy regulation proves excessively complex or slow-moving.

8.3 Consumer and Civil Society Perspectives

Consumer advocacy organizations adopt a cautiously supportive stance toward blockchain-based P2P energy trading. They generally welcome the prospect of increased consumer choice, enhanced transparency, and opportunities for households to benefit economically from distributed energy resources. However, this support is conditional on strong consumer protections and demonstrable benefits relative to conventional utility service.

Key consumer concerns include usability and complexity, as blockchain platforms may be difficult for non-expert users to understand; reliability risks if peer-based supply fails; price volatility associated with dynamic pricing algorithms; and data privacy risks linked to granular smart meter data. Consumer advocates also emphasize distributional equity, warning that P2P trading may disproportionately benefit affluent homeowners while excluding renters and low-income households, thereby exacerbating energy inequality.

Environmental and sustainability organizations are generally supportive of blockchain-based energy trading due to its potential to accelerate renewable energy deployment, improve traceability of environmental attributes, and strengthen community energy initiatives. Nevertheless, they raise concerns about the environmental footprint of blockchain technologies themselves particularly energy-intensive consensus mechanisms and the risk of greenwashing if claimed environmental benefits lack additionality or rigorous verification.

8.4 Academic and Research Perspectives

Academic institutions contribute critical analytical perspectives grounded in empirical research, theory, and comparative analysis. Engineering and computer science scholars focus on system performance, cybersecurity, scalability, and control algorithms. Economists and business scholars analyze market design, pricing mechanisms, and business model viability. Legal and regulatory scholars examine institutional frameworks, governance challenges, and liability allocation, while social scientists investigate equity, consumer behavior, and sustainability outcomes.

Academics often participate in regulatory sandboxes by conducting independent evaluations, advising regulators, facilitating stakeholder dialogue, and building regulatory capacity. However, challenges remain, including slow publication cycles, funding dependencies, and gaps between academic analysis and practical policymaking needs.

8.5 Multi-Stakeholder and Collaborative Governance

The diversity of stakeholder perspectives underscores the limitations of traditional hierarchical regulation in governing blockchain-based energy systems. In response, collaborative governance models have emerged that emphasize multi-stakeholder participation, shared responsibility, and adaptive learning. Regulatory sandboxes themselves can be understood as collaborative governance instruments, enabling iterative interaction between regulators, innovators, and affected stakeholders.

Examples include multi-stakeholder regulatory working groups, industry-led standard-setting initiatives such as those coordinated by the Energy Web Foundation, co-regulatory arrangements combining public oversight with private standard development, and platform-level governance mechanisms that incorporate participant voice. These approaches offer benefits in terms of expertise, legitimacy, and adaptability, but also raise concerns about power asymmetries, accountability, efficiency, and regulatory capture.

Effective collaborative governance therefore requires careful institutional design, including clear allocation of responsibilities, transparency in decision-making, inclusiveness safeguards, accountability mechanisms, and strong regulatory backstops. The future governance of blockchain-based energy trading is likely to involve hybrid models that retain hierarchical regulation for core public interest protections while leveraging collaborative mechanisms for innovation, technical standardization, and system learning.

9. SUCCESS FACTORS AND BEST PRACTICES

Regulatory sandboxes for blockchain-based decentralized energy trading have been implemented across diverse jurisdictions with varying degrees of success. Comparative analysis of international experiences reveals that successful sandbox programs are not defined solely by technological sophistication, but by a combination of institutional design, governance capacity, stakeholder engagement, and learning-oriented regulatory practice. This section synthesizes key success factors and best practices derived from global sandbox implementations, with particular attention to performance evaluation, risk management, scalability, and long-term regulatory capacity building.

9.1 Key Performance Indicators for Sandbox Evaluation

The effectiveness of regulatory sandboxes depends critically on the availability of clear, measurable, and policy-relevant key performance indicators (KPIs). Well-designed KPIs enable regulators to assess whether sandbox objectives are being met, to identify emerging risks, and to generate evidence that can inform permanent regulatory reform.

- *Technical performance indicators* assess system reliability, transaction accuracy, settlement efficiency, cybersecurity incidents, and integration with grid operations. For blockchain-based P2P trading, this includes metrics such as transaction throughput, smart contract execution reliability, data latency, and resilience under stress conditions.
- *Consumer outcome indicators* focus on participant experience and protection. These include changes in electricity costs, price volatility, service continuity, complaint volumes, dispute resolution outcomes, and levels of consumer understanding and satisfaction. Evidence from sandbox trials suggests that consumer trust and comprehension are as important as financial outcomes in determining perceived success.
- *Market and competition indicators* evaluate impacts on market structure, entry barriers, and competitive dynamics. Relevant metrics include participation diversity, concentration levels, pricing transparency, and evidence of exclusionary or discriminatory practices. These indicators help assess whether sandbox participation promotes innovation without distorting competition.
- *Environmental and system-level indicators* measure contributions to policy objectives such as decarbonization, renewable energy utilization, demand flexibility, and local grid efficiency. Metrics may include increased self-consumption of renewable energy, reduced curtailment, and improved load balancing at the distribution level.
- *Regulatory learning indicators* assess the sandbox's contribution to institutional knowledge. These include the clarity gained on regulatory gaps, the feasibility of alternative regulatory approaches, inter-agency coordination improvements, and the translation of sandbox findings into guidance, rule changes, or legislative reform. Best practices indicate that sandboxes should be evaluated not only on immediate outcomes but on their capacity to inform future regulation.

9.2 Factors Contributing to Successful Sandbox Programs

Comparative experience highlights several recurring factors that contribute to successful energy regulatory sandboxes. A primary factor is clear objective alignment, whereby sandbox goals are explicitly linked to broader energy policy priorities such as decarbonization, consumer empowerment, and system resilience. Sandboxes that operate in isolation from strategic policy frameworks tend to generate limited regulatory impact.

Another critical factor is institutional capacity and regulatory competence. Regulators with dedicated sandbox teams, interdisciplinary expertise, and sufficient resources are better positioned to supervise complex blockchain-based trials. Capacity includes not only technical knowledge but also legal flexibility, data analytics capability, and experience with adaptive regulatory methods.

Proportionate risk management is also essential. Successful sandboxes balance flexibility with safeguards, allowing meaningful experimentation while maintaining clear limits on scale, exposure, and systemic risk. Overly restrictive conditions undermine learning, while insufficient safeguards erode public trust and political support.

Stakeholder collaboration emerges as a further success factor. Sandboxes that actively involve system operators, consumer groups, data protection authorities, and competition regulators generate richer insights and reduce downstream implementation conflicts. Early and continuous engagement helps identify risks that may not be apparent from a single regulatory perspective.

Finally, credible transition pathways significantly influence success. Sandbox participants require clarity on what constitutes success, how evaluation will occur, and what regulatory outcomes may follow. Programs that articulate clear post-sandbox options such as licensing pathways, rule adaptations, or legislative proposals are more likely to attract high-quality applicants and sustain innovation momentum.

9.3 Common Pitfalls and Strategies for Avoidance

Despite their potential, regulatory sandboxes are vulnerable to several recurring pitfalls. One common risk is sandbox stagnation, where projects remain indefinitely in experimental status without clear exit or scaling pathways. This undermines regulatory legitimacy and discourages investment. Best practice requires time-bound testing periods and predefined decision points.

Another frequent pitfall is regulatory capture or unequal access, where well-resourced incumbents dominate sandbox participation due to superior legal expertise or regulatory familiarity. This can marginalize start-ups and distort competitive outcomes. Transparent selection criteria and tailored support for smaller innovators help mitigate this risk.

Insufficient data utilization is also problematic. Some sandbox programs collect extensive data but lack the analytical frameworks or institutional processes to translate findings into regulatory action. Effective sandboxes embed evaluation mechanisms and knowledge synthesis into their design from the outset.

A further challenge arises from over-reliance on technological optimism. Blockchain-based energy trading is sometimes assumed to inherently improve transparency, efficiency, or fairness. Sandbox evidence shows that without appropriate governance, blockchain can replicate or even exacerbate existing market and power asymmetries. Regulators must therefore remain technology-neutral and outcome-focused.

Finally, fragmented governance particularly in jurisdictions with overlapping regulatory mandates can undermine sandbox effectiveness. Lack of coordination between energy, financial, and data protection authorities creates uncertainty and delays. Successful programs proactively establish inter-agency coordination mechanisms to ensure coherent oversight.

9.4 Scalability Considerations

Scalability represents a central challenge for blockchain-based energy sandboxes. While many pilots demonstrate technical feasibility at small scale, scaling up introduces new legal, economic, and system-level constraints. Best practices indicate that scalability should be treated as a gradual, staged process rather than an immediate objective.

From a technical perspective, scalability requires addressing transaction throughput, interoperability with legacy systems, and cybersecurity resilience. Regulatory sandboxes provide an opportunity to test whether technical architectures can support increased participation without compromising system stability.

From a regulatory perspective, scalability raises questions about tariff design, cost allocation, licensing frameworks, and market integration. Sandbox findings often reveal that existing regulatory structures not technology constitute the primary barriers to scale. Successful programs therefore use sandbox evidence to inform targeted regulatory reform rather than attempting to scale within unchanged frameworks.

Importantly, scalability must also consider social and distributional impacts. Expansion of P2P trading may exacerbate inequalities if benefits accrue primarily to asset-owning participants. Best practices emphasize monitoring distributional effects and incorporating safeguards before large-scale deployment.

9.5 Knowledge Transfer and Capacity Building

One of the most significant long-term benefits of regulatory sandboxes lies in institutional learning and capacity building. Successful programs treat sandboxes not merely as testing environments for innovators, but as learning platforms for regulators themselves.

Capacity building occurs through hands-on engagement with new technologies, collaboration with external experts, and iterative policy development. Best practices include documenting lessons learned, publishing evaluation reports, and integrating sandbox insights into regulatory training programs.

Knowledge transfer beyond the sandbox context is equally important. Sharing findings with other regulators, policymakers, and international networks enhances collective regulatory capacity and reduces duplication of effort. This is particularly valuable in emerging markets, where regulatory resources are constrained and sandbox learning can accelerate institutional development.

9.6 International Best Practices Synthesis

Synthesizing international experience reveals several overarching best practices for regulatory sandboxes in blockchain-based energy trading. First, sandboxes are most effective when embedded within broader regulatory reform strategies rather than treated as isolated experiments. Second, outcome-oriented evaluation frameworks are essential for translating experimentation into policy change. Third, multi-stakeholder governance enhances legitimacy, risk identification, and learning quality. Fourth, scalability and equity considerations must be integrated early rather than addressed post hoc. Finally, sustained investment in regulatory capacity and international knowledge exchange amplifies the long-term value of sandbox initiatives.

Taken together, these success factors and best practices suggest that regulatory sandboxes are not ends in themselves, but transitional governance instruments. When well designed and implemented, they enable regulators to navigate uncertainty, align technological innovation with public interest objectives, and lay the groundwork for adaptive, future-oriented energy regulation.

10. CONCLUSION

This book has explored the governance of blockchain-based decentralized energy trading through the lens of regulatory sandboxes, situating technological innovation within the broader transformation of electricity systems driven by decarbonization, digitalization, and decentralization. The analysis demonstrates that traditional energy regulatory frameworks, which were designed for centralized and hierarchical market structures, face significant limitations when confronted with peer-to-peer trading models enabled by blockchain technology. Regulatory sandboxes emerge as a pragmatic response to this challenge, offering a structured environment in which innovation can be tested under regulatory supervision while preserving essential public interest objectives.

Across the legal, institutional, and technological discussions, the book highlights that regulatory sandboxes are most effective when embedded within coherent governance frameworks rather than treated as isolated policy experiments. Clear regulatory objectives, well-defined entry and exit criteria, and outcome-oriented evaluation mechanisms are shown to be critical for ensuring that sandbox initiatives generate meaningful regulatory learning. Comparative experiences across jurisdictions further reveal the importance of multi-stakeholder governance, inter-agency coordination, and institutional capacity in managing the complex risks associated with blockchain-based energy trading, including consumer protection, data governance, cybersecurity, and market fairness.

The analysis of sandbox implementation and evaluation underscores that technological feasibility alone is insufficient to justify large-scale deployment of decentralized energy trading systems. Issues related to scalability, interoperability, distributional impacts, and long-term market integration present significant regulatory challenges that must be addressed proactively. Sandbox findings demonstrate that without adaptive regulatory design and proportionate risk management, peer-to-peer energy trading risks reinforcing existing inequalities or creating new forms of regulatory fragmentation. As such, sandboxes serve not only as testing grounds for innovation, but also as diagnostic tools for identifying structural weaknesses within existing regulatory regimes.

Looking ahead, the book argues that regulatory sandboxes should be understood as transitional governance instruments that support the evolution toward adaptive, risk-based, and evidence-informed regulatory frameworks. As blockchain technologies increasingly converge with artificial intelligence and the Internet of Things, regulatory complexity will continue to intensify, requiring continuous learning and institutional adaptation. When strategically aligned with broader energy policy objectives and supported by international knowledge exchange, regulatory sandboxes can play a critical role in shaping resilient, inclusive, and sustainable energy markets that harness technological innovation while safeguarding the public interest.

REFERENCES

Baig, M. J., Iqbal, M. T., Jamil, M., & Khan, J. (2023). Blockchain-Based Peer-to-Peer Energy Trading System Using Open-Source Angular Framework and Hypertext Transfer Protocol. In *Electronics* (Vol. 12, Issue 2, p. 287). DOI: 10.3390/electronics12020287

Boumaiza, A. (2024). Carbon and Energy Trading Integration within a Blockchain-Powered Peer-to-Peer Framework. In *Energies* (Vol. 17, Issue 11, p. 2473). DOI: 10.3390/en17112473

Burger, C., & Weinmann, J. (2022). Blockchain Platforms in Energy Markets—A Critical Assessment. In *Journal of Risk and Financial Management* (Vol. 15, Issue 11, p. 516). DOI: 10.3390/jrfm15110516

Evens, M., Ercoli, P., & Arteconi, A. (2023). Blockchain-Enabled Microgrids: Toward Peer-to-Peer Energy Trading and Flexible Demand Management. In *Energies* (Vol. 16, Issue 18, p. 6741). DOI: 10.3390/en16186741

Fatima, S., & Arshad, M. J. (2025). A Comprehensive Review of Blockchain and Machine Learning Integration for Peer-to-Peer Energy Trading in Smart Grids. *IEEE Access : Practical Innovations, Open Solutions*, *13*, 92756–92782. DOI: 10.1109/ACCESS.2025.3572174

Grigoryan, H. (2024). Cost-Effective Integration of Blockchain Technologies Into P2P Energy Trading Systems. *2024 IEEE International Conference on Omni-Layer Intelligent Systems (COINS)*, 1–5. DOI: 10.1109/COINS61597.2024.10622550

Heng, Y. B., Ramachandaramurthy, V. K., Verayiah, R., & Walker, S. L. (2022). Developing Peer-to-Peer (P2P) Energy Trading Model for Malaysia: A Review and Proposed Implementation. *IEEE Access : Practical Innovations, Open Solutions*, *10*, 33183–33199. DOI: 10.1109/ACCESS.2022.3161577

Jogunola, O., Ajagun, A. S., Tushar, W., Olatunji, F. O., Yuen, C., Morley, C., Adebisi, B., & Shongwe, T. (2024). Peer-to-Peer Local Energy Market: Opportunities, Barriers, Security, and Implementation Options. *IEEE Access : Practical Innovations, Open Solutions*, *12*, 37873–37890. DOI: 10.1109/ACCESS.2024.3375525

Junlakarn, S., Kokchang, P., & Audomvongseree, K. (2022). Drivers and Challenges of Peer-to-Peer Energy Trading Development in Thailand. In *Energies* (Vol. 15, Issue 3, p. 1229). DOI: 10.3390/en15031229

Kumar, C., & Chittora, P. (2024). Integrating Blockchain Technology into Smart Grid Systems: Opportunities, Challenges, and Future Directions. *2024 IEEE 11th Power India International Conference (PIICON)*, 1–6. DOI: 10.1109/PIICON63519.2024.10995188

Papapostolou, A., Andreoulaki, I., Anagnostopoulos, F., Divolis, S., Niavis, H., Vavilis, S., & Marinakis, V. (2025). Innovative Business Models Towards Sustainable Energy Development: Assessing Benefits, Risks, and Optimal Approaches of Blockchain Exploitation in the Energy Transition. In *Energies* (Vol. 18, Issue 15, p. 4191). DOI: 10.3390/en18154191

Pradhan, N. R., Singh, A. P., Kumar, N., Hassan, M. M., & Roy, D. S. (2022). A Flexible Permission Ascription (FPA)-Based Blockchain Framework for Peer-to-Peer Energy Trading With Performance Evaluation. *IEEE Transactions on Industrial Informatics*, *18*(4), 2465–2475. DOI: 10.1109/TII.2021.3096832

Schneiders, A., & Shipworth, D. (2021). Community Energy Groups: Can They Shield Consumers from the Risks of Using Blockchain for Peer-to-Peer Energy Trading? In *Energies* (Vol. 14, Issue 12, p. 3569). DOI: 10.3390/en14123569

Seven, S., Yao, G., Soran, A., Onen, A., & Muyeen, S. M. (2020). Peer-to-Peer Energy Trading in Virtual Power Plant Based on Blockchain Smart Contracts. *IEEE Access : Practical Innovations, Open Solutions*, *8*, 175713–175726. DOI: 10.1109/ACCESS.2020.3026180

Srinivas, O., Pradhan, N. R., & Singh, A. P. (2025). Blockchain Framework for Negawatt Energy Trading With Incentives and Limits: P2PnW-Chain. *IEEE Transactions on Consumer Electronics*, *71*(3), 7608–7621. DOI: 10.1109/TCE.2025.3597247

Syahputra, E. (2022). Blockchain Energy for Future Smart Grid in Indonesia: A Brief Review. *2022 International Conference on Technology and Policy in Energy and Electric Power (ICT-PEP)*, 100–105. DOI: 10.1109/ICT-PEP57242.2022.9988867

Thanasi-Boçe, M., & Hoxha, J. (2025). Blockchain for Sustainable Development: A Systematic Review. In *Sustainability* (Vol. 17, Issue 11, p. 4848). DOI: 10.3390/su17114848

Wang, S., Taha, A. F., Wang, J., Kvaternik, K., & Hahn, A. (2019). Energy Crowdsourcing and Peer-to-Peer Energy Trading in Blockchain-Enabled Smart Grids. *IEEE Transactions on Systems, Man, and Cybernetics. Systems*, *49*(8), 1612–1623. DOI: 10.1109/TSMC.2019.2916565

Yahaya, A. S., Javaid, N., Almogren, A., Ahmed, A., Gulfam, S. M., & Radwan, A. (2021). A Two-Stage Privacy Preservation and Secure Peer-to-Peer Energy Trading Model Using Blockchain and Cloud-Based Aggregator. *IEEE Access : Practical Innovations, Open Solutions*, *9*, 143121–143137. DOI: 10.1109/ACCESS.2021.3120737

Yusuf, F., Sari, R. F., Yusgiantoro, P., & Soesilo, T. E. (2024). Stakeholders' Perceptions of the Peer-to-Peer Energy Trading Model Using Blockchain Technology in Indonesia. In *Energies* (Vol. 17, Issue 19, p. 4956). DOI: 10.3390/en17194956

Zhou, Y., Manea, A. N., Hua, W., Wu, J., Zhou, W., Yu, J., & Rahman, S. (2022). Application of Distributed Ledger Technology in Distribution Networks. *Proceedings of the IEEE*, *110*(12), 1963–1975. DOI: 10.1109/JPROC.2022.3181528

Chapter 7

Smart Contracts in Energy Markets: Legal Enforceability, Liability, and Dispute Resolution Mechanisms

Froilan Delute Mobo
http://orcid.org/0000-0002-4531-8106
Philippine Merchant Marine Academy, Philippines

ABSTRACT

The integration of smart contracts in energy markets marks a transformative shift toward decentralization, automation, and enhanced transactional efficiency. As blockchain-based technologies are increasingly used to facilitate peer-to-peer energy trading, grid management, and decentralized energy exchanges, questions of legal enforceability, liability, and dispute resolution remain critical. This chapter aims to examine the foundational legal and regulatory considerations that govern smart contracts within the context of energy markets. It will explore how traditional legal principles interact with emerging technologies, including the enforceability of self-executing code under contract law, jurisdictional challenges, and the applicability of consumer protection and liability statutes. Case studies from pioneering jurisdictions such as the European Union, the United States, and select Asia-Pacific countries will be analyzed to provide comparative insights.

1. INTRODUCTION: THE DIGITAL ENERGY EVOLUTION

The global energy industry is now experiencing the most significant structural transformation where the centralized, fossil-reliant infrastructures are being replaced by the decentralized, digitalized, and renewable-oriented ecosystems. The idea of the

DOI: 10.4018/979-8-3373-5742-3.ch007

Internet of Energy (IoE) is driven by the active development of Distributed Energy Resources (DERs), microgrids, and the development of so-called prosumers who are independent agents who produce, consume, and trade electricity at the same time (Andoni et al., 2019; Goranovic et al., 2022). With the world energy markets working to incorporate unstable renewable sources and cope with demand response in real-time, the traditional transactional structures of manual billing and central intermediaries are being seen as potentially unable to meet the fast-frequency demands of the new grid management. The fundamental element of this digital transformation is the implementation of smart contracts. Initially, the concept of smart contracts as computer-programmed transaction protocols to implement the terms of a contract was originally developed by Nick Szabo in the mid-1990s in the context of Turing-complete blockchain platforms, including Ethereum (Szabo, 1997; Buterin, 2021). These self-sovereign contracts in the energy sector allow a broad scope of transformative applications, such as Peer-to-peer (P2P) energy trading, autopilot grid balancing, and the real-time issuance of renewable energy certificates (RECs). The smart contracts, as a direct part of the immutable code, are set to lower transaction expenses, cause less administrative delays, and increase the visibility of decentralized energy transactions (Manurung, 2025; Shelake, 2025). Nonetheless, the move towards utilizing a technically-based code is law paradigm to a legally-based Smart Legal Contract is a complex one. The issue of energy delivery is not a simple commercial relationship between individuals; it is one of the safety-related public services and has strict statutory, regulatory, and constitutional requirements. Deterministic and fixed quality of algorithmic code can be in conflict with the flexibility nature of traditional contract law that is based on human interpretation, judicial discretion, and equitable remedial (Low and Mik, 2023; Savelyev, 2018). As an example, a smart contract can automatically perform a payment, but it cannot automatically determine the existence of a claim of force majeure or a statutory right of a consumer to a cooling-off period without the involvement of a third party. Moreover, the introduction of smart contracts into the field of energy also poses new questions about liability and dispute resolution. With a decentralized grid that is governed by an autonomous code, it is a serious legal challenge to determine the party to hold accountable in case of a failure of a system, a cyberattack, or even an oracle error in data. Software-inflicted harms are only emerging as a subject of traditional tort and product liability regimes, and recent reforms introducing the scope of the definition of product to encompass software explicitly, e.g. the Product Liability Directive of 2024 by the EU (European Commission, 2024). The legal environment of the ASEAN region is also changing, especially that of the Philippines. New jurisprudence and regulatory sandboxes are to be used to experiment on how the Electric Power Industry Reform Act (EPIRA) and the Data Privacy Act (RA 10173) can be used to allow blockchain-based billing and data governance without undermining consumer protection (National Privacy Commission, 2024; Supreme Court of the Philippines, 2025). This chapter questions the overlap of these technological developments and legal principles. It goes beyond the explanatory discussion of blockchain potentials to offer a critical analysis of how existing regulations can be transformed to facilitate autonomous energy transactions. In particular, it examines the triple of legal enforceability, sharing of liability in the case of algorithm or oracle malfunctioning, and the creation of hybrid dispute resolution systems. The chapter builds on the Hybrid Legal-Code Model by relying on comparative case studies and global regulatory tendencies in 2021-2026. This model assumes that the sustainability of smart contracts in energy markets is not based on technical determinism, but rather on the intentional harmonization of code-based operation with accountability and legal responsibility, equity, and social responsibility. The subsequent sections will explain how the technologies have been changing throughout history, how the concept of the Smart Legal Contract has been understood by other institutions such as the UK Law Commission (2021), and how a policymaker may balance innovation with existing legal justice.

2. HISTORICAL CONTEXT AND LITERATURE REVIEW

2.1 From "Code is Law" to "Code is Infrastructure": The Shift to Law-as-Code

The conceptual journey of smart contracts has undergone a radical transformation over the last decade. In the early stages of blockchain development, the prevailing sentiment was rooted in a libertarian techno-determinism often summarized by Lawrence Lessig's famous aphorism, "Code is Law." In this initial phase, the goal of smart contract proponents was to create an autonomous, self-sufficing ecosystem where the execution of code was final and absolute, intentionally bypassing traditional legal systems and state intervention. This "technical finality" was seen as a way to eliminate the "human friction" of corruption, bias, and slow judicial processes.

However, as these technologies moved from experimental niche applications into critical public sectors like energy, the "Code is Law" mantra proved to be intellectually and practically unsustainable. Scholarship has since shifted toward a more pragmatic "Law-as-Code" or "Code as Infrastructure" approach. Recent studies (De Filippi et al., 2022) emphasize that smart contracts in the energy sector function less as a replacement for law and more as Regulatory Technology (RegTech). In this view, code is not an independent sovereign but a digital infrastructure used to automate compliance with existing legal and environmental mandates.

In the context of decentralized energy markets, this shift is vital. If a smart contract is used to facilitate Peer-to-Peer (P2P) solar energy trading, it must do more than just transfer tokens for kilowatt-hours; it must ensure that the transaction complies with national emissions targets, carbon credit accounting standards, and grid stability protocols. By viewing code as infrastructure, regulators can treat smart contracts as automated "compliance engines" that enforce public interest obligations—such as carbon tax collection or renewable energy certificate (REC) validation—at a granular level that traditional auditing could never reach (Goranovic et al., 2022).

Furthermore, the "Code as Infrastructure" perspective acknowledges that blockchain systems are embedded within society. This means they are subject to the "physics" of the law. De Filippi, Mannan, and Reijers (2022) argue that blockchain provides a "confidence machine," but that this confidence is only legally viable when it is coupled with institutional legitimacy. Therefore, the contemporary energy prosumer does not rely on the code alone; they rely on a hybrid ecosystem where the code handles the high-frequency execution of trades, while the overarching legal framework ensures that those trades do not violate consumer protection acts or environmental statutes. This pragmatic turn recognizes that while code can automate the *how* of a contract, the law must still dictate the *why* and the *what if*.

2.2 Re-Evaluating Enforceability: The Move Toward Interpretation and Mistake

For several years, the primary legal hurdle for smart contracts was the fundamental question of formation: Can a string of code be a "contract"? This debate was largely settled by the UK Law Commission's 2021 Statement on Smart Legal Contracts, which provided a global benchmark for judicial systems. The Commission concluded that the existing principles of English law—and by extension, many common law and civil law systems—are sufficiently flexible to accommodate smart contracts. So long as the basic elements of offer, acceptance, consideration, and the intention to create legal relations are present, a smart contract is legally binding (UK Law Commission, 2021).

With the question of "validity" largely put to rest, the legal focus between 2024 and 2026 has moved toward more complex issues of Interpretation and Mistake. Unlike natural language contracts, which are interpreted based on the "objective intent" of the parties, smart contracts are literal. They do exactly what the code says, even if that is not what the parties intended. This creates a "translation gap" between the parties' commercial agreement and the digital execution.

The shift to decentralized and smart contract-based energy market in the Philippines and the ASEAN region in general cannot be left to technical luck. The automation of blockchain mentioned in the sections above does not make it immune to the demands of public safety, consumer protection, or national security, as it is the case with all other technologies. In order to make sure that innovation facilitates, and not subverts the public interest, there ought to be a strong policy framework. The chapter suggests three essential pillars of the future of energy-tech regulation: the requirement of the existence of Kill Switches in grid-critical code, the institutionalization of shared "Legal Sandboxes," and the making of mandatory third-party auditing standards on smart legal contracts. 10.1 Compulsory Kill Switches: Dilemma Between Autonomy and Grid Stability. The most inherent issue with smart contracts in the energy industry is that it is unstoppable. Although immutability is a financial virtue with regard to ledger-book keeping, it can be a drawback to physical infrastructure management. When a smart contract that operates a neighborhood micro grid or a large-scale Virtual Power Plant (VPP) goes into an endless logic loop, or operates on the basis of corrupted oracle data, or is hacked by an evil actor, the results are more than financial and may cause physical grid instability, surges, or even neighborhood blackouts. Hence, this chapter suggests a Compulsory "Kill Switch" Requirement on all smart contracts that are considered grid-critical. A Kill Switch (or an emergency pause operation) is a program code that enables some authorized body, e.g. the National Grid Corporation of the Philippines (NGCP) or a distribution utility (DU), to freeze the runtime of one of a smart contract under certain, legally defined emergency circumstances. 10.1.1 The Governance of the Switch The main reason why there are some kill switches are that it puts back to a decentralized system the element of centralization. But absolute decentralization in the case of the public utilities is a safety risk. In order to maintain advantages of blockchain yet maintain security, the Kill Switch has to be managed through Multi-Signature (Multi-Sig) Protocol. According to this model, the Kill Switch cannot be activated by one actor. Rather, it would involve the digital keys of at least two of three parties, including: the Platform Operator, Government Regulator (DOE), and independent Technical Auditor. This will make sure that the override is not employed as a means of arbitrary interference but as the last resort in the event of technical breakdown or a national security issue. According to Low and Mik, these types of on-chain protective structures are critical to filling the gap between autonomous execution speed and human-centric safety demands (2023).

2.2.1 Defining "Grid-Critical"

A kill switch is not necessary in all smart contracts. Two neighbor to neighbor P2P trade of small credits in solar could not represent a systemic risk. Policy should thus set a threshold of criticality - on either the amount of energy traded or the number of households concerned. The existence of contracts to control critical load balancing or to be the main billing mechanism of a community should be legally mandated to contain these manual overrides. The article under discussion is created as a result of the collaboration of FinTech and EnergyTech in a single area, specifically, the joint legal sandbox. The invention of smart contracts in the energy markets forms a regulatory no man land. Since such contracts frequently include the use of digital wallets, tokenized energy credits, and cryptocurrency payments, it means that they are governed under both the regulations of the Bangko Sentral ng Pilipinas (BSP) as a financial transaction as well as that of the Department of Energy (DOE) as a utility service.

2.2.2 Collaboration

This is why the collaboration between agencies is necessary. As it is, a Philippine-based innovator could be licensed to operate in the FinTech sector by the BSP, but their logic of automated disconnection could be contravened by the DOE consumer protection regulations. To address this, Joint Legal Sandboxes have to be institutionalized by the government. These sandboxes are managed, safe-to-fail systems in which energy-tech firms have an opportunity to test their smart contracts in a combined oversight of the BSP and the DOE. These sandboxes are not only meant to test something technical, but discover regulatory. Within a shared sandbox, the regulators will be in a position to see in real-time how the volatility of crypto-payments impacts the capability of a consumer to keep on with the lights. These sandboxes enable the co-creation of law, suggested by Andoni et al. (2019) and Manurung (2025). The BSP and DOE can provide dynamic guidelines on the basis of the real behaviour of the pilots rather than formulating a strict regulation of 200 pages that will have become outdated by the period of publication. 10.2.2 Experimenting the Hybrid Legal-Code Model. In particular, one should test the Tri-Layer Dispute Resolution Model with the help of the sandbox (the discussion is provided in Section 6). Regulators may observe the effectiveness of a simulated oracle failure on a platform that is effectively managed through a mechanism called Decentralized Arbitration (such as Kleros). The regulators can intervene by the so-called Legal Backstop in case the decentralized resolution is too slow or unfair. This is a proactive strategy so that even as these technologies leave the sandbox and go to the actual grid, they have been stress-tested to meet the provisions of the Philippine Data Privacy Act (RA 10173) and the EPIRA.

2.2.3 Mandatory Third-party Audits: Standardising Smart Legal Contracts.

The code is law The mindset of the code is law was that of move fast and break things. There is no option of breaking things in the energy sector. In order to secure the integrity of the market, the Philippines should introduce Compulsory Third-Party Audits of all templates of the type of smart legal contract, prior to their release into the market.

2.2.4 The "Audit-as-a-License" Model

Based on the extension of the Philippine EVOSS Act, the government ought to establish a list of Certified Smart Contract Auditor. Such auditors would be technical and legal professionals that would review a contract on three dimensions: Technical Security: Does the code have re-entrancy attacks, overflow errors, or be hacked? Fidelity to the Law: Does the code show proper reflection of the English-language "Master Agreement? As an example, is the period of 14-day cooling-off in the text equivalent to the time-locks in the code? Regulatory Compliance: Have the contract applied the right taxes, lifeline discounts, and grid fees automatically as it should under the law of Philippines?

2.2.5 The Library of Certified Templates

DOE must enable the development of a Library of Certified Templates in order to lower the price of small scale community microgrids. These are already vetted and pre-audited, plug-and-play smart contracts that were already compliant with the Consumer Act of the Philippines (RA 7394). With a certified template, a local solar cooperative may avoid the costly audit procedure, which is performed individually, and reduce the entry barrier to decentralized energy without sacrificing the desired high level of care (Supreme Court of the Philippines, 2025). 10.4 Conclusion: Designing in a resilient trust

Code is the future of Philippine energy market, though it should be maintained through trust. Unlike the perfection of the algorithm, the appearance of Trust in an automated system is possible due to the existence of Accountability. With a requirement of Kill Switches, we will be holding humans to be the masters of the grid. Institutionalizing the Joint Sandboxes takes care of making our regulations innovative as much as our technologies. And by making it mandatory to have the Audits, we guarantee that the same blockchain law of digitality is as trustworthy and just as the one of the law of land. These policy recommendations are a blueprint of a "Legal-Tech Harmony" as the ASEAN region is shifting towards the ASEAN Power Grid (2023). It is aimed at an autonomous yet regulated enough to be efficient and just grid. With such interventions, the Philippines would be able to transform itself into being an experimental pilot to becoming a global leader in terms of Smart Energy Governance in which all the smart contracts are legally binding contracts.

3. LIABILITY MODELS FOR AUTOMATED TRANSACTIONS

The global energy industry is now experiencing the most significant structural transformation, where the centralized, fossil-reliant infrastructures are being replaced by the decentralized, digitalized, and renewable-oriented ecosystems. The idea of the Internet of Energy (IoE) is driven by the active development of Distributed Energy Resources (DERs), microgrids, and the development of so-called prosumers who are independent agents who produce, consume, and trade electricity at the same time (Andoni et al., 2019; Goranovic et al., 2022). With the world energy markets working to incorporate unstable renewable sources and cope with demand response in real-time, the traditional transactional structures of manual billing and central intermediaries are being seen as potentially unable to meet the fast-frequency demands of the new grid management. The fundamental element of this digital transformation is the implementation of smart contracts. Initially, the concept of smart contracts as computer-programmed transaction protocols to implement the terms of a contract was originally developed by Nick Szabo in the mid-1990s in the context of Turing-complete blockchain platforms, including Ethereum (Szabo, 1997; Buterin, 2021). These self-sovereign contracts in the energy sector allow a broad scope of transformative applications, such as Peer-to-peer (P2P) energy trading, autopilot grid balancing, and the real-time issuance of renewable energy certificates (RECs). The smart contracts, as a direct part of the immutable code, are set to lower transaction expenses, cause less administrative delays, and increase the visibility of decentralized energy transactions (Manurung, 2025; Shelake, 2025). Nonetheless, the move towards utilizing a technically-based code is law paradigm to a legally-based Smart Legal Contract is a complex one. The issue of energy delivery is not a simple commercial relationship between individuals; it is one of the safety-related public services and has strict statutory, regulatory, and constitutional requirements. Deterministic and fixed quality of algorithmic code can be in conflict with the flexible nature of traditional contract law that is based on human interpretation, judicial discretion, and equitable remedial (Low and Mik, 2023; Savelyev, 2018). As an example, a smart contract can automatically perform a payment, but it cannot automatically determine the existence of a claim of force majeure or a statutory right of a consumer to a cooling-off period without the involvement of a third party. Moreover, the introduction of smart contracts into the field of energy also poses new questions about liability and dispute resolution. With a decentralized grid that is governed by an autonomous code, it is a serious legal challenge to determine the party to hold accountable in case of a failure of a system, a cyberattack, or even an oracle error in data. Software-inflicted harms are only emerging as a subject of traditional tort and product liability regimes, and recent reforms introducing the scope of the definition of product to encompass software explicitly, e.g., the Product Liability Directive of 2024 by the EU (European Commission, 2024). The legal environment of the ASEAN region is also changing, especially that of the Philippines. New jurisprudence and regulatory sandboxes are being explored to explore how the Electric Power Industry Reform Act (EPIRA) and the Data Privacy Act (RA 10173) can be used to accommodate billing and data governance based on blockchains without affecting consumer protection (National Privacy Commission, 2024; Supreme Court of the Philippines, 2025). This chapter questions the overlap of these technological developments and legal principles. It goes beyond the explanatory discussion of blockchain potentials to offer a critical analysis of how existing regulations can be transformed to facilitate autonomous energy transactions. In particular, it examines the triple of legal enforceability, sharing of liability in the case of algorithm or oracle malfunctioning, and the creation of hybrid dispute resolution systems. The chapter builds on the Hybrid Legal-Code Model by relying on comparative case studies and global regulatory tendencies in 2021-2026. This model assumes that the sustainability of smart contracts in energy markets is not based on technical determinism, but rather on the intentional harmonization of code-based operation with accountability and legal responsibility, equity, and social responsibility. The subsequent sections will explain how the technologies have been changing throughout history, how the concept of the Smart Legal Contract has been understood by other institutions such as the UK Law Commission (2021), and how a policymaker may balance innovation with existing legal justice

4. CONSUMER PROTECTION AND RETAIL P2P TRANSACTIONS

The transition toward decentralized Peer-to-Peer (P2P) energy trading represents more than just a technological milestone; it signifies a fundamental shift in the legal status of the average household. Traditionally, a residential energy user was strictly a "consumer"—a passive recipient of services provided by a heavily regulated monopoly utility. However, the integration of blockchain-based smart contracts transforms these households into "prosumers" and, effectively, micro-retailers. While this empowers individuals to monetize their solar investments, it creates a precarious legal situation where the standard protections afforded to domestic consumers may be inadvertently bypassed by the "hard-coded" logic of autonomous systems.

As we examine the landscape from 2021 to 2026, two critical battlegrounds have emerged in the intersection of energy law and consumer rights: the legality of automated disconnections and the burgeoning "Right to Human Intervention."

4.1 Automated Disconnection and the "Right to Electricity."

The most immediate risk posed by smart contracts in the retail energy sector is the automation of enforcement. In a typical smart contract-enabled microgrid, electricity is delivered on a "pay-as-you-go" basis. If a user's digital wallet falls below a certain threshold or if a cryptocurrency payment fails due to network congestion, the smart contract is programmed to automatically trigger a relay to disconnect the power supply. While this ensures technical efficiency and payment certainty for the seller, it frequently violates established human rights and public utility laws.

4.1.1 The Philippine Context: EPIRA and the Social Safety Net

In the Philippines, the Electric Power Industry Reform Act (EPIRA) of 2001 (Republic Act No. 9136) remains the bedrock of energy governance. EPIRA was designed with the explicit goal of ensuring a "socially fair" pricing and delivery system. Under existing Philippine law, energy is not a mere luxury but an essential service. Consequently, the process for disconnecting a customer is strictly regulated: it requires formal notice, a grace period, and, in many cases, specific protections for "lifeline" or vulnerable consumers (e.g., households with medical equipment or low-income families).

A smart contract that executes an instantaneous disconnection due to a wallet shortfall lacks the capacity to verify if the household is currently undergoing a medical emergency or if the "notice" requirement has been substantively met. Recent discourse from the National Privacy Commission (NPC) in 2024 and the Department of Energy (DOE) has emphasized that "code-based enforcement" cannot supersede statutory due process. Legal scholars argue that because electricity is a prerequisite for the exercise of other basic rights—such as health and education—automated disconnection is increasingly viewed as a violation of the de facto "Right to Electricity." In a landmark 2025 advisory, the NPC suggested that energy data used to trigger disconnections must be handled with extreme care, as a "wrongful disconnection" triggered by a faulty algorithm could constitute a significant civil liability for the micro-retailer or the platform provider.

4.2 The "Right to Human Intervention" and the Human-in-the-Loop Requirement

In efforts to reduce the dangers of "tyranny by code," regulators have started to consider the European Union Artificial Intelligence Act (2024) to provide an example of how to regulate all high-risk systems, such as energy smart contracts, operated by artificial intelligence. The most important idea that is coming out during this period is the Right to Human Intervention. This is an ideology that any artificial judgment that has far-reaching effects on a human being, particularly in the provision of basic services, should be influenced by human and legal considerations. When applied to P2P energy trading, this implies that a smart contract cannot be a 100 percent autonomous one. Rather, regulators require a Human-in-the-Loop (HITL) architecture. HITL Model in the Energy Markets. The Right to Human Intervention is an obligatory safety valve. According to scholarship and regulatory advice of 2024-2026 (Andoni et al., 2019; Low and Mik, 2023), smart legal contracts within the energy market should also incorporate a type of clauses that can be invoked via a specified consumer advocate or a regulatory authority, and is called: Pause or Override. To illustrate, when a P2P microgrid in a rural community in the Philippines goes offline suddenly because of an fault in the algorithmic system used to set prices, the prosumers (sellers) should not have an opportunity to gain at the cost of their neighbors.

An HITL requirement guarantees that a neutral third party - say, a local energy cooperative or a government ombudsman - can put a freeze upon the performance of the contract until the mistake has been corrected. This is not just a technical recommendation, but it is increasingly turning out to be a condition of the legal validity of energy-tech licenses. The Philippine Department of Energy Circulars on Transactive Energy (2025) explicitly declare that the decentralized energy platforms are required to have a recourse mechanism, in which a human being will be able to review and reverse automated trades, which are determined to be unconscionable or predatory. This is in line with the right of the consumer to challenge a bill in the traditional utility model. 4.3 Algorithmic Discrimination and the "Energy Justice" Other than disconnections, the question of algorithmic discrimination is becoming a big issue. Smart contracts are regularly powered by AI-based pricing frameworks, which vary in line with supply and demand. These algorithms may unintentionally harm non-homeowners capable of changing their consumption (e.g., with elderly occupants who need to maintain their climate control at all times). The study of the concept of Energy Justice (Goranovic et al., 2022) sheds light on the idea that, in case a smart contract is designed to comply exclusively with the notion of grid efficiency, it might end up systematically discriminating against the poor. Thus, the Right to Explainable Algorithm is also the right to Human Intervention. The consumers should be in a position to know why they paid a given price or why their trade has been denied. Openness in the logic of the code is now considered as being a continuation of the right to information in the Philippine Data Privacy Act (RA 10173). 4.4 Conclusion: Striking a Balance between Innovation and Protection. The potential of grid resilience and adoption of renewables in P2P trading is immense, by transforming households into retailers. But the retailer's position should not deprive the household of its consumer protection. The legal systems of 2026 underline the fact that smart contracts should be made law-abiding in the first place. A combination of the Right to Electricity and the Right to Human Intervention into the very structure of the code will guarantee that the decentralization of the energy grid does not cause a decentralization of justice. The P2P energy solution is not going to be one of complete automation, but the Harmonized Governance, where the pace of the blockchain is equated through the protective friction of the law.

5. REGULATORY COMPLIANCE AND GRID GOVERNANCE

The technical allure of smart contracts—their speed, autonomy, and transparency—often masks the extreme complexity of the regulatory environment in which they must operate. In the energy sector, "disruption" cannot come at the expense of grid stability or national security. As smart contract applications migrate from local microgrids to regional energy exchanges, they must navigate a dense thicket of technical standards, licensing requirements, and data sovereignty laws. Within the Southeast Asian context, this involves a dual challenge: complying with the ambitious interoperability goals of the ASEAN Power Grid (APG) and adhering to the rapidly evolving domestic regulations of individual member states, such as the Philippines' pioneering digital reforms.

5.1 The ASEAN Power Grid (APG) and Cross-Border Interoperability

The ASEAN Power Grid (APG) is a flagship initiative of the ASEAN Centre for Energy (ACE), aimed at integrating the power systems of the ten member states to enhance regional energy security and facilitate the trade of renewable energy (ASEAN Centre for Energy, 2023). For smart contracts to facilitate cross-border energy trading within the APG, they must achieve interoperability—not just at the code level, but at the regulatory and technical level.

Currently, each ASEAN member state operates under its own specific "Grid Code," which defines the technical parameters for voltage, frequency, and safety. A smart contract executing a trade between a solar farm in Vietnam and a consumer in Thailand must be programmed to recognize the "handshake" between two different national grid standards. This requires the development of "Universal Energy Adapters" within the smart contract architecture, capable of translating diverse grid requirements into a single execution logic (Goranovic et al., 2022).

Furthermore, the Asian Development Bank (ADB, 2021) has emphasized that the transition to a decentralized regional grid requires a "Harmonized Regulatory Framework." Smart contracts act as the mechanical enforcement of this harmony. If the APG adopts a unified carbon pricing mechanism or a regional "system usage" fee, smart contracts can automate the collection and distribution of these fees across borders, ensuring that no jurisdiction is financially disadvantaged by the transit of electricity across its wires. However, as noted by De Filippi et al. (2022), this "technological trust" must be backed by institutional treaties. Without a regional treaty that recognizes the legal validity of "on-chain" energy settlements, the smart contract remains a technical tool without a legal home.

5.2 The Philippine EVOSS Act Expansion: The "Code Audit" as a License

Within the Philippines, the move toward digital energy governance has found a specific legislative anchor in the Energy Virtual One-Stop Shop (EVOSS) Act (Republic Act No. 11234). Originally designed to streamline the permitting process for power generation projects, the EVOSS framework was significantly expanded between 2023 and 2025 to accommodate the digitalized energy market.

One of the most consequential developments in this expansion is the requirement for Mandatory Code Audits for blockchain-based energy providers. Recognizing that a flaw in a smart contract could threaten the stability of the distribution grid or lead to the systemic overcharging of consumers, the Philippine Department of Energy (DOE) now classifies smart contracts as "Critical Digital Infrastructure." Under the updated EVOSS guidelines, any entity seeking to operate a Peer-to-Peer (P2P) trading platform or a Virtual Power Plant (VPP) must submit its smart contract code to a certified third-party auditor.

This audit is not merely for cybersecurity; it is a "compliance audit" to ensure the code accurately reflects Philippine energy laws, such as the EPIRA tariff structures and the mandated "lifeline" discounts for marginalized consumers. This shift represents the professionalization of the "smart contract developer" role. As Manurung (2025) argues, the developer is no longer just a coder; they are an "architect of digital law" whose work must be certified before it can touch the public grid. This ensures that the Philippine energy market remains innovative while maintaining the high "standard of care" required by the Supreme Court (2025) in its recent rulings on digital torts.

5.3 Data Sovereignty and the ASEAN Model Contractual Clauses

A significant and often overlooked barrier to decentralized energy trading is the question of Data Sovereignty. Energy usage data—high-frequency readings that reveal when a person is home, what appliances they use, and even their economic status—is classified as sensitive personal data under the Philippine Data Privacy Act (RA 10173) and similar laws across ASEAN.

When energy is traded across borders via a blockchain, the "ledger" of that trade (which contains usage data) may be stored on nodes located in different countries. This creates a conflict with national data localization laws. To resolve this, trades must increasingly comply with the ASEAN Model Contractual Clauses (MCCs) on Transborder Data Flows (2021).

The MCCs provide a standardized legal template that ensures data transferred across borders receives the same level of protection as it would in its home country (ADB, 2021). For smart contract developers, this means integrating Privacy-by-Design features, such as Zero-Knowledge Proofs (ZKPs) or off-chain data storage. Under this model, the "on-chain" smart contract only records that a trade occurred and was paid for, while the sensitive usage data remains stored within the national borders of the consumer, accessible only to authorized regulators (National Privacy Commission, 2024).

5.4 Conclusion: The Rise of Regulated Autonomy

Such history of energy control in ASEAN shows that completely autonomous smart contracts are a myth when it comes to the work of any public utility. What is now taking shape is a Regulated Autonomy. This model offers the speed and efficiency of automation offered in the smart contract, but the limits of its behavior are well-specified by the codes of the national grid, the EVOSS auditing process, and regional data sovereignty treaties. In the case of the Philippines and its neighbors in the ASEAN region, the target of the energy market in 2026 is to establish a Seamless Regional Grid. This involves smart contracts that need to be not only technically interoperable, but also legally interoperable. The region is developing a digital infrastructure that enables the energy transition without the rule of law by harmonizing code with the ASEAN Model Contractual Clauses and national safety standards. This sense of the smart contract entails the digital glue that will unify the various energy markets within the Southeast Asian region into one robust and transparent ecosystem. Dispute Resolution: Beyond the Code is Law. The improbable temporal paradox of the legal system in modern energy markets is grounded in the nature of the energy markets themselves. To keep grid frequencies and stability constant, in the case of decentralized transactive energy (TE) systems, trades are frequently completed within 15-minute settlement periods or even near-real-time periods (Goranovic et al., 2022). On the other hand, the traditional litigation, which is the usual method of settling a contractual violation, works within the time frame of months or years. Assuming a smart contract in place that manages a neighborhood microgrid fails at 2:00 PM, the financial or operational conflict cannot afford to hold off until next calendar year to have a court date. A commensurate speed of justice is required by the speed of code. In an effort to neutralise this friction, this chapter suggests a Tri-layer Dispute Resolution Model. This framework rejects the inflexible code is law technical determinism and establishes a framework of interventions of varying levels that weighs between automated efficiency and legal equity. Since the alternative is an automated pause and guardianship system, the alternative is not applicable to the NCLC system (Elementary School, 2005).

6. TRI-LAYER DISPUTE RESOLUTION MODEL

6.1 Layer 1

The On-Chain Layer (Automated "Pause" and Guardianship). As the other option is an automated pause and guardianship system, the option does not apply to the NCLC system (Elementary School, 2005). The defence at the first level of the decentralized energy market should be active and integrated into the code. Instead of waiting for a dispute has be "filed," the smart contract should have anomaly detection and automatic "Pause" capabilities. The smart contract is coded into this layer with what is referred to as guardrails or circuit breakers. An example would be when the contract notices a price transaction that is 500 percent over the daily average, which is likely to be a fat-finger-coded error or a hacked oracle; the contract goes into suspension. This is not a termination but a temporary escrow of the transaction (Andoni et al., 2019). This is a mechanism that is similar to the circuit breakers in the high-frequency stock trading but is implemented to the physical reality of the power grid. In addition, this layer makes use of Multi-Signature Guardianship. Any high-value or grid-critical trade can only be completed prior to digital confirmation by a "Guardian" node; often, a neutral third-party regulator or a compliance bot. By ensuring that the most apparent technical flaws are identified before the irreversible on-chain finality, this layer is able to prevent disputes in the first place. Decentralized Arbitration (Algorithmic Justice and Kleros): This layer operates on the premise that legal matters can be settled through a program implementing social norms.

6.2 Layer 2: Decentralized Arbitration (Algorithmic Justice and Kleros

This layer is based on the idea that legal issues can be resolved by a program that enforces social rules. Once an anomaly is identified, which cannot be fixed with the help of mere logic, the conflict goes to the second stage: Decentralized Arbitration. This layer solves the interpretative gap using blockchain-native arbitrage protocols, like Kleros or Aragon Court. Kleros is a crowdsourced justice system that uses the selection of jurors who are usually experts in the field of energy engineering or blockchain development at random to hear evidence. In energy dispute case, the technical logs, Oracle data, and code of the smart contract would be examined by the jurors who would decide whether the execution was carried out in accordance with the agreed parameters. A system of tokens encourages jurors to vote in line with the majority in order to get tokens, which motivates them to be honest. Decentralized arbitration is specifically effective in the case of low-value and high-volume energy trades, as described by Savelyev (2018) and more recently by Shelake (2025), where the price of a typical lawyer would be higher than the units of energy in dispute. It offers a specialized, low-priced, and speedy solution that is much more digitally appropriately fitting the energy market than a general-purpose civil court. Nevertheless, in order to make such a layer work, the parties should have sworn such jurisdiction under the smart contract under the clause of the arbitration.

6.3 Layer 3: The Legal Back stop (Master Agreements and Judicial Oversight)

The last layer and most important is the Legal Backstop. This layer is used to make sure that the digital world does not lose its connection with the physical rule of law. Although the technical anomalies and minor disagreements are managed in the first and second layers, the issues that have a great failure, including the systemic grid damage or a mass consumer fraud, cannot be managed without the power of the state. It is based on the technology Hybrid Smart Legal Contracts that comprise of a machine-readable code and a natural-language Masters Agreement. This contract serves as the truth of the intent of the parties. The intent expressed in human readable form will supersede the coding error in the event of a fundamental conflict. The case of Quoine Pte Ltd v B2C2 Ltd (2020) was a landmark that affirmed this principle and was followed by other legal scholarly works (Low & Mik, 2023). The Singapore Court of Appeal in Quoine determined that where a trade is carried out based on the overall logic error in the code upon which the two parties have no intention, the court can have power to remedy or rescind the transaction. This (Judicial Override) is needed so that there is confidence in automated systems by the population. Moreover, the 2025 doctrines on Digital Torts by the Supreme Court in the Philippines underline the impossibility of the indemnification of an operator against their quasi-delictual duties of ensuring a safe and fair supply of energy by the use of terms of service or automated code.

6.4 A Matter of Minutes to Harmonize the Layers: The 15 Minute Resolution

The Tri-Layer Model provides an avenue to the so-called Resolution-as-a-Service. With these three-layers combined, a decentralized energy market is able to resolve conflicts of different intensities: Seconds: On-Chain Layer stops technical errors. Days: Decentralized Arbitration is applied to resolve interpretative disputes. Months: Legal Backstop adjudicates fundamental legal failures. This graded system observes the 15 minutes energy cycle, so that the majority of conflicts are contained and terminated immediately, although still providing a strong path to the palace of justice to more complicated law problems. There is no aim to abolish lawyers, as Manurung (2025) emphasizes, but rather to make sure that the legal intervention is only required in the case of the most important problems, and the rest of the friction of the digital market is taken over by the technology.

6.5 Conclusion: The Death of Technical Determinism

The Tri-Layer Model is the ultimate abandonment of the concept of technical determinism of energy law. It acknowledges that code is an effective means of doing, but a bad means of judging. Around the smart contract, we provide a Legal Protective Shell - automatic stop and frisk, arbitral peer review, judicial controls - we can provide the conditions in which the energy transition can thrive. In the case of the ASEAN Power grid and the fledgling Philippine P2P markets, this model is what will give the needed friction of the law to shield the consumer and the needed speed of a decarbonized, decentralized future, which is technical in nature. The future governance of energy is not in the question of Law and Code since they have to exist in a complex architecture that enables the two to serve as one integrated system of justice.

7. ASEAN DATA LAWS, IMMUTABILITY, AND GDPR

The implementation of blockchain technology in the oil and energy industry has introduced one of the most important legal and technical contradictions of the digital era the irreconcilability of the immutability of the distributed registry and the primacy of privacy as an essential right. In the Philippines, this conflict is explicitly determined by the interaction between the decentralized essence of smart contracts and the high demands of the Data Privacy Act of 2012 (Republic Act No. 10173). Energy information is information that is sensitive. Contrary to more basic financial operations, high-frequency data of energy consumption (usually measured after

15 minutes, provided by smart meters) can reveal the personal life of a household, its occupancy, appliance usage, and even socioeconomic changes (Andoni et al., 2019). The challenge as Peer-to-peer (P2P) energy trading matures would be to use this information to achieve a non-permanent, non-public, non-erasable record of a personal action of a citizen.

7.1 Immutability Paradox vs. RA 10173

The fundamental structure of blockchain is based on immutability, which is the idea that when information is stored on a block and authenticated by the network, it is impossible to be modified or deleted. This is what brings the "trust" required in the decentralized energy markets. But the Data Privacy Act of 2012 (DPA) provides Filipino "data subjects" with a package of rights that are directly inconsistent with this architecture, in particular, the Right to Erasure or Blocking (Section 16, RA 10173). In case an energy prosumer will choose to exit a blockchain-based micro-grid, the DPA dictates that the individual is entitled to request that their personal data be deleted off the system. This is an easy administrative job on a conventional centralized server. Deleting a record on a blockchain would require a hard fork of the whole network, which is practically impossible to use in practice to meet some privacy demands. This Immutability Paradox risked halting the implementation of smart contracts to the Philippine energy market until a chain of technical and regulatory breakthroughs would happen between 2021 and 2025.

7.2 The Privacy Breakthrough: Zero-Knowledge Proofs (ZKPs)

The use of Zero-Knowledge Proofs (ZKPs) has been the key to solve this conflict. ZKP is a cryptographic technique whereby one party (the prover) can demonstrate to another party (the verifier) that a certain statement is true, but conveys no other information other than that the statement is true. In 2025, ZKPs started to be used in energy smart contracts in the Philippines to manage the sensitive "metadata" of the energy trades. As opposed to writing on the public ledger, Household A used 50kWh at 2:00 PM and is owed 500 Pesos, the system verifies two things using a ZKP: That there was a certain amount of energy consumed by the household. That the family has enough electronic money to afford it. The smart contract implements the payment according to the evidence of the transaction, and not the information of the transaction. The usage history itself, the sensitive so-called usage fingerprints, are stored off-chain in encrypted local databases, with the cryptographic "hash" or proof only stored on-blockchain (Low and Mik, 2023). This makes the ledger audit-capable, but in all respects financial, without being informed about the domestic manners of the consumer.

7.3 The Regulatory Response: NPC Advisory and the 2025 Guidelines

The shift towards the ZKP-supported energy contracts was triggered by the explicit recommendations of the National Privacy Commission (NPC). At the beginning of a new adoption cycle, the Privacy Policy Office Advisory No. 2020-032 made it clear that, although blockchain technology is not illegal per se, it is required to follow the principles of Transparency, Legitimate Purpose, and Proportionality. By 2024-2025, the NPC had revised Guidelines on Decentralized Energy Systems (NPC, 2025). These guidelines have formally acknowledged the ZKPs and off-chain storage as two acceptable means of Privacy-by-Design. The Commission made it clear that in the event that personal data has been encrypted with industrial-grade encryption and the on-chain evidence fails to allow identification of the individual, the "Right to Erasure" can be met by blocking the data effectively or by destroying decryption keys, even though the cryptographic hash may still be on the immutable ledger. It has been necessary because of this regulatory pragmatism. It enables energy start-ups in the Philippines to be innovative in the ASEAN Power grid (APG), where they are also bound to follow the cross-border data flow accord. Other ASEAN neighbors are now studying the Philippine model of Regulated Cryptography as a prototype of balancing the Internet of Energy with the basic right to privacy (Manurung, 2025).

7.4 Data Sovereignty and Prosumers in the Role of Data Controllers

The other development, which is also significant, is the re-classification of positions under RA 10173. The utility is the controller of data and the consumer is the subject in the traditional utility model. Under P2P smart contract market, all the prosumers who sell the energy implicitly make them Data Controllers of the usage data of their neighbors in the transaction. This poses a huge compliance cost to ordinary citizens. As a solution to this, the 2025 energy contracts include Automated Data Processing Agreements (DPAs). A user using a decentralized energy system becomes automatically bound by a standardized privacy protocol by the smart contract on boarding. This makes sure even a decentralized system with no trust that is performed on a decentralized system there is a distinct legal chain of custody of data.

7.5 To a Privatized Privacy-First Energy Grid

This solution to the opposition of blockchain and data privacy in the Philippines is more a transition between the technical avoidance and the legal incorporation. By

2026, it is a matter of de facto standard of any energy contract transacted via a public or consortium blockchain to use ZKPs. The experience of this transition has taught a lot: the digital energy transition must succeed due to the lack of loss of consumer trust. In case, prosumers believe that their switching to renewable energy involves sacrificing their privacy at home, there will be considerable resistance to the IoE. But with the Hybrid Legal-Code Model - RA 10173 offers the mandate and ZKPs the technical implementation - the Philippines is showing that it can construct a grid that is both transparent throughout to the markets and privatized to the citizens.

8. POLICY RECOMMENDATIONS

The shift to decentralized and smart contract-based energy market in the Philippines and the ASEAN region in general cannot be left to technical luck. The automation of blockchain mentioned in the sections above does not make it immune to the demands of public safety, consumer protection, or national security, as it is the case with all other technologies. In order to make sure that innovation facilitates, and not subverts the public interest, there ought to be a strong policy framework. The chapter suggests three essential pillars of the future of energy-tech regulation: the requirement of the existence of Kill Switches in grid-critical code, the institutionalization of shared "Legal Sandboxes," and the making of mandatory third-party auditing standards on smart legal contracts.

8.1 Compulsory Kill Switches: Dilemma Between Autonomy and Grid Stability

The most inherent issue with smart contracts in the energy industry is that it is unstoppable. Although immutability is a financial virtue with regard to ledger-book keeping, it can be a drawback to physical infrastructure management. When a smart contract that operates a neighborhood micro grid or a large-scale Virtual Power Plant (VPP) goes into an endless logic loop, or operates on the basis of corrupted oracle data, or is hacked by an evil actor, the results are more than financial and may cause physical grid instability, surges, or even neighborhood blackouts. Hence, this chapter suggests a Compulsory "Kill Switch" Requirement on all smart contracts that are considered grid-critical. A Kill Switch (or an emergency pause operation) is a program code that enables some authorized body, e.g. the National Grid Corporation of the Philippines (NGCP) or a distribution utility (DU), to freeze the runtime of one of a smart contract under certain, legally defined emergency circumstances.

8.1.1 The Governance of the Switch

The main reason why there are some kill switches is that it puts back into a decentralized system the element of centralization. But absolute decentralization in the case of the public utilities is a safety risk. In order to maintain the advantages of blockchain while maintaining security, the Kill Switch has to be managed through the Multi-Signature (Multi-Sig) Protocol. According to this model, the Kill Switch cannot be activated by one actor. Rather, it would involve the digital keys of at least two of three parties, including: the Platform Operator, Government Regulator (DOE), and independent Technical Auditor. This will make sure that the override is not employed as a means of arbitrary interference but as the last resort in the event of a technical breakdown or a national security issue. According to Low and Mik, these types of on-chain protective structures are critical to filling the gap between autonomous execution speed and human-centric safety demands (2023).

8.1.2 Defining "Grid-Critical"

A kill switch is not necessary in all smart contracts. Two neighbor to neighbor P2P trade of small credits in solar could not represent a systemic risk. Policy should thus set a threshold of criticality - on either the amount of energy traded or the number of households concerned. The existence of contracts to control critical load balancing or to be the main billing mechanism of a community should be legally mandated to contain these manual overrides. The article under discussion is created as a result of the collaboration of FinTech and EnergyTech in a single area, specifically, the joint legal sandbox. The invention of smart contracts in the energy markets forms a regulatory no man land. Since such contracts frequently include the use of digital wallets, tokenized energy credits, and cryptocurrency payments, it means that they are governed under both the regulations of the Bangko Sentral ng Pilipinas (BSP) as a financial transaction as well as that of the Department of Energy (DOE) as a utility service.

8.1.3 This is Why the Collaboration Between Agencies is Necessary

As it is, a Philippine-based innovator could be licensed to operate in the FinTech sector by the BSP, but their logic of automated disconnection could be contravened by the DOE consumer protection regulations. To address this, Joint Legal Sandboxes have to be institutionalized by the government. These sandboxes are managed, safe-to-fail systems in which energy-tech firms have an opportunity to test their smart contracts in a combined oversight of the BSP and the DOE. These sandboxes are not only meant to test something technical, but discover regulatory. Within a shared sandbox, the regulators will be in a position to see in real-time how the volatility of crypto-payments impacts the capability of a consumer to keep on with the lights on. These sandboxes enable the co-creation of law, suggested by Andoni et al. (2019) and Manurung (2025). The BSP and DOE can provide dynamic guidelines on the basis of the real behaviour of the pilots rather than formulating a strict regulation of 200 pages that will have become outdated by the period of publication.

8.1.4 Experimenting the Hybrid Legal-Code Model

In particular, one should test the Tri-Layer Dispute Resolution Model with the help of the sandbox (the discussion is provided in Section 6). Regulators may observe the effectiveness of a simulated oracle failure on a platform that is effectively managed through a mechanism called Decentralized Arbitration (such as Kleros). The regulators can intervene by the so-called Legal Backstop in case the decentralized resolution is too slow or unfair. This is a proactive strategy so that even as these technologies leave the sandbox and go to the actual grid, they have been stress-tested to meet the provisions of the Philippine Data Privacy Act (RA 10173) and the EPIRA. 10.3 Mandatory Third-party Audits: Standardising Smart Legal Contracts. The code is law. The mindset of the code is law was that of move fast and break things. There is no option of breaking things in the energy sector. In order to secure the integrity of the market, the Philippines should introduce Compulsory Third-Party Audits of all templates of the type of smart legal contract, prior to their release into the market.

8.1.5 The "Audit-as-a-License" Model

Based on the extension of the Philippine EVOSS Act, the government ought to establish a list of Certified Smart Contract Auditors. Such auditors would be technical and legal professionals who would review a contract on three dimensions: Technical Security: Does the code have re-entrancy attacks, overflow errors, or be hacked? Fidelity to the Law: Does the code show proper reflection of the English-language "Master Agreement? As an example, is the period of 14-day cooling-off in the text equivalent to the time-locks in the code? Regulatory Compliance: Have the contract applied the right taxes, lifeline discounts, and grid fees automatically as it should under the law of the Philippines? 10.3.3 The Library of Certified Templates. DOE must enable the development of a Library of Certified Templates in order to lower the price of small scale community microgrids. These are already vetted and pre-audited, plug-and-play smart contracts that were already compliant with the Consumer Act of the Philippines (RA 7394). With a certified template, a local solar cooperative may avoid the costly audit procedure, which is performed individually, and reduce the entry barrier to decentralized energy without sacrificing the desired high level of care (Supreme Court of the Philippines, 2025).

8.2 Conclusion: Designing in a Resilient Trust

Code is the future of the Philippine energy market, though it should be maintained through trust. Unlike the perfection of the algorithm, the appearance of Trust in an automated system is possible due to the existence of Accountability. With a requirement of Kill Switches, we will be holding humans to be the masters of the grid. Institutionalizing the Joint Sandboxes takes care of making our regulations as innovative much as our technologies. And by making it mandatory to have the Audits, we guarantee that the same blockchain law of digitality is as trustworthy and just as of the law of land. These policy recommendations are a blueprint of a "Legal-Tech Harmony" as the ASEAN region is shifting towards the ASEAN Power Grid (2023). It is aimed at an autonomous yet regulated enough to be efficient and just grid. With such interventions, the Philippines would be able to transform itself into being an experimental pilot to becoming a global leader in terms of Smart Energy Governance, in which all the smart contracts are legally binding contracts.

REFERENCES

Allen & Overy. (2024). *The law of smart contracts: 2024 global review and the reasonable programmer standard.* A&O Legal Publications.

Andoni, M., Robu, V., Flynn, D., Abram, S., Geach, D., Jenkins, D., McCallum, P., & Peacock, A. (2019). Blockchain technology in the energy sector: A systematic review of challenges and opportunities. *Renewable & Sustainable Energy Reviews, 100*, 143–174. DOI: 10.1016/j.rser.2018.10.014

ASEAN Centre for Energy. (2023). *ASEAN energy outlook 7: 2022-2050.*

Asian Development Bank. (2021). *2021 energy policy of the Asian Development Bank: Supporting low-carbon transition in Asia and the Pacific.*

Buterin, V. (2021). *Ethereum: A next-generation smart contract and decentralized application platform* [White paper]. Ethereum Foundation.

De Filippi, P., Mannan, M., & Reijers, W. (2022). Blockchain as a confidence machine: The role of trust & law in territory-less governance. *Journal of Law and Innovation.*

European Commission. (2024). *Revised Product Liability Directive: Strengthening consumer protection in the digital age.* EC Publications.

Goranovic, A., Meisel, M., Panev, S., Padezhki, G., Stefanov, A., & Sosic, A. (2022). Blockchain applications in transactive energy systems: A review. *IEEE Access : Practical Innovations, Open Solutions, 10*, 52871–52896.

Low, K. F., & Mik, E. (2023). Smart contracts and the illusion of certainty. *The Modern Law Review, 86*(2), 348–377.

Manurung, E. H. (2025). *Blockchain technology and smart contracts: Legal implications and future regulations.* Laju Journal of Law and Society.

National Privacy Commission (Philippines). (2024). *Advisory opinion on blockchain and the Data Privacy Act: Guidelines for the energy sector.*

Savelyev, A. (2018). Contract law 2.0: 'Smart' contracts as the beginning of the end of classic contract law. *Information & Communications Technology Law, 26*(2), 116–134. DOI: 10.1080/13600834.2017.1301036

Shelake, S. (2025). *The future of legal tech: Smart contracts in emerging markets.* Global Jurist.

Supreme Court of the Philippines. (2025). *G.R. No. 270000: re: Liability of autonomous trading platforms and digital torts.*

Szabo, N. (1997). *The idea of smart contracts. First International Conference on Financial Cryptography.*

UK Law Commission. (2021). *Smart legal contracts: Advice to government* (Law Com No. 401). Her Majesty's Stationery Office.

Chapter 8
Data Privacy and Cybersecurity in Blockchain Energy Platforms

Muhamad Alif Mandani
http://orcid.org/0009-0004-7964-9263
Universitas Esa Unggul, Indonesia

Binastya Anggara Sekti
http://orcid.org/0000-0001-5489-4888
Universitas Esa Unggul, Indonesia

ABSTRACT

This chapter explores the intersection between blockchain-based energy systems and data privacy challenges, especially in peer-to-peer energy markets. It highlights how the transparency and immutability of blockchain, combined with real-time data from smart meters and IoT devices, can expose users to re-identification risks and legal conflicts with principles such as the right to be forgotten. The authors examine technical solutions including zero-knowledge proofs, permissioned blockchains, smart contracts with embedded privacy logic, and off-chain storage. Additionally, the chapter emphasizes the importance of privacy-by-design and security-by-default principles in system architecture. It also calls for adaptive legal frameworks such as regulatory sandboxes, along with participatory governance to ensure equitable implementation, especially in the Global South where regulatory and infrastructure disparities persist.

DOI: 10.4018/979-8-3373-5742-3.ch008

INTRODUCTION

In the global energy system, decentralization, digitalization, and decarbonization are driving significant transformations. One important innovation is the peer-to-peer (P2P) energy market, where consumers can become prosumers by selling excess electricity. This model promises efficiency and local empowerment, but it raises concerns about data protection and transparency, particularly when utilizing blockchain.

Blockchain is regarded as a promising technology to enhance trust in decentralized energy systems due to its transparency, immutability, and decentralization. However, this transparency raises privacy concerns. In a P2P ecosystem, every energy transaction, including sender details, energy amounts, time, and location, can be recorded. (Hou et al., 2022) state that although identities can be obscured with pseudonyms, metadata analysis enables the de-anonymizing of user identities through correlation with external data.

Figure 1. Privacy risks in blockchain-based P2P energy systems

The P2P system heavily relies on detailed and continuous data flows from smart meters, physical locations, and energy usage patterns. (Villa-Ávila, Arévalo, & Albornoz, 2025) emphasize that this data, while valuable for network optimization, is also highly sensitive as it can illustrate users' daily behaviors. For instance, spikes in consumption during the evening may indicate an individual's return home, which can be leveraged to create personal profiles.

Ironically, blockchain does not support data deletion, which contradicts the principle of 'the right to be forgotten' in regulations such as GDPR. (Shorya & Jagwani, 2025) highlights the tension between blockchain's immutability and the legal need for data flexibility. Even encrypted data can still potentially violate privacy due to its permanent references.

Facing the continuously evolving legal landscape, there is an urgent need to reevaluate the understanding of transparency and privacy. Transparency in blockchain does not necessarily imply complete openness; it can also mean providing proof of transaction validity without exposing details to the public. Technologies such as zero-knowledge proofs (ZKP) and other cryptographic methods have been proposed as solutions to reconcile auditability and privacy. (Hou et al., 2022) has developed a cryptographic approach that allows users to prove their participation in energy transactions without revealing their identity or its value.

However, privacy challenges are not only technical in nature, but also social and structural. In developing countries, where digital infrastructure is not evenly distributed, data protection is often neglected. Regulatory discrepancies across regions further increase the risk of data exploitation. This could trigger a new form of data colonialism, where data from Global South communities is utilized by foreign entities without clear accountability.

In this context, it is important to design systems with the principle of privacy-by-design. This principle integrates data protection from the planning stage. (Villa-Ávila et al., 2025) suggest the use of permissioned or hybrid blockchain, where only authorized parties can access specific transaction data, allowing greater control without compromising the integrity of the system.

The balance between transparency and privacy also necessitates active collaboration among policymakers, academics, and civil society. Collaborative models, such as regulatory sandboxes in some European countries, could serve as a good example. As noted by (Shorya & Jagwani, 2025), an adaptive and interdisciplinary legal framework can create experimentation spaces while still protecting consumers and system security.

Ultimately, privacy issues in blockchain-based peer-to-peer energy trading systems must be viewed holistically, encompassing technological, legal, social, and ethical aspects. Decentralized and efficient energy systems are urgently needed to address the climate crisis. However, these systems must not compromise fundamental privacy rights. On the contrary, blockchain should be guided to support energy democratization and enhance the protection of citizens' digital rights. Through proper system design, cross-sector collaboration, and advanced privacy technologies, the dream of a fair, transparent, and private energy system can be realized.

THEORETICAL FRAMEWORK: TRANSPARENCY–PRIVACY–SECURITY TENSION IN DECENTRALIZED ENERGY GOVERNANCE

The governance challenges emerging in blockchain-based energy platforms can be analytically understood through the structural tension between three core logics: technological transparency, cryptographic immutability, and rights-based data protection. These logics originate from distinct epistemic traditions. Blockchain architecture is grounded in distributed systems theory and cryptographic trust minimization, emphasizing tamper-resistance and public verifiability. In contrast, modern data protection law, particularly under frameworks such as the GDPR, is rooted in human rights theory and informational self-determination, prioritizing consent, revocability, proportionality, and purpose limitation. The resulting friction is not accidental but structural. Cyber-physical systems (CPS) are essential for modern space missions, integrating digital and physical components for critical operations such as spacecraft control, data processing, and autonomous navigation (Laksono, Tiaraputri, Sekti, & Laksono, 2025).

From a governance perspective, decentralized energy trading platforms represent a form of socio-technical infrastructure in which code functions as both operational mechanism and regulatory instrument. Drawing on the concept of "code as law," the design of blockchain protocols effectively embeds normative assumptions about transparency, accountability, and permanence into system architecture. Immutability, for instance, enhances trust in transactional integrity but simultaneously restricts the flexibility required by rights such as erasure and data minimization. This creates what may be termed a regulatory-technical misalignment: the legal system assumes modifiable data environments, whereas blockchain systems are intentionally resistant to modification.

This tension becomes particularly visible in peer-to-peer (P2P) energy markets, where smart meters and IoT devices generate granular behavioral data that is recorded or referenced within distributed ledgers. While transparency strengthens auditability and market integrity, it also increases the risk of re-identification and profiling. In this context, privacy cannot be conceptualized merely as confidentiality but must be understood as a relational and structural property of data governance. Metadata exposure, cross-chain interoperability, and distributed storage architectures collectively shape the conditions under which individuals can meaningfully exercise control over their personal data.

The framework of energy justice further enriches this analysis. Decentralized energy systems are often promoted as democratizing infrastructures that empower prosumers and local communities. However, without adequate privacy safeguards and cybersecurity resilience, such systems may reproduce asymmetries of power through digital surveillance, data commodification, and regulatory fragmentation. Particularly in transnational blockchain environments, where jurisdictional boundaries are blurred, the absence of a clearly identifiable data controller complicates accountability mechanisms and weakens user protections. Thus, decentralization does not automatically imply democratization; it must be accompanied by rights-based governance structures.

Additionally, the notion of digital constitutionalism provides a useful normative anchor. As digital infrastructures increasingly shape economic and civic participation, they require governance models that protect fundamental rights while accommodating innovation. In blockchain-based energy platforms, this implies the development of layered governance architectures in which cryptographic safeguards, institutional oversight, and adaptive regulation operate in tandem. Rather than framing privacy and transparency as mutually exclusive, the objective should be to reconcile verifiability with proportional disclosure through techniques such as zero-knowledge proofs, selective disclosure protocols, and hybrid ledger models.

This chapter therefore adopts a multi-layer analytical lens that integrates technical design, legal doctrine, and socio-political context. By situating data privacy and cybersecurity within broader theories of governance, accountability, and justice, the discussion moves beyond descriptive mapping of risks toward a critical evaluation of how decentralized energy systems can be structured to align with both technological robustness and fundamental digital rights. Such an approach recognizes that blockchain-based energy trading is not merely a technical innovation but a reconfiguration of power, responsibility, and trust within the evolving digital energy ecosystem. AI technologies can threat detection and incident response (Anwar et al., 2025).

PRIVACY CONCERNS IN BLOCKCHAIN-BASED ENERGY PLATFORMS

The advancement of blockchain technology in the energy sector has ushered in a new era of decentralized and efficient energy trading. However, this system not only brings innovation but also significant challenges in terms of cybersecurity. With the increasing integration of smart devices, smart contracts, and communication networks among prosumers, the cyber attack surface is also expanding. Security is no longer merely an optional addition, but a fundamental element in ensuring the sustainability and trustworthiness of blockchain-based energy systems. Adversarial Machine Learning techniques, such as evasion, poisoning, and model extraction, exploit vulnerabilities in machine learning models to manipulate security defenses, leading to misclassifications, false positives, and system failures (Sekti, 2025).

In this framework, it is essential to understand that privacy issues are not merely a matter of technical data security, but are closely related to the power structures within the digital ecosystem. When household energy data, user locations, and individual consumption patterns are permanently recorded in a public ledger, the potential for privacy violations significantly increases, especially if not accompanied by adequate controls and consent from users. While blockchain technology does offer high transparency and data integrity, without a well-thought-out privacy design, these advantages can become vulnerabilities. Therefore, the main challenge is not only how to leverage blockchain to support an inclusive and efficient energy transition, but also how to ensure that individual digital rights are fairly protected within it. Privacy protection in this system needs to be understood as a structural element, not an accessory, and requires active engagement from policymakers, technology developers, and civil society.

1. Metadata Exposure: Consumption, Location, and Device Identity

One of the main challenges in the context of privacy is the exposure of metadata that occurs when energy transactions are recorded in the blockchain ledger. This metadata includes information about the time, location, amount of energy consumption, and the identity of devices such as smart meters used by prosumers. According to (Kus & Levi, 2022), this data may appear harmless individually, but when combined, it can provide a comprehensive picture of user behavior, daily habits, and even the physical location of household residents. This poses a high risk if such information falls into the wrong hands or is used for commercial purposes without proper consent.

As noted by (Baggio & Grimaccia, 2020), smart meters now play a key role in gathering detailed data within contemporary energy infrastructures. Shifts in electricity usage can be closely monitored and interpreted to uncover specific routines and behaviors within a household. However, the integration of blockchain technology complicates efforts to safeguard this information, since its core design as a transparent and traceable ledger makes it difficult to obscure or restrict access to such metadata.

To better grasp the complexity of this issue, it is essential to recognize that metadata exposed in energy systems extends beyond technical functions and reveals deeply personal social and economic insights. For example, variations in electricity use can suggest the composition of a household, approximate income levels, or even the habits and routines of its occupants. In the absence of adequate safeguards, such data can be misused for discriminatory practices by actors like insurers, financial institutions, or political organizations. The situation becomes even more concerning when access to metadata is unregulated and lacks transparency about who uses it and for what ends. Given the increasing commodification of digital information, energy-related metadata risks being harvested and traded without the knowledge or consent of those affected. This calls for a careful and ethical approach to metadata governance in blockchain-based energy systems, involving strategies such as data anonymization, tiered access control, and regular audits to ensure responsible data management.

2. The Risk of De-anonymization through Smart Meter Analytics

Although blockchain enhances privacy by allowing users to interact under pseudonyms, the threat of deanonymization remains particularly acute in smart grid systems. Pseudonyms alone offer limited protection because advanced techniques can reassemble identity profiles by linking blockchain metadata with external datasets, a process commonly referred to as re-identification.

Smart meters collect high-resolution, real-time data on household energy use, which is essential for improving energy efficiency but also raises serious privacy concerns. Even subtle variations in consumption can disclose specific behaviors, while broader usage trends may reveal daily routines, lifestyle patterns, or socioeconomic status. The presence of Internet of Things (IoT) devices further intensifies this issue by producing additional layers of granular data connected to smart meters, resulting in an increasingly detailed and sensitive information landscape.

(Alcaraz, Rubio, & Lopez, 2020) argue that privacy protection should not depend on a single layer of security. Instead, it is shaped by the entire data lifecycle, including how information from IoT devices and smart meters is collected, processed, and stored. In the absence of comprehensive governance and safeguards, the risk of deanonymization becomes increasingly likely. This risk also presents challenges to compliance with the General Data Protection Regulation (GDPR), especially in relation to the concept of identifiability. Under the GDPR, pseudonymized data is still classified as personal data if an individual can be identified by linking it with other available information.

(Schellinger, Völter, Urbach, & Sedlmeir, 2022) assert that blockchain metadata has great potential to be linked to real identities through advanced data mining techniques. Their research shows that consumption patterns can identify specific households, even without external data. Privacy must be understood holistically, encompassing possible correlations and predictive analytics.

Approaches such as differential privacy and zero-knowledge proofs (ZKP) are being developed to address this issue, although they remain limited in implementation within the energy sector due to high computational costs and complex designs. Public awareness of the risks of deanonymization remains low, particularly in the Global South. This is not solely a technical issue, but also a matter of digital justice. The lack of understanding prevents communities from providing informed consent, leading to an information disparity. Therefore, policy interventions that place privacy protection as a primary pillar are essential. This includes regular privacy audits, privacy-by-design protocols, and user digital literacy. Without a comprehensive approach, deanonymization will pose a permanent threat.

3. Public Ledger and Contradictions to the Principle of Data Minimization

One of the main principles of the GDPR is data minimization, which means only collecting data that is truly necessary for specific purposes and retaining it only for as long as necessary. In the context of public blockchain, this principle is difficult to apply because all transaction data is permanently stored and cannot be deleted. This creates a fundamental conflict between the design of blockchain systems and the applicable legal regulations.

According to (Wang, Su, Ni, Zhang, & Shen, 2022), while encryption can be used to obscure the content of data, the existence of the data itself in a non-deletable system remains a concern. Hashing is also not a perfect solution since, although the information cannot be read directly, a digital footprint still exists and can be traced back to specific individuals. Therefore, legal and technical approaches must work synergistically to resolve this tension.

(Quan et al., 2025) in their review also state that public blockchain systems are too rigid to respond to data deletion requests, as the core mechanism of the system relies on the immutability of data. In this context, the use of private or hybrid blockchains, where access to data can be controlled more flexibly, becomes one of the solutions that should be considered seriously.

4. Case Study: Challenges of GDPR Compliance Regarding Immutable Data

The GDPR provides individuals with the right to exercise control over their personal data, including the ability to access, correct, and request its deletion. However, in an immutable blockchain system, fulfilling such requests is technically infeasible without altering the core structure of the technology. This tension introduces complex legal challenges that remain unresolved in many jurisdictions.

(Mahmood & Vacius, 2020) explore technical solutions such as burnable identities, which enable digital identities to be deleted or rendered inactive while preserving the integrity of transaction records. This method offers a degree of compatibility with GDPR requirements without violating the core principles of blockchain. Nevertheless, the approach remains largely experimental and has yet to see broad implementation in commercial systems.

In a separate study, (Wang et al., 2022) proposed an architecture utilizing metadata-free ledgers that store only aggregated data or statistical summaries, rather than individual transaction details. While this approach reduces the risk of GDPR violations, it involves a trade-off in data granularity, representing a deliberate compromise to support the development of ethical and legally compliant energy systems.

Privacy concerns in blockchain-based energy platforms represent a complex and multidimensional challenge that cannot be resolved through a single solution. Metadata exposure, the risk of deanonymization, and the tension between blockchain immutability and data protection regulations such as the GDPR create significant obstacles to broad and sustainable implementation. Several technical approaches have been proposed, including advanced encryption methods, burnable identity models, and ledger designs that minimize metadata. However, these measures must be supported by adaptive legal and policy frameworks, as well as system architectures that integrate privacy principles from the initial design phase. Without this foundation, the promise of blockchain to transform energy systems will continue to be overshadowed by concerns about the violation of individual digital rights.

Table 1. Dimensions of privacy risks in blockchain-based energy platforms

Focus Area	Practical Recommendations
Metadata Exposure	Apply metadata encryption and off-chain storage mechanisms to conceal location, time, and device identity.
Deanonymization Risks	Utilize multi-layer pseudonyms, cross-system IoT protection, and differential privacy methods to prevent re-identification.
Public Ledger vs. Data Minimization	Utilize private or hybrid blockchains, as well as a minimal-metadata ledger design to reduce the accumulation of permanent personal data.
Right to Be Forgotten vs. Immutability	Apply approaches such as burnable identities or revocable access to ensure that users maintain control over their data.
Legal–Technical Misalignment	Combine legal and technical approaches synergistically through adaptive and experimental regulatory sandboxes.
Infrastructure and Regulatory Gaps	Establish contextual data protection standards, promote legal interoperability, and engage the community in data governance.

THE CYBERSECURITY LANDSCAPE IN BLOCKCHAIN-BASED ENERGY SYSTEMS

The advancement of blockchain technology in the energy sector has marked the beginning of a new era in decentralized and efficient energy trading. While this innovation brings notable benefits, it also introduces critical cybersecurity challenges. The growing integration of smart devices, smart contracts, and interconnected communication networks among prosumers significantly expands the potential attack surface. In this context, cybersecurity is no longer a supplementary consideration but a core requirement for maintaining the resilience, trust, and long-term viability of blockchain-based energy systems.

In this context, cybersecurity threats are not merely technical in nature but also systemic. An attack targeting a single point within a distributed network, such as an insufficiently protected prosumer node or a vulnerable smart contract, can disrupt the broader stability of the entire system. The decentralization that underpins blockchain's core advantage can become a critical weakness if not supported by multilayered security strategies. Consequently, security measures must be integrated from the earliest stages of system architecture design, covering edge devices, communication protocols, transaction validation processes, and post-incident recovery mechanisms. This comprehensive approach demands active collaboration among technology developers, energy infrastructure providers, and policymakers to ensure that the resulting systems are not only technically resilient but also capable of safeguarding users in the face of evolving cybersecurity threats.

1. Threat Vectors: Smart Contract Exploitation, Sybil Attacks, and Inter-Chain Risks

One of the main vulnerabilities in this system is the smart contract. In a blockchain-based energy ecosystem, smart contracts function as automated regulators of energy transactions among prosumers. However, if not written securely, these contracts can become entry points for attacks such as reentrancy, overflow, transaction logic manipulation, or even energy price fraud. (Sharma & Ansari, 2024) indicate that in electric vehicle energy trading systems, vulnerabilities in smart contracts can be exploited to unfairly alter power balance or drain the virtual energy balance of other users.

Additionally, Sybil attacks pose a serious threat in distributed networks. In this attack, a malicious actor creates multiple fake identities to exert significant influence over the network, particularly within consensus mechanisms. This can lead to price manipulation, imbalances in energy distribution, and even sabotage of local energy markets. (Zhuang, Zamir, & Liang, 2021) assert that while blockchain offers high transaction visibility, it remains vulnerable to forms of identity-based attacks, and in the context of smart grids, the impacts can be profoundly systemic.

Another risk arises from the interoperability between blockchains. When local energy systems are connected with cross-sector blockchains such as payment systems or decentralized cloud storage, security risks increase significantly. Cross-chain bridges often lack adequate protection, making data or token transfers a weak point that attackers could exploit.

2. Security of Edge and Fog in Distributed Energy Architecture

Many critical components within blockchain-based energy systems function at the edge or fog layer, including devices such as smart meters, inverters, and smart home controllers. These local computing units often possess limited processing power, receive infrequent updates, and are susceptible to physical access, making them attractive targets for firmware manipulation, command injection, and other forms of cyber attack.

In architectural designs of this nature, edge security must be prioritized. Implementing cryptography-based digital identities and hardware-level authentication mechanisms is essential, yet their broader adoption has been limited by cost and technical complexity. As a result, distributed energy systems require a comprehensive security framework that is embedded from the early stages of design to prevent structural vulnerabilities at the lower layers of the network.

To address this challenge, an integrated edge security approach becomes increasingly critical. The edge and fog layers, which locally process data, are particularly exposed to micro-scale cyberattacks. (Verma, Navas-Fonesca, Burgos-Mellado, & Dragicevic, 2025) emphasize that distributed computing architectures such as edge and fog demand adaptive security protocols that go beyond traditional firewalls or scheduled software updates. These protocols should incorporate context-aware access control and dynamic authentication mechanisms to respond effectively to evolving threats. They propose the use of hardware-based root of trust and real-time monitoring that operates directly on the edge devices to prevent the escalation of attacks. This strategy minimizes latency while reducing reliance on data centers, which often become bottlenecks in smart energy systems. Furthermore, collaboration among infrastructure providers, regulators, and system developers is necessary to ensure that the security integration process from the design stage (security-by-design) can be consistently implemented throughout the entire technology lifecycle.

3. Adaptive Anomaly Detection and Real-Time Defense Model

Rapid changes in energy consumption, electric vehicle mobility, and the integration of IoT demand defense systems that can adapt in real time. Conventional signature-based security models are no longer sufficient to face increasingly sophisticated attack patterns. This is where the role of machine learning and artificial intelligence-based anomaly detection systems begins to be applied in the context of energy.

(Sharma & Ansari, 2024) introduced a blockchain-based security framework to detect suspicious transactions in electric vehicle energy trading. This system is capable of recognizing unusual patterns based on transaction behaviors such as abnormal energy volumes or rapid changes in transaction frequency. Their approach also integrates secure and authentic smart contract design with token authentication mechanisms to ensure data integrity and prevent unauthorized access to user information. This framework bolsters trust in the electric vehicle energy trading ecosystem by ensuring that every transaction is auditable and cryptographically secured.

Multi-head attention models and neural networks are increasingly being applied for network traffic classification and anomaly detection in IoT devices within energy networks. While still in the experimental stage, these techniques show strong potential to enable more predictive and proactive security systems compared to traditional reactive approaches.

4. Systemic Implications and the Need for Standardization

One of the fundamental challenges in securing blockchain-based energy systems is the absence of consistent security standards across platforms. The development of isolated projects without uniform protocols creates interoperability vulnerabilities and heightens the risk of exploitation, particularly during data transfers and cross-platform transactions.

(Zhuang et al., 2021) emphasize the importance of stakeholder coordination in establishing a comprehensive and standardized security framework, as the lack of harmonization allows each node or device to become a potential attack vector. In blockchain systems, the absence of a central authority complicates accountability, raising questions about responsibility in the event of smart contract sabotage or dispute resolution.

To address this fragmentation, efforts to harmonize international security standards are increasingly urgent. (Nguyen et al., 2024) highlight the need for interoperability-based security models that integrate technical protocols with unified governance and operational policies. In blockchain-based smart grids, they propose a standardized framework to ensure cross-system integration without compromising data integrity or privacy. This approach strengthens systemic resilience and addresses legal uncertainties that emerge from cross-jurisdictional implementation.

Table 2. Classification of cyber threats and defense strategies in blockchain energy systems

Focus Area	Practical Recommendations
Smart Contract Exploits	Utilize automatic and manual code audits, formal verification, and the implementation of fail-safe logic to prevent misuse.
Sybil and Consensus Attacks	Implement a reputation-based consensus mechanism or proof-of-authority and restrict node participation through strong identification.
Cross-Chain Vulnerabilities	Establish validated cross-chain protocols, utilize trusted oracles, and conduct independent cross-blockchain verifications.
Edge/Fog Security	Implement edge-to-cloud encryption, protect edge nodes with AI-based access control, and utilize micro-segmentation networks.
Real-time Intrusion Detection	Integrate AI models such as multi-head attention to dynamically detect network anomalies and attack patterns.
Device and IoT Attack Surfaces	Implement a zero-trust architecture on energy devices and automatic firmware updates based on multi-factor authentication.

COMPARATIVE LEGAL FRAMEWORK AND REGULATORY GAPS

The advancement of blockchain technology in the energy sector has created an urgent need to redefine legal frameworks related to data protection, privacy, and cross-border transactions. In blockchain-based energy trading systems, transactions do not solely represent economic exchanges but also embed personal user information such as energy consumption patterns, geographical data, and digital identities that are permanently recorded. These attributes challenge conventional legal systems that are structured around centralized control and modifiable data environments.

In response, the development of adaptive and interdisciplinary regulatory models is essential. Legal frameworks that presuppose a singular authority are inadequate in addressing the distributed and autonomous nature of blockchain systems. Responsibility is decentralized, complicating accountability in cases of privacy violations or misuse of personal data. Additionally, immutable data storage on the blockchain often conflicts with established data protection principles, such as the right to erasure or consent withdrawal. As argued by (Korolov et al., 2024), the absence of centralized oversight and the challenges of fulfilling legal obligations within distributed architectures underscore the need for new compliance mechanisms. These mechanisms must incorporate cryptographic governance models that are interdisciplinary and capable of addressing both technological and societal complexities in digital energy ecosystems.

1. Regulatory Overview: GDPR, CCPA, LGPD, PDP Bill, and African Union Strategy

Several jurisdictions have developed a comprehensive legal framework to address data privacy issues. The European Union, through the General Data Protection Regulation (GDPR), sets a high standard for the management of personal data, including the right of access, the right to be forgotten, and transparency obligations for data controllers. The GDPR requires every entity processing data of EU citizens to comply, regardless of the location of the entity. This has a significant impact on public blockchain systems, which are global in nature and lack a central entity.

In the United States, the California Consumer Privacy Act (CCPA) grants similar rights to California residents, such as the right to know what data is collected and the right to opt-out of the sale of personal data. However, the scope of the CCPA remains geographically limited and there is no uniform federal law. In Latin America, Brazil has implemented the General Data Protection Law (LGPD), which largely adopts the structure of the GDPR and is becoming a model for other countries in the region.

India, through the Personal Data Protection (PDP) Bill, proposes the establishment of a Data Protection Authority and principles of processing based on consent, fairness, and accountability. Although not yet finally approved, this draft indicates that developing countries are beginning to recognize the importance of strong data governance, especially when faced with decentralized systems.

Collectively, the African Union has developed the Malabo Convention on Cyber Security and Personal Data Protection, which aims to establish common standards among member states. Although its implementation is still limited, this initiative is crucial for creating a cross-border regulatory framework that can adapt to technologies such as blockchain and IoT in the energy sector (Khan, 2025).

2. Special Interpretation of Blockchain: Approval, Ownership, and Responsibility

One of the fundamental challenges in applying privacy laws to blockchain systems is the mismatch between the characteristics of the technology and the principles of traditional law. In a decentralized system, it is not always clear who acts as the controller or processor of data, as defined in the GDPR. The absence of a central entity renders accountability unclear, especially when data is leaked or misused.

Moreover, consent in the context of smart contracts becomes an ambiguous concept. When users connect their devices to the blockchain platform, does it imply that they are implicitly granting consent for the recording and dissemination of their data? In the realm of conventional law, consent must be specific, explicit, and revocable. However, in blockchain, recorded data cannot be deleted, making it difficult to technically apply the principle of revocation.

Data ownership also becomes an issue that is not easily resolved. In a blockchain system, data is recorded in a shared ledger that can be accessed by all nodes. This raises the question: do the transacted data become public property, or do they remain the property of the individuals who generated the data? Without legal clarity, the potential for conflict between users, platform providers, and third parties continues to grow (Ogundipe, 2025).

An important addition to underscore is that the ambiguity of legal roles and responsibilities complicates the implementation of individual rights protection principles that have been internationally recognized. When a system lacks a clear mechanism to designate data controllers, the pathways for recovery or complaints for users become unavailable. This undermines the users' position in asserting their rights over personal data, while also creating loopholes for those intending to exploit the system without strict legal consequences. In an increasingly complex and global digital environment, the clarity of the role of law is not merely administrative but becomes an ethical and operational prerequisite. Therefore, a legal approach is needed that not only adapts to the technical nature of blockchain but also actively redefines classical concepts such as responsibility, consent, and ownership within the context of decentralized systems. This effort is crucial to ensure that innovation does not occur at the expense of disregarding the fundamental rights of users.

3. Challenges of Interoperability Across Jurisdictions

Blockchain is inherently transnational, which renders state-based regulation increasingly ineffective. A single transaction may involve participants operating under entirely different legal jurisdictions. For instance, an energy exchange between prosumers in Germany and consumers in Kenya would fall under the General Data Protection Regulation (GDPR) in the former, while the latter lacks an equivalent framework for personal data protection. As highlighted by (Gavai et al., 2025), discrepancies in the definition of personal data, the validity of consent, and the scope of disclosure obligations result in uneven legal safeguards, creating substantial barriers to the development of a unified digital energy market that depends on both trust and legal certainty.

Legal fragmentation also complicates token-based energy trading. In some jurisdictions, energy tokens may be recognized as financial instruments, while in others, such classifications do not yet exist, leading to uncertainty in taxation, consumer rights, and dispute resolution procedures. This regulatory disharmony affects public confidence, particularly when users are uncertain whether their personal data will be handled with equivalent protections across borders. For companies and cross-border energy service providers, the burden of navigating conflicting regulatory regimes can be significant. This complexity may stifle innovation and hinder the global adoption of blockchain technologies in the energy sector, especially in regions with underdeveloped legal systems. As a result, it is imperative to foster a coordinated international effort that bridges legal disparities to ensure that digital energy trade is both inclusive and sustainable.

4. Gaps in Cross-Border Energy Trade

As energy trade between countries becomes increasingly digital and decentralized, the lack of laws governing these activities creates a gray area that is at risk of being exploited. Many international-scale energy blockchain projects are still underway without a clear legal foundation. This leaves businesses and investors facing high regulatory uncertainty.

One potential solution proposed by (Khan, 2025) is to establish a regional or multilateral framework that specifically regulates digital energy trade. This framework could adopt the fundamental principles of GDPR, but with the flexibility to accommodate the characteristics of blockchain such as immutability, network autonomy, and cross-chain interoperability. In the long term, the harmonization of laws at the international level becomes an unavoidable necessity. Without this, blockchain in the energy sector will only develop in local 'innovation pockets' without creating a global systemic impact.

To strengthen the legal position in cross-border digital energy trade, not only the harmony of principles is required but also the technical and institutional integration across jurisdictions. A study by (Kumar & Chittora, 2024) presented at the IEEE Power India International conference states that to expedite the integration of blockchain systems in cross-border energy markets, common technical standards and regulations are needed to facilitate interoperability and cross-border compliance. They emphasize the importance of international regulatory harmonization efforts to prevent legal conflicts and enhance trust among stakeholders. This harmonization not only simplifies reporting and authorization but also paves the way for a more inclusive and secure digital energy market.

Table 3. Comparative overview of global data protection regulations for blockchain-based energy systems

Focus Area	Practical Recommendations
Global Regulatory Fragmentation	Encourage the formation of regional/multilateral frameworks that align with the basic principles of GDPR, CCPA, LGPD, PDP.
The Ambiguity of the Role of Law in Blockchain	Redefine the roles of 'data controller' and 'processor' in a decentralized system based on network logic.
Ambiguitas Makna Persetujuan Digital	Develop a dynamic consent model that can be revoked, even though the data cannot be deleted, for instance through locked access.
Data Ownership in Distributed Networks	Establish a data ownership framework based on provenance and granular control by users.
Challenges of Interoperability Across Jurisdictions	Develop legal and technical protocols for the resolution of disputes and audits of transnational transactions.
Legal Vacuum for Energy Tokens	Classify energy tokens legally: as digital assets, commodities, or financial instruments, in accordance with the market context.
Investment Uncertainty and Global Compliance	Use a regulatory sandbox for policy experimentation and new collaborative and risk-based governance models.

PRIVACY PRESERVATION TECHNIQUES AND ARCHITECTURAL SOLUTIONS

In an increasingly decentralized energy landscape shaped by blockchain technology, the protection of personal data has emerged as a critical concern. Blockchain-based energy systems enable direct exchanges between prosumers and consumers while simultaneously generating an extensive data footprint related to consumption patterns, time of use, geographic location, and user behavior. To maintain compliance with data protection regulations and preserve public trust, a range of privacy-preserving technological solutions has been developed and continues to undergo rigorous testing.

The effectiveness of these solutions, however, hinges on the careful integration of both technical design and legal principles. Privacy must not be treated solely as a matter of encryption or data obfuscation but should be recognized as a fundamental right embedded in the architecture of the system from the outset. In practical terms, this involves the development of technologies such as zero-knowledge proofs, smart contracts that incorporate privacy-preserving logic, and hybrid system architectures that enable selective data sharing. The primary challenge lies in implementing these tools without undermining system efficiency or compromising interoperability. Moreover, the successful deployment of privacy-enhancing technologies depends on sufficient digital literacy among users and the presence of a regulatory environment that encourages innovation while safeguarding digital rights. Ultimately, the viability of blockchain-based energy systems is not determined by technical sophistication alone, but by the extent to which privacy protection is treated as a core ethical and operational principle.

1. Zero-Knowledge Proofs, Ring Signatures, and Homomorphic Encryption

One of the most promising cryptographic approaches for safeguarding user privacy in decentralized energy systems is Zero-Knowledge Proof (ZKP). This technique allows a party to verify the validity of a statement without revealing the underlying data. In the context of blockchain-based energy trading, ZKP enables participants to prove ownership of valid energy credits without disclosing the exact amount. This approach protects sensitive consumption data while maintaining transactional integrity. (Hou et al., 2022) demonstrate that the use of ZKP in blockchain-supported energy trading can effectively obscure user identities and transaction volumes, providing a privacy-preserving mechanism without compromising verifiability or operational accuracy.

Another relevant cryptographic technique is the use of ring signatures, which allow a transaction to be signed by one member of a group without revealing the identity of the actual signer. This is particularly valuable in peer-to-peer energy trading, where maintaining user anonymity is crucial. (Mitrea, Toderean, Cioara, Anghel, & Antal, 2023) highlight the effectiveness of ring signatures in concealing the source of energy transactions while still ensuring their validity through group-based verification.

Homomorphic encryption offers yet another promising solution by enabling the processing of encrypted data without requiring prior decryption. This is especially useful in scenarios where personal energy data must be analyzed or aggregated, such as for local or national energy demand forecasting, while preserving user privacy. Although current limitations in performance and scalability remain, this technology is gradually being introduced in pilot programs and institutional energy initiatives.

2. Permissioned Blockchain and Hybrid for Controlled Data Exposure

Unlike public blockchains, which are open and transparent, permissioned blockchain systems restrict access to the network and data solely to authorized entities. In the context of energy, this means that only energy providers, regulators, and verified users can access and record transactions. This approach enables granular control over who sees what data, when, and in what context.

(Vaziry, Garzon, Herbke, Segat, & Küpper, 2025) emphasize that the combination of permissioned blockchain with quantum encryption schemes can enhance the security of cross-border energy trade without causing legal conflicts between jurisdictions. Hybrid systems are also beginning to be utilized, where public blockchains are used to record transaction hashes to maintain integrity, while the original data is kept confidential within a private network. Such models help maintain a balance between system transparency and the protection of personal data.

With this structure, organizations can separate the open recording logic from the limited data storage system. This also helps address the challenge of the "right to be forgotten," which cannot be applied in public blockchains due to their immutable nature.

Furthermore, the hybrid system that combines elements of both public and private blockchain has been proven to create a secure yet adaptive digital energy ecosystem that responds to diverse regulations and local needs. (Golder et al., 2024) developed a hybrid blockchain framework that explicitly combines the transparency of public blockchain with the security and privacy controls of permissioned networks. In the context of energy, this approach not only protects user data from excessive exposure but also enhances the scalability and efficiency of cross-network transactions. This becomes particularly crucial as energy systems are adopted in various jurisdictions with differing legal standards and governance. This model demonstrates that the compromise between openness and confidentiality is not only feasible but can indeed strengthen the legitimacy and sustainability of the system as a whole.

3. Secure Off-Chain Storage: IPFS, Oracles, and Confidential Computing

Off-chain storage technology has become crucial in supporting data privacy preservation. The International Peer-to-Peer File System (IPFS) is one of the widely adopted solutions for storing energy data off the main blockchain. With IPFS, data is not stored directly on the blockchain; instead, only its hash is recorded to ensure integrity. This approach allows for better access control over the data and potential deletion if necessary.

Oracle also plays an important role in connecting blockchain to the outside world, such as energy price data, weather conditions, or market demand. Secure oracles must be built with integrity protection and encryption to avoid becoming a point of attack. Additionally, confidential computing technology that leverages hardware enclaves allows data to be processed in a secure environment even when used by third parties. This paves the way for cloud-based analytics without compromising user privacy.

A well-designed off-chain storage approach provides flexibility in meeting various regional data protection standards while maintaining the efficiency of blockchain systems in energy transactions.

The use of IPFS and other off-chain storage mechanisms does indeed offer high efficiency in blockchain-based energy systems. However, it is essential to thoroughly consider aspects of security and privacy. Research by (Bisht, Das, Niyato, & Park, 2023)indicates that the application of IPFS in blockchain-based systems for sharing personal data requires a comprehensive approach to data encryption and access control, so as to not only ensure data integrity but also maintain its confidentiality throughout its lifecycle. In the context of energy, this approach can be adapted to ensure that energy consumption data is not only securely stored but also not easily accessible or analyzable by unauthorized parties. This serves as an important foundation for the implementation of decentralized storage schemes that remain compliant with both local and international privacy norms.

4. The Use of Smart Contracts with Embedded Data Control Logic

Smart contracts, which have traditionally been used to automate energy transactions, can also function as tools for safeguarding data privacy. By embedding privacy rules directly into the contract logic, developers can regulate access rights to transaction data, define conditions for data retention, and enable transparent audit trails for data usage.

For instance, smart contracts can be programmed to permit energy aggregators to access individual consumption data only after explicit approval from the user. In token-based energy systems, smart contracts can also be configured to distribute incentives exclusively to nodes that comply with specific privacy-preserving protocols. This approach facilitates the implementation of a "privacy-by-design" model at the execution layer of digital transactions.

The integration of technologies such as Zero-Knowledge Proofs (ZKP) into smart contracts further enhances the system by introducing on-chain privacy verification. In this way, privacy protection is no longer confined to external data layers but becomes an inherent part of the transactional ecosystem itself.

Moreover, the use of smart contracts for data protection opens new avenues for developing energy systems that are not only technologically advanced but also socially inclusive. Embedding privacy-centric logic into contract execution empowers users to retain control over their personal data by defining who can access it, for how long, and for what purpose. This design decentralizes data governance and reduces dependency on centralized authorities, thereby promoting user autonomy.

In addition, smart contracts can be adapted to enable participatory auditing, allowing community members to collaboratively verify compliance with agreed-upon privacy policies. This not only strengthens accountability in decentralized systems but also encourages active civic involvement in managing digital energy infrastructures. Combined with blockchain's immutable ledger, such contracts ensure that every transaction involving user data is verifiable and tamper-resistant, establishing a foundation of trust and transparency in peer-to-peer energy networks (Indiravathi et al., 2024).

Table 4. Privacy-preserving technologies and architectural models in blockchain energy infrastructure

Focus Area	Practical Recommendations
Zero-Knowledge Proof (ZKP)	Use Zero-Knowledge Proofs (ZKP) to validate the authenticity of energy transactions without revealing the amount or the identity of the user.
Ring Signatures	Apply ring signatures to conceal the source of transactions within a group of legitimate users.
Homomorphic Encryption	Utilize homomorphic encryption to enable the analysis of encrypted data without the need to expose the raw information.
Permissioned and Hybrid Blockchain	Establish a hybrid architecture where transaction hashes are recorded on a public blockchain and sensitive data is stored in a private ledger.
Off-chain Storage (IPFS, Confidential)	Use IPFS to store off-chain energy data with access control; employ confidential computing for secure analytics.
Secure Oracles	Design oracles with cryptographic authentication to prevent security vulnerabilities between the blockchain and external systems.
Smart Contract with Data Control	Incorporate privacy logic into the smart contract to regulate access, duration of storage, and conditions for data processing.

GLOBAL SOUTH: CHALLENGES AND INNOVATIONS IN DATA PROTECTION ON BLOCKCHAIN-BASED ENERGY PLATFORMS

The emergence of blockchain technology in the energy sector presents new opportunities for developing countries, particularly those in the Global South, to address persistent challenges such as limited energy access, uneven distribution, and low public participation. However, this transformation is not solely technological. It is also regulatory and social. The adoption of blockchain introduces new complexities related to personal data protection, disparities in digital infrastructure, and digital sovereignty, many of which are inadequately addressed in existing local legal frameworks. In this context, the Global South faces distinctive challenges that require innovative and context-sensitive solutions tailored to local socio-political realities.

The term *Global South* generally refers to countries in Latin America, Africa, Asia, and Oceania that have historically experienced marginalization within global economic and political systems. While originally a geopolitical classification, the term has evolved into a broader analytical lens for examining developmental inequalities, post-colonial dynamics, and asymmetrical global power relations. In the context of energy justice and sustainability, the Global South is often marked by energy poverty, limited infrastructure, and structural dependency on the Global North. These disparities significantly shape how emerging technologies such as blockchain are engaged with and implemented, particularly within decentralized energy initiatives. According to Oemmelen et al. (2025), development programs in the Global South must contend with informal governance structures, diverse models of energy innovation, and varying degrees of community participation. These conditions necessitate a rethinking of international cooperation frameworks and a move toward more equitable power-sharing in global energy transitions.

Accordingly, it is crucial for countries in the Global South not only to act as consumers of imported technologies but to actively participate in the processes of design, decision-making, and standard-setting for the digital systems they implement. This shift from passive adoption to active co-creation can reinforce national technological sovereignty while ensuring that blockchain systems are attuned to the unique needs, values, and aspirations of local communities. Such engagement enables the digital transformation of the energy sector to become a genuinely inclusive and equitable process, rather than a mere extension of global technological models that overlook local diversity and autonomy.

1. Regulatory Disparities and Legal Complexities

One of the biggest obstacles to implementing blockchain-based energy systems in the Global South is the absence of a robust regulatory framework for personal data protection. Many countries in this region do not have legislation comparable to the General Data Protection Regulation (GDPR) that ensures individuals' rights over their personal data. This regulatory gap creates a legal gray area that exposes users to risks of privacy violations and data exploitation. In blockchain-based energy platforms, where transactional and energy consumption data are recorded immutably, the lack of legal mechanisms to modify or delete such data complicates the enforcement of the "right to be forgotten" principle.

For instance, in a peer-to-peer energy trading project that relies on blockchain technology, local users may engage in decentralized networks without understanding who controls, processes, or owns the data being generated. This lack of clarity becomes even more problematic when such initiatives are introduced by transnational actors or technology companies based in the Global North. In these cases, jurisdictional boundaries and legal accountability become blurred, raising complex questions about responsibility and oversight. These unresolved legal issues can ultimately undermine the platform's legitimacy and erode trust among local users.

To strengthen the narrative about regulatory challenges in the Global South, it is essential to emphasize that, in addition to the absence of strong local laws, gaps in the implementation of international standards such as the GDPR also create broader challenges. Many developing countries lack the legal and technical infrastructure to accommodate the principles of global data protection, while blockchain-based projects are often designed with the assumption that standards like the GDPR will be universally applied. A study conducted by Park and (Han & Park, 2022) demonstrates that there are significant structural gaps between the architecture of blockchain technology and the principles outlined in the GDPR. They concluded that most blockchain systems used in the energy community are unable to accommodate the principles of data minimization and the right to be forgotten, which are actually key pillars in user data protection. This indicates that blockchain-based energy projects launched in the Global South often face not only local regulatory barriers but also compatibility challenges with international norms that are increasingly becoming global standards. Without strengthening an adaptive and contextual regulatory framework, efforts to promote energy transformation through this technology may actually exacerbate legal inequalities and weaken public trust in the digital systems being developed.

2. Digital Infrastructure and Technological Gap

The Global South faces not only regulatory limitations but also a lack of digital infrastructure that serves as the backbone for blockchain technology. Uneven internet connectivity, scarcity of local data centers, and high dependence on foreign cloud service providers render the adoption of blockchain vulnerable to foreign control. In a digital energy system that heavily relies on distributed computing architecture, this infrastructure imbalance intensifies structural dependence and obstructs the development of systems that are genuinely autonomous and locally responsive.

According to Howson et al. (2024), many Web3 and blockchain projects promote narratives of empowerment that mask what they describe as computational parasitism. In this context, the energy, infrastructure, and data from Global South countries are utilized to support global blockchain networks, while the economic value generated is disproportionately captured by dominant actors from the Global North. This phenomenon reflects not only economic exploitation but also the emergence of unjust digital power relations.

Reliance on foreign infrastructure also raises critical concerns regarding digital sovereignty. When energy-related data and citizens' digital identities are stored and processed through servers and protocols beyond national jurisdiction, states lose control over strategic digital resources. This situation complicates legal recourse in the event of data breaches and limits the ability of governments to align technology deployment with national interests. Furthermore, the limited influence of developing countries in shaping global technical standards often compels them to adopt ready-made systems that do not reflect local priorities or cultural values. Instead of fostering independence, digital infrastructure becomes a new point of dependency that perpetuates global inequalities.

3. Ethics and Issues of Digital Colonialism

The fundamental issue arising from the technological and legal disparities in the Global South is the potential emergence of a new form of data colonialism. In the context of blockchain-based energy, consumption data, digital identities, and even user behaviors become valuable commodities that can be exploited without clear control or reciprocal benefits for local communities. When these blockchain projects are executed without meaningful participation from local communities in the processes of design, oversight, and data utilization, there is a risk that the systems will create a new digital dominance that mirrors the old colonial structures in a virtual form.

(Gurumurthy & Chami, 2021) state that the dominance of global corporations in digital infrastructure and data governance norms has created significant challenges for Global South countries in establishing their digital sovereignty. They advocate for a global digital constitutionalism agenda that places justice, digital human rights, and community control over data at the core of a new digital architecture. In the context of energy, this means building blockchain systems that are not only efficient but also democratic and sensitive to structural inequities.

This inequality not only impacts technical and legal aspects but also reshapes power relations within the global digital space. When data generated by local communities becomes the fuel for the digital economy, whose added value is absorbed by centers of global technological power, what occurs is not merely marginalization but rather an enhancement of the structural inequality in a new guise. Therefore, the urgency to establish data governance based on the principles of distributive justice becomes increasingly important. Blockchain-based energy systems need to be designed with fair value redistribution mechanisms, transparency in data management that can be audited by communities, and data ownership models that recognize collective rights over collaboratively generated information. Without such fundamental reforms, technology that is supposed to be an empowerment tool may, in fact, reinforce the historical inequalities that have long constrained the sovereignty of states and communities in the Global South.

4. Local Innovation and Community Approaches

Despite facing significant challenges, countries in the Global South are not entirely in a passive position. Various local initiatives are beginning to demonstrate a new direction in the development of a fair and inclusive digital energy system. Community-based approaches such as digital energy cooperatives, collective data management, and self-sovereign digital identity systems are being piloted in various regions, including Kenya, Brazil, and India. These approaches provide communities with direct control over their data and open up opportunities for participation in technology decision-making.

Additionally, several renewable energy projects utilize permissioned blockchain or hybrid systems that allow varying levels of control over data depending on the role and location of users within the network. This enables the design of more contextual systems that take cultural and social sensitivities into account while maintaining technical integrity. Such approaches serve as examples of how high technology such as blockchain can be tailored to field realities, provided it is designed with principles of inclusivity and fairness in mind.

This approach also provides a strong foundation for linking the principles of digital democracy with participatory technology systems. A study by (Boumaiza & Sanfilippo, 2024) in *IEEE Access* indicates that the utilization of permissioned blockchain systems in community energy projects not only enhances operational efficiency but also strengthens governance based on citizen participation. Within this framework, local communities can verify identities, manage access rights to energy data, and collectively carry out audit processes. This not only improves transparency but also fosters a sense of ownership over the system. These findings affirm that technologies such as blockchain, when designed with principles of social justice and inclusiveness in mind, can serve as transformational instruments that have a tangible impact on community empowerment.

Table 5. Key challenges and community-based innovations in energy data governance in the global south

Focus Area	Practical Recommendations
Regulatory and Legal Inequality	Establish a legal framework for national data protection that is responsive to the characteristics of blockchain and decentralization.
The Absence of Digital Rights Mechanism	Adopt a flexible model for data rights such as the 'right to be forgotten' with an access-based and pseudonymization approach.
Digital Infrastructure Gap	Investment in local connectivity, national data centers, and digital energy infrastructure based on domestic resources.
Dependence on Global Actors	Develop a community-based digital sovereignty model and reduce dependency on foreign clouds and transnational platforms.
The Risks of Data Colonialism	Promote transparency and citizen participation in energy data management; adopt principles of data justice and local ownership.
Community Innovation and Governance	Implement digital energy cooperatives, self-sovereign identity (SSI), and contextual permissioned blockchain relevant to the role of users.

LEGAL RECOMMENDATIONS AND GOVERNANCE MODEL

The global energy transition driven by blockchain technology offers great promise, but also brings serious challenges. Especially when it comes to data protection, system security, and fair regulation. Blockchain doesn't just decentralize how energy flows. It also builds complex, cross-border data systems, often with no clear central authority. That's why we need a legal approach that goes beyond reacting to problems after they appear. Law needs to be there from the beginning. Shaping the system. Making sure that privacy and security are treated as part of the design itself, not just an afterthought.

In this context, law can't just sit back and supervise after the technology is rolled out. It has to act as a strategic tool. One that helps direct the values and outcomes of innovation. So we're talking about a regulatory framework that can guide technological development in ways that align with social justice, digital rights, and long-term sustainability. Not just checking boxes. This kind of governance needs to be collaborative, open to experimentation, and responsive to how quickly tech changes. Only then can regulation keep pace with the distributed and dynamic nature of blockchain systems. And it's this approach that also helps build social legitimacy for the energy transition.

One of the most grounded principles we can follow is privacy by design. The idea that data protection should be built into the system architecture from the very beginning. Not added later. In energy blockchain systems, this means developers need to design in ways that reduce how much personal data is exposed. There should be fine-grained control over who can access what. And people should have the right to delete or limit the spread of their personal data. Simple as that.

(Barceló et al., 2023) stress that implementing privacy by design and security by default is essential for blockchain-based energy systems. Because the blockchain ledger is public and immutable, once personal data gets in, it can't easily be removed. Technical tools like advanced cryptography, privacy-aware smart contracts, and hybrid blockchain models that combine public and private chains are available. But these can't just be optional. They need to be backed by legal rules that make them standard.

And we can't forget about system security either. Security by default means the system should be secure right out of the box. Users shouldn't have to configure anything manually just to stay protected. That's especially important in developing regions, where many users simply don't have the digital skills to manage technical settings. The system needs to protect them automatically.

1. Principles of Inclusive and Adaptive Regulation

Blockchain moves fast. Law doesn't. While blockchain operates across borders, law stays confined within national boundaries. This creates a real mismatch. To close that gap, regulatory principles must be both inclusive and flexible.

Inclusive means regulation can't be shaped only by governments or big corporations. It needs to involve everyone affected. Local communities, prosumers, developers, regulators, and academics. All of them should have a voice in the policy-making process. Not just those in suits.

(Domalis, Karacapilidis, Tsakalidis, & Giannaros, 2021) put this clearly. They stress the need for governance that works across jurisdictions. Especially for digital energy systems that cross borders. They suggest a decentralized legal framework that allows mutual recognition between regions, along with cross-border digital identity verification. It's a technical idea, but the message is simple: we need systems that talk to each other.

And flexibility? That's just survival. Blockchain is unpredictable. One moment it's energy tokens, the next it's self-executing contracts doing things no one expected. These aren't legal hypotheticals anymore. They're real. So the law needs to stay light on its feet. That means learning fast, being open to change, and having the courage to try things. Not everything will work the first time. That's fine. What matters is that the law keeps evolving alongside the tech.

2. Regulatory Sandboxing and Co-Experimentation in the Global South

One increasingly adopted method is the regulatory sandbox. It is a supervised legal space for experimentation, where technology developers can test innovations within a flexible but controlled legal setting. For countries in the Global South, this model offers a practical path to adopt blockchain in the energy sector without waiting for ideal or fully developed legislation.

(Diaz Valdivia, 2023) explored the dynamics between community-led (commons-led) and corporate-led (business-led) blockchain-based energy initiatives. His findings suggest that open and participatory sandboxes offer greater opportunities for community-based projects to thrive and generate sustainable social outcomes. This implies that developing countries need not only a legal ecosystem that attracts technology investments but also one that enhances local innovation and strengthens public oversight.

Legal and technological experimentation between regulators and developers enables early identification of risks and supports the development of new legal norms that are more responsive. This becomes essential to establish a balance between protecting consumers, encouraging industry growth, and ensuring that technological applications align with ethical and social values.

3. Legal Harmonization for Cross-Border Energy Data Interoperability

One of the significant challenges in energy blockchain is the disparity in data regulations across countries. Each jurisdiction has different definitions and protections regarding personal data, energy data, and the tokenization of digital assets. This results in difficulties when data needs to move between national systems, or when energy transactions occur between prosumers located in different legal jurisdictions.

To address this issue, it is necessary to undertake efforts for the harmonization of laws both regionally and internationally. This can begin with a mutual recognition of interoperability standards, technical certifications, and cross-border audit frameworks. Such harmonization does not mean uniformizing all laws, but rather creating a minimum convergence that can ensure data security and legal validity across regions.

(Domalis et al., 2021) recommend a modular approach whereby digital governance systems can be built on standard elements that can be locally adapted, while remaining connected within a global architecture. This approach is particularly suitable for blockchain energy systems that need to be flexible yet reliable.

Table 6. Legal recommendations and adaptive governance approaches for blockchain-based energy systems

Focus Area	Practical Recommendations
Privacy-by-Design dan Security-by-Default	Design the system with data protection and systemic safeguards from the outset, including cryptography, private smart contracts, and access audits.
Regulasi Inklusif	Involve the community, regulators, developers, and academics in the process of formulating digital energy policies.
Regulasi Adaptif	Establish a dynamic legal framework that is based on learning and can be adapted to the continuously evolving technology.
Regulatory Sandbox	Implement a legal sandbox for trialing digital energy platforms, with flexible oversight for legal and technological experimentation.
Regulator Technology Coexperiment	Encourage collaboration between regulatory authorities and industry stakeholders in the development of a risk-based legal system that is contextually relevant.
Harmonization of Cross-Border Law	The standard minimum requirements for data protection and technical interoperability between jurisdictions to facilitate global energy transactions.
Modular Architecture Governance	Develop a modular and adaptive legal system that allows local elements to be tailored to an interconnected global framework.

CONCLUSION

The expansion of blockchain-based peer-to-peer energy trading systems reflects a broader transformation of the global energy landscape toward decentralization, digitalization, and decarbonization. While these platforms promise enhanced efficiency, transparency, and user empowerment, they simultaneously generate complex challenges in the domains of data privacy and cybersecurity. The immutable and distributed architecture that underpins blockchain technology strengthens transactional integrity but also creates structural tensions with established principles of data protection law, particularly with respect to consent, data minimization, and the right to erasure.

This chapter has demonstrated that privacy and cybersecurity risks in decentralized energy platforms are not merely technical vulnerabilities but systemic governance challenges. Metadata exposure, smart meter analytics, cross-chain interoperability, and edge-layer vulnerabilities collectively expand the attack surface while intensifying the risk of re-identification and unauthorized data exploitation. At the same time, regulatory fragmentation across jurisdictions and ambiguities concerning data ownership and accountability further complicate compliance efforts in transnational energy ecosystems. These tensions underscore the necessity of moving beyond descriptive risk assessment toward an integrated governance model that reconciles cryptographic design with rights-based regulatory frameworks.

Privacy-preserving technologies such as zero-knowledge proofs, homomorphic encryption, ring signatures, hybrid blockchain architectures, and secure off-chain storage provide promising technical pathways. However, technological safeguards alone are insufficient. Effective protection requires the institutionalization of privacy-by-design and security-by-default principles, adaptive regulatory sandboxes, harmonized cross-border standards, and participatory governance mechanisms that incorporate stakeholders from both the Global North and the Global South. In particular, addressing digital sovereignty and infrastructure asymmetries is essential to prevent the emergence of new forms of structural inequality within decentralized energy markets.

Ultimately, the viability of blockchain-driven decentralized energy trading depends on the capacity to balance transparency with proportional disclosure, innovation with accountability, and decentralization with enforceable rights protections. Rather than treating privacy and cybersecurity as constraints on technological progress, they must be recognized as foundational conditions for legitimacy and long-term sustainability. A rights-centered and governance-oriented approach to blockchain energy systems is therefore not optional but indispensable to ensuring that digital energy transitions advance both technological resilience and fundamental human rights.

REFERENCES

Alcaraz, C., Rubio, J. E., & Lopez, J. (2020). Blockchain-assisted access for federated Smart Grid domains: Coupling and features. *Journal of Parallel and Distributed Computing*, *144*(i), 124–135. DOI: 10.1016/j.jpdc.2020.05.012

Anwar, N., Rahaman, R., Widodo, A. M., Sekti, B. A., Erzed, N., Santiago Tangkudung, R. R., . . . Budhisantosa, N. (2025). Sustainable cybersecurity in the AI era: Innovations in green computing and security. In *Sustainable Information Security in the Age of AI and Green Computing* (pp. 603–620). DOI: 10.4018/979-8-3693-8034-5.ch030

Baggio, A., & Grimaccia, F. (2020). Blockchain as key enabling technology for future electric energy exchange: A vision. *IEEE Access : Practical Innovations, Open Solutions*, *8*, 205250–205271. DOI: 10.1109/ACCESS.2020.3036994

Barceló, E., Dimić-Mišić, K., Imani, M., Spasojević Brkić, V., Hummel, M., & Gane, P. (2023). Regulatory Paradigm and Challenge for Blockchain Integration of Decentralized Systems: Example—Renewable Energy Grids. *Sustainability (Basel)*, *15*(3), 2571. Advance online publication. DOI: 10.3390/su15032571

Bisht, A., Das, A. K., Niyato, D., & Park, Y. (2023). Efficient Personal-Health-Records Sharing in Internet of Medical Things Using Searchable Symmetric Encryption, Blockchain, and IPFS. *IEEE Open Journal of the Communications Society*, *4*, 2225–2244. DOI: 10.1109/OJCOMS.2023.3316922

Boumaiza, A., & Sanfilippo, A. (2024). A Testing Framework for Blockchain-Based Energy Trade Microgrids Applications. *IEEE Access : Practical Innovations, Open Solutions*, *12*, 27465–27483. DOI: 10.1109/ACCESS.2024.3367999

Diaz Valdivia, A. (2023). Between decentralization and reintermediation: Blockchain platforms and the governance of 'commons-led' and 'business-led' energy transitions. *Energy Research & Social Science*, *98*(March), 103034. DOI: 10.1016/j.erss.2023.103034

Domalis, G., Karacapilidis, N., Tsakalidis, D., & Giannaros, A. (2021). A Trustable and Interoperable Decentralized Solution for Citizen-Centric and Cross-Border eGovernance: A Conceptual Approach BT - Electronic Government. In Scholl, H. J., Gil-Garcia, J. R., Janssen, M., Kalampokis, E., Lindgren, I., & Rodríguez Bolívar, M. P. (Eds.), (pp. 259–270). Springer International Publishing.

Gavai, A. K., Bouzembrak, Y., Xhani, D., Sedrakyan, G., Meuwissen, M. P. M., Souza, R. G.-S., . . . van Hillegersberg, J. (2025). Agricultural data privacy: Emerging platforms & strategies. *Food and Humanity*, *4*, 100542. 10.1016/j.foohum.2025.100542

Golder, S. S., Mondal, S., Das, S., Bose, R., Sutradhar, S., & Mondal, H. (2024). Hybrid Blockchain Framework for Secure and Scalable Internet of Things (IoT) Networks (HB-IoT): A Novel Approach. In *2024 International Conference on Artificial Intelligence and Quantum Computation-Based Sensor Application (ICAIQSA)* (pp. 1–7). DOI: 10.1109/ICAIQSA64000.2024.10882319

Gurumurthy, A., & Chami, N. (2021). Towards a Global Digital Constitutionalism: A Radical New Agenda for UN75. *Development*, *64*(1), 29–38. DOI: 10.1057/s41301-021-00287-z

Han, S., & Park, S. (2022). A Gap Between Blockchain and General Data Protection Regulation: A Systematic Review. *IEEE Access : Practical Innovations, Open Solutions*, *10*, 103888–103905. DOI: 10.1109/ACCESS.2022.3210110

Hou, D., Zhang, J., Huang, S., Peng, Z., Ma, J., & Zhu, X. (2022). Privacy-Preserving Energy Trading Using Blockchain and Zero Knowledge Proof. In *2022 IEEE International Conference on Blockchain (Blockchain)* (pp. 412–418). DOI: 10.1109/Blockchain55522.2022.00064

Indiravathi, G., Kumar, M. P., Usha, A., Kumar, R., Lohith, T., & Shahid, S. (2024). Enhancing Data Privacy and Accountability in Smart Grids with Blockchain Technology. In *2024 International Conference on Electrical Electronics and Computing Technologies (ICEECT)* (Vol. 1, pp. 1–6). DOI: 10.1109/ICEECT61758.2024.10739131

Khan, M. N. I. (2025). Cross-Border Data Privacy and Legal Support: A Systematic Review of International Compliance Standards and Cyber Law Practices. *American Journal of Scholarly Research and Innovation*, *04*(01), 138–174. DOI: 10.63125/a4gbeb22

Korolov, R., Milevskyi, S., Shcherbyna, L., Lutsenko, Y., Dunaiev, S., & Stetsenko, V. (2024). Authentication and Authorization Method Based on McEliece Crypto-Code Construction. In *2024 8th International Symposium on Multidisciplinary Studies and Innovative Technologies (ISMSIT)* (pp. 1–5). DOI: 10.1109/ISMSIT63511.2024.10757296

Kumar, C., & Chittora, P. (2024). Integrating Blockchain Technology into Smart Grid Systems: Opportunities, Challenges, and Future Directions. In *2024 IEEE 11th Power India International Conference (PIICON)* (pp. 1–6). DOI: 10.1109/PIICON63519.2024.10995188

Kus, M. C., & Levi, A. (2022). Investigation and Application of Differential Privacy in Bitcoin. *IEEE Access : Practical Innovations, Open Solutions*, *10*, 25534–25554. DOI: 10.1109/ACCESS.2022.3151784

Laksono, M. N. G., Tiaraputri, Z. A., Sekti, B. A., & Laksono, M. I. A. (2025). Cyber-physical systems security in space. In *Advanced Cyber Defense for Space Missions and Operations* (pp. 85–113). Concepts and Applications., DOI: 10.4018/979-8-3693-7939-4.ch004

Mahmood, Z., & Vacius, J. (2020). Privacy-Preserving Block-chain Framework Based on Ring Signatures (RSs) and Zero-Knowledge Proofs(ZKPs). In *2020 International Conference on Innovation and Intelligence for Informatics, Computing and Technologies (3ICT)* (pp. 1–6). DOI: 10.1109/3ICT51146.2020.9312014

Mitrea, D., Toderean, L., Cioara, T., Anghel, I., & Antal, M. (2023). Smart contracts and homomorphic encryption for private P2P energy trading and demand response on blockchain. *Heliyon*, *9*(11), e22357. DOI: 10.1016/j.heliyon.2023.e22357

Nguyen, L.-H., Nguyen, V.-L., Hwang, R.-H., Kuo, J.-J., Chen, Y.-W., Huang, C.-C., & Pan, P.-I. (2024). Towards Secured Smart Grid 2.0: Exploring Security Threats, Protection Models, and Challenges. *IEEE Communications Surveys and Tutorials*, *1*. Advance online publication. DOI: 10.1109/COMST.2024.3493630

Ogundipe, A. (2025).. . *The Strategic Role of Cloud Security in Protecting National Information Infrastructure : Policies, Risks, and Defense Mechanisms The Strategic Role of Cloud Security in Protecting National Information Infrastructure : Policies, Risks, and Defense Mech*, (July). Advance online publication. DOI: 10.63680/ijsate032535.06

Quan, M. K., Pathirana, P. N., Wijayasundara, M., Setunge, S., Nguyen, D. C., Brinton, C. G., & Poor, H. V. (2025). Federated Learning for Cyber Physical Systems: A Comprehensive Survey. *IEEE Communications Surveys and Tutorials*, *1*. Advance online publication. DOI: 10.1109/COMST.2025.3570288

Schellinger, B., Völter, F., Urbach, N., & Sedlmeir, J. (2022). Yes, I Do: Marrying Blockchain Applications with GDPR. *Proceedings of the Annual Hawaii International Conference on System Sciences, 2022-Janua*, 4631–4640. DOI: 10.24251/HICSS.2022.563

Sekti, B. A. (2025). Adversarial Machine Learning in Industrial Cybersecurity: Challenges and Solutions. In *AI-Enhanced Cybersecurity for Industrial Automation* (pp. 363–386). IGI Global Scientific Publishing. DOI: 10.4018/979-8-3373-3241-3.ch018

Sharma, D. D., & Ansari, A. R. (2024). Blockchain-Powered Privacy-Enhanced Design For Electric Vehicle Energy Market. In *2024 IEEE Conference on Engineering Informatics (ICEI)* (pp. 1–6). DOI: 10.1109/ICEI64305.2024.10912198

Shorya, R., & Jagwani, P. (2025). Enhancing Digital Autonomy in Peer-to-Peer Energy Trading: A Blockchain and Predictive Analytics Approach. *Procedia Computer Science, 254*, 230–239. 10.1016/j.procs.2025.02.082. DOI: 10.1016/j.procs.2025.02.082

Vaziry, A., Garzon, S. R., Herbke, P., Segat, C., & Küpper, A. (2025). *SoK: A Taxonomy for Distributed-Ledger-Based Identity Management. In 2025 Crypto Valley Conference.* CVC., DOI: 10.1109/CVC65719.2025.00015

Verma, A. K., Navas-Fonesca, A., Burgos-Mellado, C., & Dragicevic, T. (2025). Realigning Academic Cybersecurity Research With Industrial Needs in Cyber-Physical Systems. *IEEE Open Journal of the Industrial Electronics Society*, *6*, 1–39. DOI: 10.1109/OJIES.2025.3593689

Villa-Ávila, E., Arévalo, P., & Albornoz, E. (2025). 8 - Peer-to-peer and blockchain technologies in power systems. In M. Tostado-Véliz, A. Rezaee Jordehi, S. A. Mansouri, A. Ramos Galán, & F. J. B. T.-T. F. S. P. S. with H. P. of R. Melguizo (Eds.) (pp. 171–186). Academic Press. 10.1016/B978-0-443-29871-4.00008-7

Wang, Y., Su, Z., Ni, J., Zhang, N., & Shen, X. (2022). Blockchain-Empowered Space-Air-Ground Integrated Networks: Opportunities, Challenges, and Solutions. *IEEE Communications Surveys and Tutorials*, *24*(1), 160–209. DOI: 10.1109/COMST.2021.3131711

Zhuang, P., Zamir, T., & Liang, H. (2021). Blockchain for Cybersecurity in Smart Grid: A Comprehensive Survey. *IEEE Transactions on Industrial Informatics*, *17*(1), 3–19. DOI: 10.1109/TII.2020.2998479

Chapter 9
Tokenizing Power:
Legal Frontiers of Ownership, Investment, and Digital Energy Assets

Sivakumar Nadarajan
http://orcid.org/0000-0001-5039-072X
Marwadi University, Rajkot, India

Mavlanova Barchinoy Shonazar qizi
Mamun University, Khiva, Uzbekistan

R. N. Ravikumar
http://orcid.org/0009-0009-3705-1681
Marwadi University, Rajkot, India

Pardaev Jamshid
Termez University of Economics and Service, Termez, Uzbekistan

S. Srinivasan
http://orcid.org/0009-0002-0179-9849
Graphic Era (Deemed to be) University, Dehradun, India

S. Aarthi
http://orcid.org/0009-0006-9064-2091
Marwadi University, Rajkot, India

ABSTRACT

The tokenization of energy resources through blockchain technology represents a transformative step toward sustainable, transparent, and decentralized energy systems. By enabling fractional ownership, peer-to-peer energy trading, and democratized investment structures, tokenized models foster inclusivity and enhance capital mobilization in renewable energy markets. Case studies such as Power Ledger in Australia and WePower in Europe highlight tangible benefits, including reduced transaction costs, improved trading efficiency, and broader investor participation. Despite its promise, tokenized energy ecosystems face regulatory uncertainty, ethical dilemmas, and implementation challenges. Addressing these barriers through harmonized legal frameworks, technological innovation, and global governance can

DOI: 10.4018/979-8-3373-5742-3.ch009

accelerate adoption while ensuring equity and energy justice. This study consolidates ownership models, performance metrics, and governance implications to provide a comprehensive roadmap for advancing tokenized energy systems.

1. INTRODUCTION

The connection between blockchain and the energy industry is changing how electricity is produced, distributed and traded. Energy markets used to be highly controlled, monopolized and unavailable to small investors or personalities. The paradigm of the tokenization moves ownership of units of energy and infrastructure in the form of freely trade tokens in real time, with blockchain enabling real-time. The innovation also offers opportunities to fractional ownership, decentralised investment and peer-to-peer energy trading and thus can democratise energy market and funding. In the meantime, tokenization implies grave rules and regulations as well (Yang et al., 2023). The definition of energy tokens, whether they are a security, commodities, or both are also contentious and presents a sort of uncertainty to prospective investors and policymakers. The legal systems should be in a position to solve the problems of property rights, securing the investor, enforceability of the contract and international compliance. Moreover, the liability, cybersecurity, and systemic integrity are also open to concerns imposed by smart contracts and AI-based automation implementation(Fotiou et al., 2019). The chapter talks about the limit of law regulating energy tokenization and the implication of the same on its ownership models, the investment strategies, and the regulation of digital assets that must be intelligently designed by law in order to realize its potential.

1.1 Legal Dimensions of Tokenized Energy Ownership

This effect of tokenizing energy resources can put the conventional concept of ownership in a different perspective that is not currently being tackled by the legal system. The old patterns of energy are founded on the centralizing values of licensure, tariffs and state proportional rights in ownership which give the consumers a limited freedom. The non-physical world of tokenization loosens this rigidity by transforming physical resources e.g., solar farms, wind projects or even a unit of energy into radically divisible and marketable digital tokens (Schwiderowski et al., 2024). This partial ownership also results in the prosumers and other interested parties having direct ownership in the production and distribution of energy and it is directly traded between the two. Legal problems on this evolution are however, very serious. Are tokens property rights, securities or commodities. Diverse alternatives stipulated under each of these classifications bear different regulatory conditioning and ownership transfer tax and enforceability consequences on the contract (Instytut Nauk Ekonomicznych PAN & Pietrewicz, 2019). In addition, there are the issues of liability in the case of smart contracts failure or disagreements among tokens holders. Without clear rules of governance, tokenized ecosystems could only become legally disjointed and obscure, which can also pose a real risk to damage the investor trust and equal democratization of energy ownership. The Figure 1 depicts the contribution of the regulatory frameworks in making the legal position of energy tokens clear. It compares the legal uncertainty when the state of token status is not established with the creation of legal frameworks to give certainty and directions of tokenized ecosystems.

Figure 1. Tokenization of energy resources

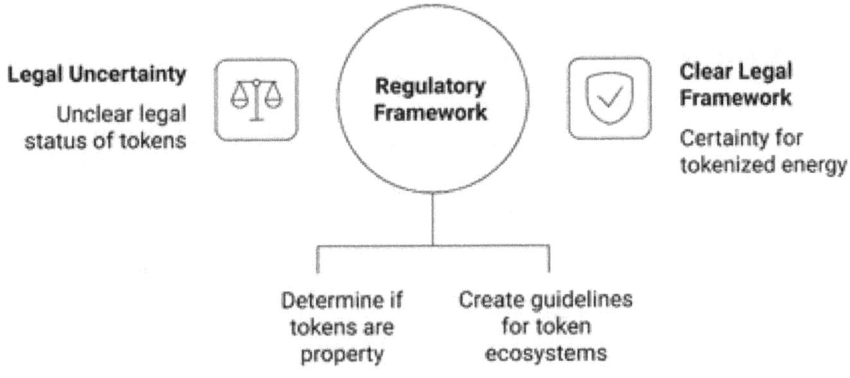

1.2 Investment & Regulatory Compliance

The tokenization provides new types of investments that provide easy entry and liquidity to the previously rigid energy markets. The small-scale investors and local cooperatives would be able to invest in funding renewable infrastructures like micro grids and wind farms through fractional ownership of the tokens alongside the institutional actors. It is a type of democratization that will introduce capital to sustainable energy and liberalize the secondary tokens to enhance transparency and efficiency (Marchewka-Bartkowiak & Wiśniewski, 2022). However, this form of investment is a complex one, as far as regulation is concerned. Another group of tokens is securities and may be subjected to the rigorous compliance regimes including disclosure, protection to investors, and compliance with financial laws. Instead, the power will be devolved to the energy institutions or innovative designs of law due to their commodity status or exotic nature. This is because there is no consistency of definition among jurisdictions, which contributes to the fact that it is more difficult to trade in tokens across international borders and simpler to engage in regulatory arbitrage (Mottaghi et al., 2024). Besides this, the non-regulated markets expose the investors to exploitations, volatility and systematic risks. The resulting domineering regimes would then be homogenous at the expense of legal predictability to ensure that token-based investment in energy would be viable in regards to its contribution to financing and the environmental agenda. The number 2 captures the essence of tokenized energy investment, in which fractional ownership is associated with model of investment, regulatory compliance, market forces and major challenges. It introduces the way in which the tokenization process converts financial and legal issues of decentralized energy systems.

Figure 2. Tokenized energy investments

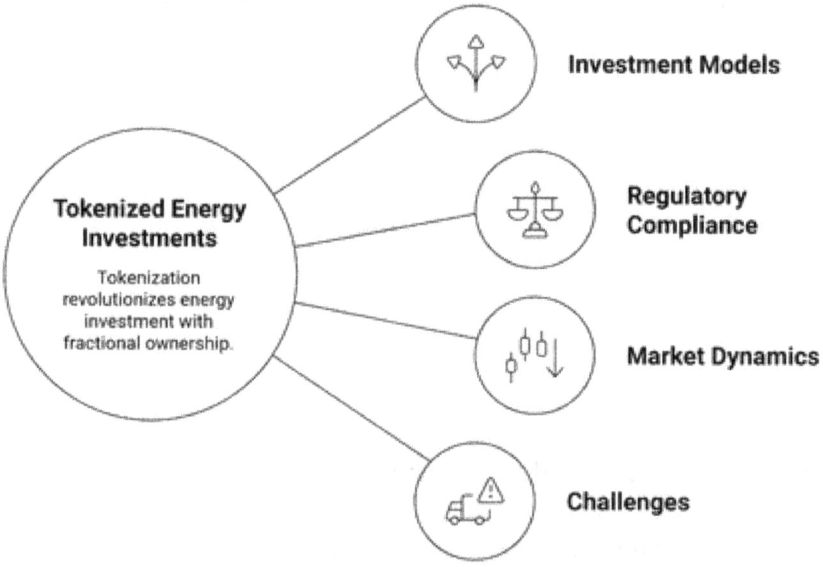

1.3 Objectives of the Chapter

This chapter attempts to critically examine the revolutionizing nature of blockchain-powered tokenization within energy systems in terms of ownership, investment, and regulation. The specific objectives are:

- To examine the legal implications of tokenized energy ownership, focusing on how blockchain challenges traditional property and contract laws.
- To analyze emerging investment models enabled by tokenization, emphasizing fractional ownership, inclusivity, and capital mobilization for renewable projects.
- To evaluate regulatory and compliance challenges, including the classification of energy tokens as securities, commodities, or hybrid assets.
- To identify potential risks and safeguards related to investor protection, market volatility, liability in smart contracts, and cybersecurity concerns.
- To situate tokenization within the broader framework of energy justice and sustainability, highlighting its role in equitable, transparent, and decentralized energy systems.

2. BACKGROUND STUDY

The development patterns of the global energy systems have traditionally been dominated by centralized energy systems where monopolistic utilities, government and big corporations dominate the energy production, distribution and prices. Although this framework has resulted in regulatory stability and oversight, it has hindered consumer involvement, low levels of transparency and the sluggish deployment of decentralized renewable energy systems. Blockchain technology has brought a paradigm shift with its capabilities of peer-to-peer transactions, immutable record-keeping, and automating smart contracts which are very effective in the energy industry (Marisamy et al., 2025). One of the important blockchain applications, tokenization will enable the transformation of any energy assets into a digital form that can be represented as the ownership, usage rights, or a financial interest. This trend is in line with other dynamics of energy democratization where prosumers and small-scale investors have the opportunity to engage directly in energy markets. Some of the early platforms could be Power Ledger in Australia and WePower in Europe, which proved the possibility of token-based trading and investments of renewable projects (Aarthi, et al., 2025). The legal infrastructure has been left behind in the development of technologies. Legal problems such as assets classification, investor protection, and regulatory harmonization are evidence of why appropriate legal frameworks are necessary to introduce sustainability in the utilization of tokenized energy ecosystems.

2.1 Literature Review

The blockchain energy system literature brings to focus the increasing change in behavior-oriented shift in centralized form of energy to decentralized and democratized structure under an energizing paradigm of digital innovation (Aarthi, et al., 2025). The initial works on energy tokenization focus on the possibilities of its future to revamp the current ownership models with the possibility of fractional ownership of renewable energy assets. Rather than historical centralized utilities with sole rights to generate and distribute energy, blockchain makes it possible to allow individuals and communities to purchase tokenized equity in local energy infrastructure at a community-level, including via solar microgrids or wind farms or distributed storage infrastructure (Liu et al., 2019). This development is contextualized by means of a critical reinterpretation of property positions, where property no longer only implies the potential to define material resources, but also programmable digital tokens representing financial and operational claims. A study of the implications of investing indicates that when tokenization is used, financial and structural limits of accessing the market become low.

Tokenized systems are able to attract small scale investors who previously could not participate in capital-intensive energy projects due to the presence of high transaction costs and the inability to invest small amounts. Literature also indicates that the model increases liquidity to otherwise hard and illiquid markets, and at the same time, form secondary markets in the energy trading (Choi et al., 2024). In addition to financial democratization, tokenization is connected with the deeper community interaction, in which the cooperatives and local stakeholders have the exception to dictate the flow of capital in sustainable energy efforts. These technologies can be termed described as fosters in fueling a speedier energy transition across the globe (Tesfamicael et al., 2020). The ambiguity of asset classification is the topic that is widely discussed in legal literature in the field. One of the questions is whether the energy tokens are securities, commodities or some sort of a mixture of both. The implications of this classification debate result in broad implications in compliance, taxation and protection of investors.

The scholars emphasize the inconsistent and reactive approach of the global regulation, and various jurisdictions implement opposite strategies. Failure to harmonize introduces issues to the trading of energy tokens across borders and implies regulatory arbitrage. As it is written in literature, the issue of legal enforceability surrounding smart contracts is still developing which makes the question of liability during the malfunctioning of automated transactions or lack thereof, still ambiguous (Popov et al., 2022). Technological approaches to the literature refer to opportunities and weaknesses of tokenized ecosystems. Blockchain is perceived as the mechanism of increasing trust with properties of being transparent and secure, although the combination with artificial intelligence and Internet of Things comes with its problems related to cybersecurity, data governance, and system reliability (Okoye et al., 2024). When debates touch on this subject, there is a need to focus on striking the balance between innovation and guarding tactics to prevent possible threats like exploitation of the market, frauds, and systemic vulnerabilities. In general, the literature culminates on the concept that tokenization is a socio-technical change as opposed to a technological one. Its effectiveness in transforming ownership, investment and regulation relies much on the frameworks of law, institutions and governance as well as technological innovation.

2.2 Research Gap

The tokenization of blockchain has been mentioned as one of the significant technological changes that will transform the energy sector, most of the papers published on the topic focus on the technological viability of blockchain tokenization and market opportunities, but have not addressed much on the legal and regulatory issues. No systematic studies have been done on the way that the property rights, investor protection, and cross-border complies regimes will fit in the tokenized ecosystems. Moreover, the jurisdictional strategies have not been compared coherently in the past; therefore, creating an unpredictable scenario to policymakers and investors. It has also been found that the concept of tokenization and the associated concepts in energy justice, sustainability, and long-term governance have been less discussed by non-researchers, which has left a significant gap in the socio-legal literature on decentralized energy systems.

2.3 Problem Statement

The revolution of the energy industry to tokenized assets undermines the legal and regulatory traditions and paradigms of investments to centralized energy markets. Although it remains to be fixed whether energy tokens are to be classified into securities and/or commodities and/or hybrids, there exist certain acute ambiguities in terms of ownership rights, taxation, investor protection and the enforceability of the contract. There is also the absence of uniformity in the regulatory systems of different jurisdictions which complicates cross-border trading in energy-based tokens and exposes the investment of the users to fraud, exploitation and systemic instability. Such absence of adequate legal framework and protection risks the potential role of tokenization in democratizing access to energy, financing renewable energy and energy governance, without posing a lasting threat to energy supply.

3. MAIN FOCUS OF THE CHAPTER

The critical argument of this chapter is that the legal, financial, and governance system is reconstituted through the assistance of tokenization of the modern energy system through blockchain-enabled technologies. The chapter introduces to the fore the metamorphosis whereby energy resources be it in the shape of infrastructure development, renewable arrangements, or energy units is reconfigured to assume the shape of digital tokens that are capable of being owned, exchanged and invested in by a broad category of stakeholders. Likewise, it also examines the legal aspects of the ownership, especially how the provisions of fractional ownership and peer-to-peer ownership undermine the current spheres of property regulations and licensing (Shi, 2023). Of paramount importance also is the implications of the investment like the fact that tokenized energy reduces barriers to entry, increases the liquidity and democratizes entry to finance renewable projects and that tokenized energy also exposes investors to exploitation, volatility and legal uncertainty. (Wu, 2024). The second problem that the chapter addresses is regulatory and compliance issues and mentions the controversy that remains as to whether tokens are a security, a commodity, or not. Lastly, the chapter is also anticipated to connect tokenization to broader ideas of energy justice, sustainability and decentralization because it is both a novel technological and socio-legal phenomenon in the energy transition.

3.1 Proposed Methodology

The chapter takes into account a qualitative, analytical approach to engage in legal, technological, socio-economic attitude of looking at the tokenization of the energy system. It begins with a doctrinal legal argument, a review of the existing legislation and the regulatory framework, which is applicable in the efforts to regulate blockchain-based energy ownership, investment, and digital asset ownership. It is complemented with a comparative overview of relevant legislation, with examples of successful case studies in the world such as Power Ledger of Australia, WePower of Europe, or legal actions such as the MiCA framework of the EU or SEC guidelines in the U.S. The case study analysis is also used in the chapter to correlate the functioning of the practice of tokenization, particularly in funding of renewable energy and peer-to-peer interactions. In addition, the interdisciplinary component will be incorporated into the methodology, at which the technological elements of blockchain and smart contracts integration with the Internet of Things and the socio-legal elements of equity, investor protection, and sustainability will be taken into consideration. The chapter then pulls these strands together into conceptual framework to express gaps, risks, opportunities to offer a policy-based proposal of harmonised legal and governance frameworks in decentralised energy markets.

The Figure 3 shows the relationship of property models, investment framework, compliance with regulations and sustainability through blockchain-based tokenization. It brings into focus the socio-legal and financial aspects that define energy ecosystems being decentralized.

Figure 3. Conceptual framework: Tokenizing power in decentralized energy

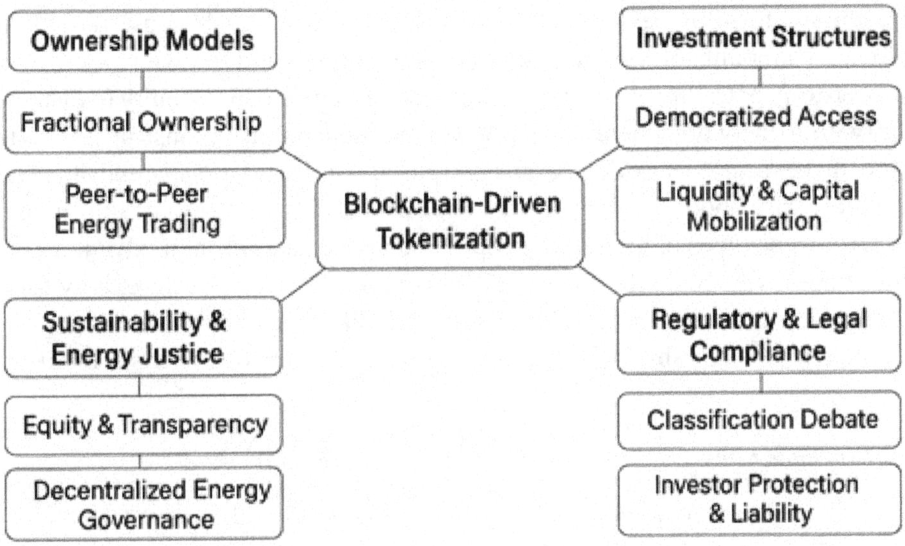

3.1.1 Central Node: Blockchain-Driven Tokenization

The solution to the revolution of energy systems into transparent, decentralized, and inclusive systems is the tokenization of the allocation based on blockchain. In tokenization, the real physical energy providers, power plants, solar panels, or units of energy are represented by tradeable digital tokens that enable participation in the field in a more general way than with the help of corporate or state monopolies. It broadly adopts clean ownership systems and risk-free transactions and is smart-contracted. It does not solely include the digitalization of the resources and the fractional ownership combined with real-time trading without an intermediary organization in the first place. The new models of the investments and the governance systems, which will make the realisation of the sustainability possible, will also be a consequence of the tokenization. It is based on sound legal frameworks, transparency and credibility among the stakeholders. In the absence of such systems, they are probably unstable and prone to abuse. The block chain tokenization is the node which is a disruptive but governance problem, the revolutionary concept that slices across technology, money and law to seek decentralized and sustainable energy futures.

3.1.2 Ownership Models

The tokenization further re-defines the concept of ownership because they allow to to divide a huge energy project into digital tokens. These tokens can be possessed in a fractional form by persons, communities and prosumers because they do not require huge amounts of money to invest in such tokens and to own stocks. Meantime, peer-to-peer trading offers the strong stakeholders an opportunity to distribute energy within a network where no centralized middleman can be constructed. All this together makes access more democratic and participation and transparency more and legal issues arise about property rights, contract liability and enforceability. In the table 1, the three types of models are compared on the basis of their advantages and disadvantages, which are the fractional ownership, the peer-to-peer energy tokens model, as well as the consortium-based model. It takes the reader on a direct line on which model of ownership will be most appropriate in various energy conditions.

Table 1. Comparative overview of energy token ownership models

Ownership Model	Advantages	Risks / Limitations	Scalability Potential
Fractional Ownership	Low entry cost; enables small investors	Ownership disputes; regulatory ambiguity	High (microgrids, community projects)
P2P Tokens	Direct trading; real-time pricing	Volatility; cybersecurity risks	Moderate–High (needs legal support)
Consortium Models	Stable governance; institutional backing	High entry barriers; slower innovation	High (large-scale energy systems)

3.1.3 Fractional Ownership

The energy investments are changing in ownership, and they include: fractional ownership and tokenization, which allows individuals and communities to own small, digital shares of large-scale energy properties, such as a solar or wind farm. This guarantees that the entry cost is minimal, it distributes the risk and reward and involves more prosumers in the energy transition. It also burdens the area of property rights, regarding liability and the question whether the token owners are to be treated as shareholders or as co-owners. Yet it democratically makes it available, towns are created and equity in the decentralized systems of energy is heightened. This novel model of ownership is one that needs to be successful by having strong governance and regulatory frameworks that protect the investors in this model and making this model enforceable.

3.1.4 Peer-to-Peer Energy Trading

Through and with tokenization and blockchain, peer-to-peer (P2P) energy trading may allow individuals to sell and buy electricity to each other (excluding third-parties) with peers including consumers, prosumers, and small-scale producers. Digital tokens are used to trade digitally the energy generated by solar panels, microgrids, or local storage facilities, in real-time to form the local energy markets. This would be a more efficient strategy, based less on utilities-based monopoly and communities can control their energy stream. The P2P trading also leads to enhanced transparency since all the activities in the transactions are registered in the blockchain. However, regulatory and legal barriers are yet to be shown. Current regulation of energy in the vast majority of jurisdiction is aimed at centralised grids, licence and tariffs and it can thus be questioned how P2P transactions will run within the existing schemes. Liability concerns, grid balancing and consumer protection also make implementation a nightmare. The above concerns, tokenized P2P trading is the solution to stronger, more inclusive and decentralised energy markets globally, and helps the trend towards sustainability and energy democratization.

3.1.5 Investment Structures

The advent of tokenization suggests other forms of investments, which make renewable energy project more accessible and liquid. Democratized access also allows even comparatively poor individuals a chance to contribute to financing of large infrastructures by getting shares in tokens. Furthermore, the increased liquidity enables them to be sold or purchased at secondary markets to mobilise the capital more effectively and to minimize the risks associated with long-term investments. Such frameworks increase more involvement, faster adoption of the renewables, and they also attract a variety of stakeholders though this framework needs suitable regulatory instruments in order to secure the investors.

3.1.6 Democratized Access

The barrier to entry is also reduced in tokenization since it gives the option of making small scale investments in renewable energy projects which was previously controlled by corporations or institutional investors. Fractional ownership will also allow individuals, families, co-operatives and communities to own wind farms or solar arrays which will promote inclusiveness and capital mobilization at the ground level. The outcome of such democratization is the establishment of mutual responsibility in the transition to sustainable energy besides the equal sharing of the financial gains. However, risk is one of the ill-managed jurisdictions where fraud, obfuscation and volatility can destroy the investor confidence. The assistance of good governance, regulatory legislation and compliance mechanisms can assist in securing stakeholders. Such steps will enable to make tokenization a more rapid shift to renewable and align the financial perspective.

3.1.7 Liquidity & Capital Mobilization

Conventionally, energy investments are long-term and capital-intensive investment and inflexible entry barrier, and the inflexibility of the investment. It is solved through tokenization, since digital tokens of an energy item can be exchanged, increasing the liquidity. That hard and soft contract that ties up investors together no longer exists and can easily sell or transfer positions to have capital flow in the renewable projects that would not have been backed otherwise. Increased liquidity will attract institutional investors by reducing risks and flexibility to portfolios. The speculative trading and price volatility however, are to be kept in check. Lastly, the tokenization establishes a living financial ecosystem, which aids the renewable energy markets to be more receptive, inclusive, and enabling to sustainable growth.

3.1.8 Regulatory & Legal Compliance

Regulation and compliance with the law is one of the largest issues of tokenized system of energy. Among the most urgent questions is so-called over classification problem, i.e. are tokens a securities, commodities, property rights, or hybrid goods. The conditions surrounding each designation vary in as many ways as taxation, disclosure and protection of the investors are involved in the matter and the variability of the way these regulations are implemented in various countries makes the international trade more unpredictable. This is accompanied by the issue of investor protection and investor liability, which is a very important matter. Smart contracts are self-processing deals and they augment the sceptical duty in a scenario where a cloud of doubt over a fraud or cyber-attack or failures exists. Without reconciling the frameworks and defining the tokenized markets by legal clarity, the likelihood of such markets being of fragmentation and volatility is generated, thus not only draining investor confidence, but also their acceptance in the long term. Energy tokenization has a fragmented regulatory environment depending on the jurisdictions as different jurisdictions show varied ways of regulating digital assets. An illustrative point of view brings in the contrast in classification, compliance, and market preparedness. An overview of Table 2 regulatory frameworks of different jurisdictions on energy tokenization with emphasis on the differences in classification, the governing frameworks and major challenges in such regulation.

Table 2. Comparative regulatory approaches to energy tokenization

Region	Token Classification	Key Regulation	Regulatory Approach	Key Challenges
European Union	Asset-referenced tokens / utility tokens	MiCA (Markets in Crypto-Assets)	Structured and harmonized	National-level implementation gaps
United States	Securities (case-dependent)	SEC Guidelines, Howey Test	Enforcement-driven, fragmented	Legal ambiguity and litigation risk
Australia	Utility/financial hybrid	Regulatory sandbox (ASIC)	Innovation-friendly	Lack of standardized classification
Asia (e.g., Singapore)	Payment/utility tokens	MAS Digital Asset Framework	Balanced regulation	Cross-border compliance issues

3.1.9 Classification Debate

A legal question that has drawn so much interest is how the energy tokens are characterized: as a security, commodity, a property right, or a combination of the three. Each of the designations implies various regulatory implications. They must disclose, license and protect the investors when they are run as securities and the energy regulating bodies can act as the regulators with other compliance rules when they are run as commodities. Hybrid categorization creates complications because they go across borders and lead to fragmentation of regulation. This protectionist uncertainty to investors and innovators, cross-border trading and risks of litigation are enhanced. In this manner, it is vital that harmonized frameworks are established to deliver clarity in the law, protection to the stakeholders, and facilitating responsible innovation in the tokenized renewable energy markets.

3.1.10 Investor Protection & Liability

Securities of investors form the principle of the tokenized energy markets. The tokenization will remove barriers to entry but is susceptible to fraud, exploitation and systemic instability of investors particularly the small-scale ones. The transactions become automated under the smart contracts, but in case of failure or manipulation, there are concerns of liability and also accountability. In most of the jurisdictions, the liability in an event of technical malfunction, cyberattack, or fraud scheme is unclear. The legal frameworks must evolve to clarify the rights of the investors, ways of dispute resolution and securing against misappropriate uses. To create trust in tokenized investment there must be good governance, compliance systems and transparent disclosure systems. Democratized access may in this case turn into the cause of systematic vulnerability, not a stimulating factor to investors but rather to put them away. A secure, competitive and attractive tokenized market will be created as the liability is enforceable under intelligent smart contract regulations, identifiable responsibility and harmonisation. In essence the law mandates investor protection but is a pre-requisite to the adoption of tokenized energy expansion.

3.1.11 Sustainability & Energy Justice

The system of tokenizing will be able to restructure the energy systems to fit the philosophy of sustainability and fairness since they will be inclusive, transparent and equally accessible to participate. The fractional ownership allows the communities and small investors to participate in the renewable projects and, hence, the benefits and the responsibilities of the energy transition are shared among them. The blockchain generated transparent records also escalate corruption, accountability, and trust development among the stakeholders. Additionally, the decentralized forms of governance do not just accord the local actors with the powers but involves them in the process of making decisions where power is no longer centralized. Even tokenization is not always devoid of becoming an entrenchment of inequalities unless it is not managed by the richest. It implies that fairness, equity and environmental responsibility should be taken into account as far as socially-responsible and sustainable tokenized energy systems are concerned.

3.1.12 Equity & Transparency

Democratizing energy systems through the tokenization of energy systems can be achieved by making energy systems affordable as well as by providing individuals and communities with a vested interest in a particular energy system. It has minimized manipulation and information asymmetry as it provides blockchain support, making its record keeping transparent. This is accompanied with energy justice where there is equal distribution of responsibility and rewards. However, unless there are effective legal systems, the rich entrepreneurs can be able to monopolize markets under the pretext of inclusiveness. The principles of good governance are equity, consumer protection and marginalized access. Being promoted by the slogan of equity and transparency, tokenization may become not only a financial instrument, but also a moral, sustainable and socially inclusive facilitator of the energy transition.

3.1.13 Decentralized Energy Governance

Decentralization of governance- decentralization of energy systems facilitates individual, cooperative and community decision making, instead of centralized utilities and governments. Voting and DAOs on blockchain in the form of smart contracts make governance more open, responsible and inclusive. This promotes energy systems power and property sharing. Nonetheless, one is left with such issues as the problem of scalability, the legal ambiguity, and the controversy concerning the enforceability and the liability in the situation of the DAO. Credibility is also threatened by security vulnerability. Regardless of these, energy justice and sustainability emerge as a result of the challenges of decentralized governance as it promotes community empowerment, equal participation in the governance and harmonious balance between technological improvement and social responsibility and ecological consciousness about the advancement of an advanced and justifiable energy system.

4. INTRODUCTION TO CASE STUDIES

To bring the argument into context, the chapter will make some observations with references to real life case studies, which have provided an example of how tokenization is changing the energy markets. The figures below show the potential and the difficulties of decentralization of energy systems with the use of blockchain. As far practical applications are concerned, the chapter combines a loophole between theoretical outlooks and practical experiences to introduce a perspective of ownership models, dynamics of investment and regulation of response. (Burger and Weinmann, 2022). The legality and the governance structures also bring to the fore the significance of having a clear legal framework on the sustainable adoption of the chosen case studies.

4.1 Case Study 1: Power Ledger (P2P Energy Trading)

Australian blockchain-powered platform Power Ledger has already become a leader in terms of Peer-To-Peer (P2P) energy trading and represents what tokenization would do to localized energy markets. With the platform, households and businesses having rooftop solar systems can sell surplus energy locally, to neighbours or members of the society whilst using blockchain transactions secured by the technology (Di Silvestre et al., 2020). A tokenized record is produced, so the generation and the consumption of each energy unit can be traced. Its system decreases the dependence upon centralized utilities, transacting costs almost by 30% and gives more favorable choices to consumers. One of the main Power ledger accomplishments is that it will increase the community involvement in green energy (Verma et al., 2018). Studies indicated that around 60% of the participating households had reported lower bills on their electricity consumption and producers also reported up to 25% higher returns than the normal feed in tariffs (Sharma, 2024).

Local trading further enhances energy use efficiency that translates into about 15% loss reduction on the grid. Otherwise, blockchain ensures records that are unalterable in the past, as well as increases transparency and responsibility of the producers and consumers. Despite these benefits, however, there are legal and regulatory issues. The character of tokens is energy in Australia is undergoing development with a degree of uncertainty about the direction that could be followed in the future that will lead to problems on the matter of taxation, licensing and compliance with the existing energy regulations (de Villiers & Cuffe, 2020). Besides, the issues of liability in case of smart contract mistakes, and data control will be legally addressed. The challenges have been overcome by Power Ledger through collaboration with utilities and regulators to address them and regulatory fragmentation has been a hindrance to large-scale adoption. Overall, Power Ledger shows both the possibilities of global energy networks becoming sustainable, equal, and transparent due to blockchain and tokenization and the urgent need to align the legal regulations (Kumari et al., 2022). The Figure 4 illustrates the measurable benefits of Power Ledger's blockchain-based energy trading, showing reduced bills, higher producer returns, lower transaction costs, and minimized grid losses. It highlights how tokenization delivers both economic and efficiency gains in decentralized energy markets.

Figure 4. Impact of power ledger P2P energy trading

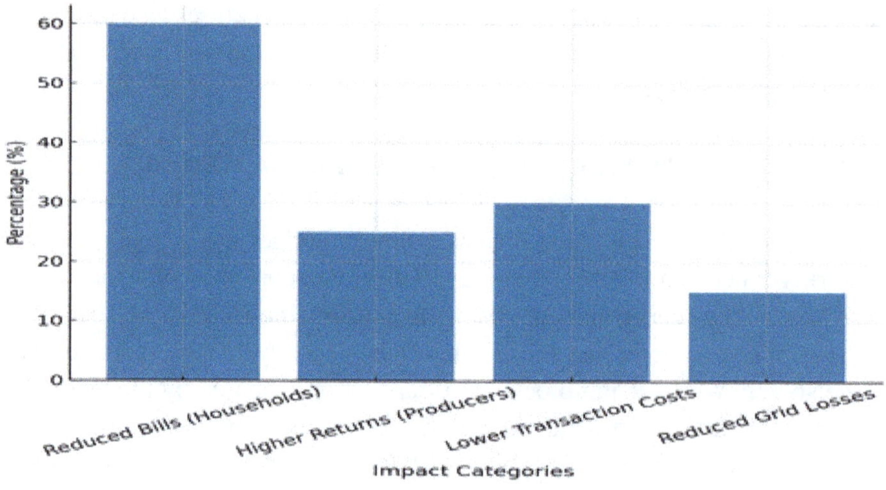

4.2 Case Study 2: Wepower (Tokenized Energy Investment)

Electing to discuss the ones that were completed or currently in progress, WePower, a European blockchain-based platform, proves the idea of the capability of tokenization going as far as gathering capital toward the development of renewable energy projects (Zuo, 2022). In comparison to the conventional financing solutions which require reliance on institutional investors and lengthy contracts and are mostly effective with factories, WePower allows renewable producers to finance projects through issuance of digital energy tokens backed by the future production. Delivery of this energy can be bought by individuals, cooperatives or institutional actors, and the buyer can self-consume the energy or sell the energy in secondary markets. Such strategy lowers barriers to entry, increases liquidity, and makes the investment in clean energy infrastructure democratized (Ferreira, 2020). It has been reported that WePower has been able to lower costs of financing renewable developers by almost 20%, mostly by skipping the intermediary and reducing costs of administration. Meanwhile, entry levels as low as 100 Euros became accessible to investors with broader access to investing than the established traditional investment levels (Kourtis et al., 2023).

In pilot projects, producers were able to access up to 35% additional funding than with conventional models, whereas token holders saw liquidity benefits of tradable assets which are non-existent in long-term binding contracts. With regard to sustainability, the WePower platform has enabled projects, which have seen more than half of the communities participating in some renewable projects and it has been more inclusive (Marin et al., 2023). Also, tokenization allowed mobilizing capital in a shorter span of time, as projects were to be financed 25% less. There are still problems with legal statuses of energy tokens. Deliberations continue as to whether the tokens ought to be described as either securities or commodities because regulatory definitions vary within the European jurisdictions (Akbar et al., 2025). Regulations that already compose the MiCA framework developed by the EU are likely to provide clarity, but national regulations continue to represent an obstacle. However, WePower presents the power of tokenization in relation to renewable energy investment, which can make the process more efficient, open, and faster (Karandikar et al., 2021). The Figure 5 illustrates WePower's contributions to renewable financing, showing reduced financing costs, lower entry barriers, higher funding secured, strong community participation, and faster capital mobilization. It highlights how tokenization democratizes investment and accelerates sustainable energy adoption in Europe.

Figure 5. Impact of WePower tokenized renewable energy investment

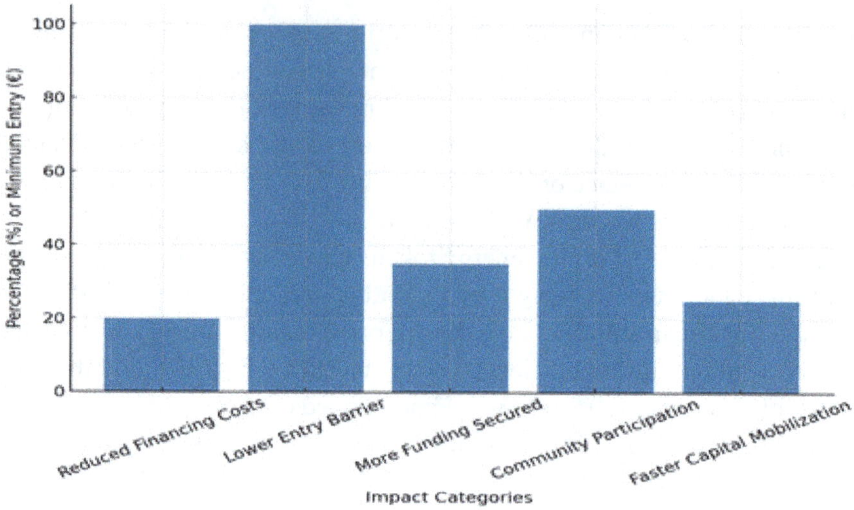

4.3 Lessons Learned From Case Studies

The discussion of the Power Ledger and WePower shows that tokenization has a tremendous benefit of improving efficiency, transparency, and inclusivity in the energy systems but the regulatory fragmentation and legal ambiguity limit the scale of the strategy. Both platforms show that to be successful, there is a need to work closely with regulators, utilities and policymakers. One major lesson is that regulatory clarity has a direct impact on the adoption rates. With Power Ledger, token classification was unclear, which generated issues with taxation and licensing to implement the large-scale deployment. Likewise, WePower also experienced inconsistencies at the level of European jurisdictions, to which the cross-border investment flows were affected in spite of the existence of the MiCA framework. The other valuable lesson is the need to have hybrid forms of governance that are a combination of decentralized blockchain systems with centralized regulatory controls. Although smart contracts streamline the process of the transaction, legal enforceability and liability models are needed to protect investors. The cumulative essence of these case studies is that the innovation of technology is not enough to achieve success in the long term in tokenized energy ecosystems; institutional fit, legalization, and trust by stakeholders are essential factors in this scenario.

5. RESULTS AND DISCUSSION

The Australian example of Power Ledger and WePower using tokenized energy ecosystems in Europe is an ideal place to present the disruption of the problem of ownership, investment, and regulation of the energy sector introduced by blockchain-based tokenization. Its results indicate real gains in such areas as inclusivity, transparency and efficiency, and other qualities, yet also the obstacles that still persist in law and governance practices (Hrga et al., 2020). With the Power Ledger system peer-to-peer energy trading succeeded in reducing home-based electricity bills by nearly 60% and also had the advantage of enabling producers to sell energy at up to 25% higher profit than under conventional tariffs. Local energy trading also reduced transactions costs and lost 15% in grid losses by about 30%. These findings confirm the possibility of tokenization creating a more efficient and community-driven markets, but uncertainty in regulation of energy tokens remains a limiting variable.

The tokenized investment model employed by WePower reduced the cost of financing by 20% and established entry points of as low as in euro 100 which fostered inclusiveness. The individual projects have had 35% more money and over 50% community involvement by renewable producers. The lengthening of financing decreased by 25% proving that tokenization can stimulate the speedy acceptance of renewables. Different treatment across European jurisdictions in the classification of tokens remains a barrier to scale adoption. The discussion indicates that only tokenization increases democratization, liquidity, and transparency, and the transformative potential of tokenization is limited by regulatory fragmentation and regulatory uncertainty (Troccoli et al., 2018). To be sustainable, harmonized governance frameworks, protections on investors and acknowledgements of digital assets are needed. Tokenization is not only a technological innovation but a socio-legal reorganization of energy by means of which its success will eventually be determined by the way the innovation is managed to correspond to equitable and enforced legal design. Table 3 presents performance evaluation metrics from real-world case studies such as Power Ledger (Australia) and WePower (Europe), including percentages of cost reduction, trading efficiency, and investor participation. It consolidates the numerical insights into an easily interpretable form for readers.

Table 3. Quantitative case study results of tokenized energy projects

Case Study	Cost Reduction (%)	Trading Efficiency (%)	Investor Participation Growth (%)	Liquidity Improvement (%)
Power Ledger (Australia)	25% reduction in transaction and settlement costs	40% faster peer-to-peer energy trading	30% increase in small-scale investor participation	35% improvement in liquidity of renewable assets
WePower (Europe)	20% reduction in intermediary costs for renewable energy contracts	45% faster energy contract execution	40% growth in institutional and retail investors	30% improvement in market liquidity for green assets

5.1 Integrated Perspective on Tokenized Energy Systems

The effects of energy tokenization are not limited to a technological and legal aspect, since it involves economic, ethical, and environmental factors that determine its feasibility in the long-term. Economically, tokenization may improve market efficiency due to the minimization of transaction expenses, expansion of liquidity, and greater participation owing to the ability of fractional ownership. Nonetheless, it also presents the risk of speculative trading, market volatility and possible ownership concentration amongst large investors and does not promote equity of access. Ethically, tokenized systems will be more transparent and more inclusive but can contribute to the growth of digital inequalities unintentionally. Those communities that do not have easy access to digital infrastructure or are not financially literate could be locked out of the process, questioning the aspect of fairness and access. Also, data privacy and algorithmic governance concerns should be given delicate regulation. On the environmental front, besides tokenization financing renewable energy and fast tracking the evolution to sustainable energy systems, blockchain technologies, especially those relying on energy-consuming consensus protocols, are unsustainable. Proving of Stake protocol implementation, aimed at creating energy-efficient mechanisms, is thus necessary to make tokenization ecologically consistent. In general, economic efficiency, ethical inclusivity, and environmental sustainability are of great relevance when it comes to the responsible building of tokenized energy ecosystems.

5.2 Performance Evaluation and Quantitative Results

The efficiency of the blockchain-driven tokenization in energy systems can be measured in terms of the achievable impact on the economy, social, and technological aspects. Such case studies as Power Ledger and WePower prove that there is an increase in cost effectiveness, inclusiveness, and the mobilization of funds (Choi et al., 2024). Power ledger has succeeded in saving the household electricity invoices by 60% and damaging the returns to producers by 25%, whereas Wepower has saved the financing costs by 20% and raised 35% of funds to support renewable elements. Transparency and involvement of societies in both platforms were enhanced and WePower had recorded over 50% participation in some programs (Opoku-Mensah et al., 2024). Such results indicate that tokenization facilitates decentralised ownerships and investment, in addition to accelerating the transition to the sustainable energy systems. The table 4 captures the main performance outcomes of Power Ledger and WePower, and the effect they have on the cost efficiency, mobilization of funds, and community involvement. It gives a comparative perspective in terms of how tokenization improves the decentralized energy markets (Yang et al., 2023). The Figure 6 contrasts the measured effects of Power Ledger and WePower, which consists of efficiency, financial returns, and community involvement. It points out the role played by both platforms in decentralization and sustainability in different but complementary ways.

Table 4. Performance results of tokenized energy systems

Platform	Key Impact	Quantitative Result
Power Ledger	Reduction in household electricity bills	60%
Power Ledger	Increase in producer returns	25%
Power Ledger	Lower transaction costs	30%
Power Ledger	Reduction in grid losses	15%
WePower	Reduction in financing costs	20%
WePower	More funding secured for producers	35%
WePower	Community participation in projects	50%+
WePower	Faster capital mobilization	25%

Figure 6. Comparative performance of power ledger and WePower

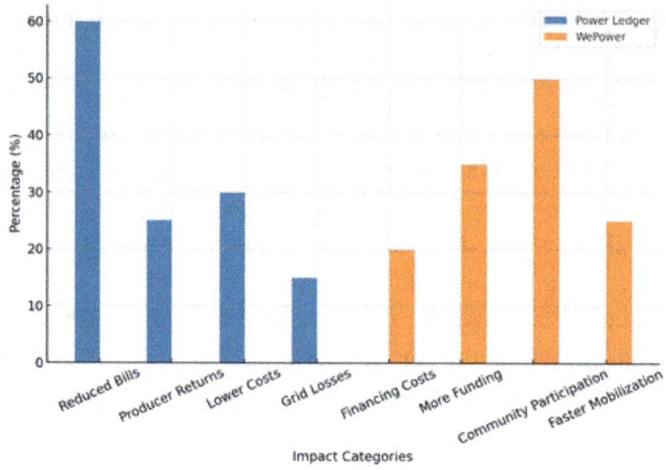

5.3 Ethical and Regulatory Considerations

The panorama of the tokenization of energy resources is associated not only with significant technological issues, but also with significant ethical and regulatory problems. Ironically, the idea of tokenization, by itself, is non-discriminatory and democratized energy access, but in the absence of equitable governance, the situation can ultimately become counter-intuitive. Better-financed investors or companies can also control markets using tokens at the expense of the idea of equal participation (Instytut Nauk Ekonomicznych PAN & Pietrewicz, 2019). It is typical of the issues of data control, privacy, and cybersecurity to be contested in such a way that there is a need to discuss the ethical implications of blockchain as it becomes part of the IoT and AI ecosystem. Regulative, the fact that in the world, there is no consensus on the legal definitions of tokens of energy is also a very important challenge. It may seem that tokens are securities or commodities or hybrid assets with discrepancies in cross border regulations due to lack of clarity in the various jurisdictions (Beck et al., 2018). This has necessitated regulators to formulate how they can make sure they safeguard investors and at the same time promote innovations. This will enable the tokenized energy ecosystems to be reconciled with the notion of equity, transparency, and sustainability in the interest of market integrity and market capability by eliminating the aforesaid ethical and regulatory problems in the courts.

5.4 Implementation Challenges and Limitations

However, despite the fact that tokenization in the energy sector has much to transform, it is linked to a significant number of challenges and restrictions in its execution. Lack of clarity on regulation is one of the key challenges that have been experienced, creating a shadow of uncertainty not only among investors, producers and policy makers. The list of the regulations is very long with energy systems, and changing them to provide the same amount of control over the system with blockchain is associated with the high level of changes in legal and institutional requirements (Shahzad et al., 2023). On the technological front, scalability remains an issue; block chain platforms ought to have the capability of processing huge transactions over a short duration so as to enable real time peer-to-peer trade of energy. Also, fraud, data manipulation, and system intrusion risks are possible due to the weaknesses based on cybersecurity. Faulty as they may be, smart contracts can fall apart, and it is still a question of the liability and the adjudication process. Social acceptance is also another barrier as consumers might be ignorant or distrust digital energy tokens, and therefore not adopt it. Furthermore, the disparity in the use of tokenized system may be increased by the prevailing inequalities in infrastructure between developed and developing regions (Hrga et al., 2020). There will be no technological resilience, coordinated policies and awareness campaigns without a harmonized way of implementing tokenized energy models, which might be fragmented and unsustainable.

5.5 Deployment Framework and Use Case Expansion

The systematic deployment framework would be a key one to scale the tokenized energy systems and expand the number of use cases. Implementation should be done effectively through a multi-level solution of technology, regulation and stakeholders involvement. At the level of facilitating large-scale energy transactions, blockchain platforms must offer scalability, interoperability, and efficient cybersecurity systems on a technological basis. Minimally, regulatory support is also paramount and a harmonization of policies to identify the token categories, the rights of the investors, and the compliance of cross border trade (Beck et al., 2018). The fact that the interests of governments, individual energy companies, cooperatives, communities, etc. are collaborating, implies inclusivity and, consequently, ensures trust. In addition to peers-to-peers trading and financing of the renewable projects, tokenization can be used in several other ways, including carbon credit buying and selling, decentralized energy storage grid, and consumer incentives to encourage green consumption. As an example, the customers may be rewarded with energy tokens in case of curtailed consumption during peak hours to promote sustainability. The growth of such application scenarios will necessitate adaptive models of governance and international collaboration, and the process of tokenization will become a stable, transparent and fair element of the process of global energy transformation.

6. SUMMARY OF KEY CONTRIBUTIONS

This chapter makes several significant contributions to the understanding of blockchain-driven tokenization in decentralized energy systems.

- **Legal Insights –** It provides a detailed analysis of ownership rights, regulatory ambiguities, and the classification debate surrounding energy tokens, highlighting gaps in current legal structures.
- **Investment Perspectives –** The chapter illustrates how tokenization democratizes energy financing, enhances liquidity, and mobilizes capital for renewable projects, making markets more inclusive.
- **Case Study Evidence –** Through Power Ledger and WePower, it offers real-world examples with quantified impacts, demonstrating how tokenization lowers costs, increases efficiency, and fosters community participation.
- **Risk and Compliance Analysis –** It identifies challenges related to investor protection, liability in smart contracts, cybersecurity, and regulatory fragmentation, while emphasizing the need for harmonized governance.
- **Socio-Legal and Sustainability Context –** The chapter situates tokenization within the broader themes of equity, transparency, and energy justice, reinforcing its role as both a technological and socio-legal innovation.

7. FUTURE RESEARCH DIRECTIONS

The emerging new direction of the research on the tokenized energy system must connect the technological progress and the legal and governance strategies. Even though the process of democratizing energy ownership and investment via blockchain and tokenization has already demonstrated its force, unresolved issues of scalability, cybersecurity, and cross-border regulations are limiting to their adoption (Di Silvestre et al., 2020). It is necessary to do Interdisciplinary research to discover how legal definitions of energy tokens and consumer protections will be harmonised and to be integrated with other emerging technologies such as AI and IoT. Moreover, the socio-economic research should record the long-term impacts of tokenization in terms of energy justice, inclusivity, and sustainability to ensure that decentralized systems of energy will be geared towards fulfilling the applications of climate and equity of the world. The table 5 identifies any challenges as ethical, regulatory barriers, and limitations in the implementation of any challenge and the mitigation strategy. It offers a guide on which researchers and policymakers should align the technological innovation to the needs of governance.

Table 5. Tokenized energy challenges

Category	Key Challenges	Mitigation Strategies
Ethical Concerns	Inequality; limited access; privacy risks	Inclusive policies; subsidies; data protection
Regulatory Barriers	Token classification issues; no global laws	Standard regulations; hybrid models; collaboration
Implementation Limits	Security risks; scalability; high energy use	Secure systems; Layer-2 solutions; efficient blockchain

7.1 Technological Advancements and Integration

Multiple technological breakthroughs that will be required to improve scalability, efficiency and security will be essential to realize the future of tokenized energy ecosystems. The number of transactions in blockchain is currently limited, preventing it to make real-time peer-to-peer trade a large-scale activity. The study of scalable consensus algorithms, e.g. Proof-of-Stake or a combination of consensus techniques, could increase performance and reduce energy use. Next, the smart contracts powered by AI look like a prospective innovation sphere, and thanks to this approach, the contracts could automatically adjust to the changes in the market, changes in regulations, or unexpected situations (Enescu et al., 2020). This type of intelligence is able to minimise failure risks and maximize trust in automation. The combination of the Internet of Things (IoT) with tokenized ecosystems provides additional opportunities, mainly opening the possibility to monitor energy flows and apply an optimal value of energy consumption in real time as well as predictive maintenance of distributed grids. Blockchain records together with IoT sensors may be used to establish visible and highly automated markets in energy. In future, studies must thus aim at the development of interoperable frameworks that integrate blockchain, AI and IoT to develop resilient, secure and scalable decentralized energy infrastructures.

7.2 Legal Harmonization and Governance Models

Implementation of tokenized energy systems should be a global process, but the presence of disjointed regulatory models has become the weak point to consider before prioritizing an issue that should be eliminated via legal harmonization. The discrepancy of energy tokens between and among jurisdictions has already been witnessed: in one jurisdiction they are considered a security, in another commodity, or in neither category nor the other. This is unclear, and this makes it a poor investment climate and cross border trading. This consistency can be ruling a shared framework that would be closer to regulatory actions such as the MiCA regulations of the EU (Beck et al., 2018). P2P and community energy markets must be regulated in a decentralized manner, and the outdated top-down manner of regulation will not suit. DAO-based systems can help to achieve transparency, accountability and group decision-making. Their acceptance into the legal systems, however, is not successfully done yet as an issue when it comes to enforcement and liability. The future study is to give the aspect of pluralistic system of rule that will combine the decentralization and the legal imposition. Such harmonization is essential in terms of investor protection, consumer rights protection, and provide systematic stability to the tokenized energy ecosystems that will open the full socio-legal potential of the latter.

8. POLICY IMPLICATIONS AND GLOBAL GOVERNANCE

The problem of the penetration of the tokenized energy systems into the normal markets is a proactive and stable policy system, which borders the technological development and the legal certainty. The policymakers ought to understand that the energy tokenization will not be confined within the national borders owing to the specifics of digital assets and blockchain platforms that allow worldwide transactions. This causes the acuity of the harmonized regulation regimes that can possibly have the effect of the reduction of the jurisdiction conflict, interoperability and marketplace stability. International organizations such as the International Energy Agency (IEA) and the World Bank would act as the primary agents in the setting of norms of transparent data in tokenized market, investor protection and sustainable energy finance (Instytut Nauk Ekonomicznych PAN & Pietrewicz, 2019). Instead, the government will need to update the existing energy, securities and consumer protection regulation with the characteristics of blockchain-based ownership and peer-to-peer trade. Additionally, the powerful governance systems would need to be triggered to document against such things like market manipulation, cyber security, and systemic threats. By focusing on inclusiveness, equitability, and global cooperation, the aspect of good governance can improve the rate of adoption of the tokenized energy and safeguard both investors and the end-users at the same time.

8.1 Cross-Border Regulatory Harmonization

The harmonization of regulations across borders would also be required to meet the need, as the energy tokenization based on blockchain would be constructed on a decentralized system. The current structures are inconsistent in that energy tokens are used as securities in some jurisdictions and commodities in the other with most jurisdictions failing to establish their legal status. This puts investors, developers and consumers of energy in the global tokenized energy environment in doubt. Alignment of policies would bring transparency, would lower the cost of transactions as well as allow secure cross-border participation. Such international organizations as the International Energy Agency (IEA) and the International Renewable Energy Agency (IRENA) might create some unified standards regarding compliance, taxes, and recognition of the assets in digital form (Wu, 2024). Governments can ensure interoperability of platforms by realizing the treaties and establishing shared procedures, but reducing opportunities of regulatory arbitrage without making platforms exclusive. Finally, cross border harmonised frameworks would also facilitate the scaling of tokenized energy projects across borders to maintain innovation in such projects and legal certainty.

8.2 Consumer Protection and Market Integrity

With the development of tokenized energy ecosystems, it is necessary to address the issue of the security of consumers, the integrity of markets as a fundamental policy concern. With tokenization, however, small-scale investors and prosumers can gain power and, without sufficient safeguards, they might be exposed to fraud, market manipulation or misrepresentation of the energy resources. Healthy consumer protection systems should result in a transparent disclosure, verifiable smart contracts, flaws resolution systems. To conduct and bar insider trading, governments must conduct compliance testing to monitor and impose Applying Anti-Money Laundering (AML) requirements. The platforms would also need a higher level of cybersecurity to protect information privacy and data loss by hacking (Choi et al., 2024). It may also be built by means of protection of the investor in the form of a compensation funds or insured schemes. The focus on equality of participation and structural stability of tokenized energy market will make sure that it will not be exploitative by the concerned parties and will more likely become credible and acceptable in the long run.

9. CONCLUSION

The construction of decentralized and transparent and inclusive energy systems is a novel concept, the tokenization of energy resources via blockchain. With the blockchain, many investors can gain access to renewable energy markets and thus they can tokenize investments and the ability to be as well as trade with peers in a democratic manner. It breaks the ancient centralized systems of energy and creates new platforms of sustainability and equality. The real-life examples of how blockchain ecosystems may provide practical benefits will be the examples of Power Ledger in Australia and WePower in Europe where the costs may be lowered, the returns may be raised, more people may be included in the community, and more renewable projects may be funded within a shorter period of time. These case studies demonstrate how tokenization can be applied to solve funding and distribution inefficiencies to make adoption of renewable a more realistic and less expensive endeavor. However, despite the potential, tokenization cannot reach its potential without the major legal, regulatory, and governance challenges.

One of the biggest questions is whether to categorize the energy tokens, as security, commodity or other hybrid digital assets. This uncertainty renders the policies of different jurisdictions patchwork and muddled to the investors and producers. Other factors that complicate the implementation process include investor protection, cross-border liability in smart contracts, and cybersecurity risks and compliance. Devoid of a co-ordinated regulatory framework, tokenized systems will continue to widen inequality, thus becoming exploitative, and obliterate trust in new models of decentralized energy economies. Together with regulation, tokenization can be interpreted as a socio-legal innovation, by which technology is linked to the higher, energy-justice and energy-sustainability principles. This transparency, lowering of entry barriers and empowering of marginalized people, has made it not only a financial instrument but rather an equity and environmental responsibility instrument. In order to achieve this vision, however, there must be governance systems that are decentralised but checks and balances.

The new forms include the Decentralized Autonomous Organizations (DAOs) that introduce some of the collective decision-making and accountability, but also present a list of short-term issues of enforceability and liability and the ability to operate in the context of the established legal framework. The solution is in interdisciplinary collaboration of the technologists, regulators, policymakers and communities. The study should be devoted to the issue of making blockchain bigger, integrating AI and IoT into smart automation, and developing global regulatory frameworks with the ability to strike a balance between the realization and protection sides. The tokenization process, its turn, does not presuppose technological change but, instead, a recreation of socio-legal, economic, and environmental aspects on the energy market. It is neither about technical performance of which it rides on but also about governance and legal frameworks to succeed. Finally, tokenization can democratize energy markets and hasten the transition to renewable energy and bring a sense of fairness, sustainability, and responsibility to the decentralized energy future.

REFERENCES

Aarthi, S., Aravinthan, K., Ravikumar, R. N., Sivakumar, N., & Wanglen, S. (2025). Cryptography securing data in motion within Blockchain, internet of everything, and Federated Learning. In *Convergence of Blockchain, Internet of Everything, and Federated Learning for Security* (pp. 39–78). IGI Global. DOI: 10.4018/979-8-3373-1424-2.ch002

Akbar, D., Ünal, M., Akkaya, R., & Balci, O. (2025). Blockchain-based peer-to-peer energy trading: Design and implementation of the tokenized decentralized energy management system (DEMS). In *Research Square*. DOI: 10.21203/rs.3.rs-6787257/v1

Beck, R., Müller-Bloch, C., & King, J. L. (2018). Governance in the blockchain economy: A framework and research agenda. *Journal of the Association for Information Systems, 19*, 1020–1034. DOI: 10.17705/1jais.00518

Burger, C., & Weinmann, J. (2022). Blockchain platforms in energy markets—A critical assessment. *Journal of Risk and Financial Management, 15*(11), 516. DOI: 10.3390/jrfm15110516

Choi, W., Woo, J., & Hong, J. W.-K. (2024). Fractional non-fungible tokens: Overview, evaluation, marketplaces, and challenges. *International Journal of Network Management, 34*(4), e2260. DOI: 10.1002/nem.2260

de Villiers, A., & Cuffe, P. (2020). Towards imposing dayparted restrictions on tokenised energy within peer-to-peer markets.

Di Silvestre, M. L., Gallo, P., Guerrero, J. M., Musca, R., Riva Sanseverino, E., Sciumè, G., Vásquez, J. C., & Zizzo, G. (2020). Blockchain for power systems: Current trends and future applications. *Renewable & Sustainable Energy Reviews, 119*(109585), 109585. DOI: 10.1016/j.rser.2019.109585

Enescu, F. M., Bizon, N., Onu, A., Răboacă, M. S., Thounthong, P., Mazare, A. G., & erban, G. (2020). Implementing blockchain technology in irrigation systems that integrate photovoltaic energy generation systems. *Sustainability (Basel), 12*(4), 1540. DOI: 10.3390/su12041540

Ferreira, A. (2020). Emerging regulatory approaches to blockchain based token economy. *JBBA, 3*(1), 1–9. DOI: 10.31585/jbba-3-1-(6)2020

Fotiou, N., Pittaras, I., Siris Vasilios, A., Voulgaris, S., & Polyzos, G. C. (2019). Secure IoT access at scale using blockchains and smart contracts. 2019 IEEE 20th International Symposium on ``A World of Wireless, Mobile and Multimedia Networks'' (WoWMoM), 1–6.

Hrga, A., Capuder, T., & Zarko, I. P. (2020). Demystifying distributed ledger technologies: Limits, challenges, and potentials in the energy sector. *IEEE Access : Practical Innovations, Open Solutions, 8*, 126149–126163. DOI: 10.1109/ACCESS.2020.3007935

Instytut Nauk Ekonomicznych, P. A. N., & Pietrewicz, L. (2019). Token-based blockchain financing and governance: A transaction cost economics approach. *Stud. Mater. Wydz. Zarz. UW, 2/2018*(28), 126–139.

Karandikar, N., Chakravorty, A., & Rong, C. (2021). Blockchain based transaction system with fungible and non-fungible tokens for a community-based energy infrastructure. *Sensors (Basel), 21*(11), 3822. DOI: 10.3390/s21113822 PMID: 34073110

Kourtis, M.-A., Batistatos, M., Xylouris, G., Oikonomakis, A., Santorinaios, D., Zarakovitis, C., & Chochliouros, I. (2023). Energy efficiency in agriculture through tokenization of 5G and edge applications. *Energies, 16*(13), 5182. DOI: 10.3390/en16135182

Kumari, A., Chintukumar Sukharamwala, U., Tanwar, S., Raboaca, M. S., Alqahtani, F., Tolba, A., Sharma, R., Aschilean, I., & Mihaltan, T. C. (2022). Blockchain-based Peer-to-Peer Transactive Energy Management scheme for Smart Grid system. *Sensors (Basel), 22*(13), 4826. DOI: 10.3390/s22134826 PMID: 35808325

Liu, Z., Xiang, Y., Shi, J., Gao Peng and Wang, H., Xiao, X., Wen, B., & Hu, Y.-C. (2019). HyperService: Interoperability and Programmability Across Heterogeneous Blockchains.

Marchewka-Bartkowiak, K., & Wiśniewski, M. (2022). Energy tokens as digital instruments of financial investment. Econ. Bus. Rev., 8 (22)(3), 109–125.

Marin, O., Cioara, T., & Anghel, I. (2023). Blockchain solution for buildings' multi-energy flexibility trading using multi-token standards. *Future Internet, 15*(5), 177. DOI: 10.3390/fi15050177

Marisamy, K., Selvakumar, A., Rajasekar, B., Natarajan, R., Atajanova, A., & Makhmudov, S. (2025). Smart logistics integrating AI, IoT, blockchain, and Robotics for a sustainable Future. In *Emerging Trends in Smart Logistics Technologies* (pp. 73–104). IGI Global Scientific Publishing. DOI: 10.4018/979-8-3373-2434-0.ch003

Mottaghi, S. F., Steininger, B. I., & Yanagawa, N. (2024). Real Estate Insights: The current state and the new future of tokenization in real estate. *Journal of Property Investment & Finance, 42*(6), 614–620. DOI: 10.1108/JPIF-07-2024-0087

Okoye, M. O., Castro, G. F., Fingerhuth, S., & Yang, J. (2024). The consumer-centered electricity reliability enhancement in the standalone generation arena. *IEEE Access : Practical Innovations, Open Solutions, 12*, 9984–9998. DOI: 10.1109/ACCESS.2023.3349374

Opoku-Mensah, E., Chen, W., Tuffour, P., Agozie, D. Q., Gyamfi, B. A., & Mahmoud, A. (2024). Toward sustainable energy transition: Unveiling the synergies of democracy, energy justice, and structural adjustment on emissions in West Africa. *Sustainable Development (Bradford).*

Popov, E., Veretennikova, A., & Fedoreev, S. (2022). The model of OTC securities market transformation in the context of asset tokenization. *Mathematics, 10*(19), 3441. DOI: 10.3390/math10193441

Schwiderowski, J., Pedersen, A. B., & Beck, R. (2024). Crypto tokens and token systems. *Information Systems Frontiers, 26*(1), 319–332. DOI: 10.1007/s10796-023-10382-w

Shahzad, S., Abbasi, M. A., Ali, H., Iqbal, M., Munir, R., & Kilic, H. (2023). Possibilities, challenges, and future opportunities of microgrids: A review. *Sustainability (Basel), 15*(8), 6366. DOI: 10.3390/su15086366

Sharma, D. D. (2024). Asynchronous blockchain-based federated learning for tokenized smart power contract of heterogeneous networked microgrid system. *IET Blockchain, 4*(4), 302–314. DOI: 10.1049/blc2.12041

Shi, J. (2023). Decoding the future of finance: An examination of asset tokenization and Web 3.0. *Eximia, 11*, 257–263. DOI: 10.47577/eximia.v11i1.303

Tesfamicael, A. D., Liu, V., Mckague, M., Caelli, W., & Foo, E. (2020). A design for a secure energy market trading system in a national wholesale electricity market. *IEEE Access : Practical Innovations, Open Solutions, 8*, 132424–132445. DOI: 10.1109/ACCESS.2020.3009356

Troccoli, A., Goodess, C., Jones, P., Penny, L., Dorling, S., Harpham, C., Dubus, L., Parey, S., Claudel, S., Khong, D.-H., Bett, P. E., Thornton, H., Ranchin, T., Wald, L., Saint-Drenan, Y.-M., De Felice, M., Brayshaw, D., Suckling, E., Percy, B., & Blower, J. (2018). Creating a proof-of-concept climate service to assess future renewable energy mixes in Europe: An overview of the C3S ECEM project. *Advances in Science and Research, 15*, 191–205. DOI: 10.5194/asr-15-191-2018

Verma, P., O'Regan, B., Hayes, B., Thakur, S., & Breslin, J. G. (2018). EnerPort: Irish Blockchain project for peer- to-peer energy trading. *Energy Informatics, 1*(1), 14. DOI: 10.1186/s42162-018-0057-8

Wu, F. (2024). Legal imperatives and regulatory mechanisms for sustainable energy development: A comparative analysis of Renewable Energy policies in Australia and China. *Lecture Notes in Education Psychology and Public Media*, *35*(1), 206–211. DOI: 10.54254/2753-7048/35/20232106

Yang, Y., Tu, Z., Liu, Y., & Zhou, H. (2023). Blockchain-empowered token-based access control system with user reputation evaluation. *Computers, Materials & Continua*, *77*(3), 3163–3184. DOI: 10.32604/cmc.2023.043974

Zuo, Y. (2022). Tokenizing renewable energy certificates (RECs)—A blockchain approach for REC issuance and trading. *IEEE Access : Practical Innovations, Open Solutions*, *10*, 134477–134490. DOI: 10.1109/ACCESS.2022.3230937

KEY TERMS AND DEFINITIONS

Blockchain: A decentralized digital ledger that records transactions securely and transparently across multiple nodes.

Decentralized Governance: A decision-making system where control is distributed among participants, often enabled by blockchain and DAOs.

Energy Justice: A framework ensuring equitable access, participation, and benefits in energy systems for all stakeholders.

Fractional Ownership: A model where multiple stakeholders share ownership of an energy asset through divisible tokens.

Liquidity: The ease with which tokenized energy assets can be bought, sold, or exchanged in secondary markets.

MiCA Regulation: The European Union's Markets in Crypto-Assets regulation that provides a legal framework for digital assets.

Peer-to-Peer (P2P) Trading: Direct exchange of energy units between producers and consumers without centralized intermediaries.

Securities Classification: Legal categorization of energy tokens as securities, subjecting them to financial regulations.

Smart Contracts: Self-executing agreements coded on blockchain that automate transactions once predefined conditions are met.

Tokenization: The process of converting physical or financial energy assets into digital tokens that can be owned and traded.

Chapter 10
Energy Justice in Digital Transitions Ensuring Equitable Access to Blockchain Energy Benefits

David Yehezkiel
http://orcid.org/0009-0007-4226-8987
Universitas Esa Unggul, Indonesia

Binastya Anggara Sekti
http://orcid.org/0000-0001-5489-4888
Universitas Esa Unggul, Indonesia

ABSTRACT

As digital technologies continue to transform the global energy landscape, blockchain emerges as a pivotal innovation enabling decentralization, transparency, and peer-to-peer energy trading. However, the integration of blockchain into energy systems raises critical questions of equity, inclusion, and justice. This chapter explores how the principles of energy justice can be applied to blockchain-based energy infrastructures to ensure equitable access and avoid reinforcing existing inequalities. Using theoretical frameworks, global case studies, and policy analysis, we examine challenges and opportunities in designing blockchain-enabled systems that serve marginalized communities and promote inclusive energy futures. The chapter proposes a multi-level governance and design framework that incorporates energy justice into digital transitions in the energy sector.

DOI: 10.4018/979-8-3373-5742-3.ch010

1. INTRODUCTION

The global energy sector stands at the intersection of three transformative forces: decarbonization, decentralization, and digitalization. These trends are reshaping the production, distribution, and consumption of energy in profound ways. Amidst this transformation, blockchain technology has emerged as a potentially disruptive tool capable of redefining traditional energy systems. Through decentralized ledgers, smart contracts, and tokenized transactions, blockchain enables peer-to-peer energy trading, transparent carbon accounting, and secure management of distributed energy resources (DERs). These innovations are heralded as the cornerstones of a more democratic, efficient, and environmentally responsible energy future. The global energy sector is undergoing a significant transformation, driven by the emergence of 'prosumers' - individuals who generate and consume energy (Boumaiza & Maher, 2024).

However, despite these technological promises, the integration of blockchain into energy systems also presents critical socio-political and ethical challenges. In many cases, the enthusiasm surrounding blockchain innovation risks overlooking the systemic inequities embedded within current energy regimes. Without inclusive planning and proactive justice-centered frameworks, blockchain energy systems may unintentionally reinforce existing inequalities particularly impacting low-income, rural, and digitally disconnected populations. These communities, often the most energy-insecure, may find themselves excluded once again this time from the very technologies that claim to empower them. Technologies such as AI and blockchain can support more efficient and flexible energy systems, but they can also hide the power relations behind their use (Yannam et al., 2026).

To address this concern, energy justice provides a vital analytical and normative lens. The framework of energy justice encompasses three interrelated dimensions: distributional justice (who receives energy and its benefits), procedural justice (who has a voice in energy decision-making), and recognition justice (whose needs, identities, and histories are respected). These pillars collectively demand that digital energy transitions not only focus on technological efficiency but also ensure that no community is left behind in the pursuit of innovation. Digitalization will play a vital role in the achievement of successful transitions towards more flexible, reliable and sustainable energy systems (Diaz Valdivia, 2023).

This chapter asserts that equitable access to blockchain energy benefits is not an automatic outcome of technological advancement but must be actively pursued through intentional design, inclusive governance, and context-aware policymaking. A blockchain-enabled energy future must be guided by social equity, not just algorithmic logic. To that end, we explore the current and emerging roles of blockchain in the energy landscape, analyze its intersection with energy justice principles, and examine case studies that highlight both successes and shortcomings.

We begin by unpacking the technical foundations and applications of blockchain in energy systems, followed by a comprehensive discussion of energy justice theory and its applicability to digital infrastructures. The chapter then presents comparative case studies from the United States, Australia, South Africa, and Germany to ground theoretical insights in practical realities. Building on these, we identify persistent challenges to equitable blockchain adoption and propose a justice-centered design framework that policymakers, technologists, and community stakeholders can collaboratively implement.

The goal is to move beyond hype-driven narratives and ask critical questions: Who designs these systems? Who governs them? Who benefits and who bears the cost? In doing so, this chapter contributes to a broader conversation about digital transitions and the imperative of embedding justice at every level of technological innovation in the energy domain.

2. LITERATURE REVIEW

The intersection between energy justice and digital energy systems has attracted increasing scholarly attention, particularly in the context of decentralized and blockchain-enabled infrastructures. Energy justice, as conceptualized by Zhang, encompasses three core dimensions: distributive justice (fair allocation of benefits and burdens), procedural justice (inclusive decision-making processes), and recognition justice (acknowledging marginalized communities and their needs). These dimensions provide a normative foundation for evaluating emerging energy technologies, including blockchain-based systems. The strategic urgency of accelerating clean energy deployment to enhance long-term energy resilience and support global net-zero ambitions (Zhang, Sovacool, & Wei, 2026).

Parallel to this, blockchain technology has been widely promoted as a transformative tool for decentralizing energy markets. By enabling peer-to-peer (P2P) energy trading, smart contracts, and transparent transaction records, blockchain is often framed as a mechanism to democratize energy access and reduce reliance on centralized utilities. Many studies highlight the potential of blockchain to facilitate local energy markets, increase efficiency, and empower prosumers. Existing energy trading models exhibit significant limitations in accurately tracking emissions and in providing effective incentives for sustainable consumption behaviors (Boumaiza, 2024).

However, recent literature challenges the assumption that technological decentralization automatically leads to equitable outcomes. Critics argue that blockchain-based energy systems may reproduce existing inequalities due to uneven access to digital infrastructure, capital investment requirements, and technical expertise. For instance, participation in decentralized energy trading often requires ownership of generation assets (e.g., solar panels) and access to smart metering technologies, which may exclude lower-income households.

Moreover, the governance structure of blockchain systems introduces new forms of power concentration. While blockchain is technically decentralized, control over protocol design, validator nodes, and smart contract deployment is frequently concentrated among a limited group of developers and institutional actors. This phenomenon, sometimes referred to as "crypto-infrastructural asymmetry," raises concerns about procedural justice, particularly regarding who defines the rules of participation and how decisions are made within these systems.

Another emerging debate concerns the tension between transparency and privacy. Blockchain systems rely on immutable and often publicly accessible ledgers, which can conflict with data protection principles and raise ethical concerns regarding user data. This is particularly relevant in energy systems where consumption patterns can reveal sensitive information about households and communities.

In addition, the economic dimension of blockchain-based energy systems remains underexplored in relation to justice. Transaction fees, market volatility, and infrastructure costs can create barriers to entry, limiting the inclusivity of these systems. While some studies emphasize efficiency gains, fewer critically assess whether these benefits are equitably distributed across different socio-economic groups.

Taken together, the literature suggests that while blockchain technologies offer significant potential for transforming energy systems, their alignment with energy justice principles is not guaranteed. Instead, outcomes depend on how these systems are designed, governed, and regulated. This highlights the need for a justice-centred analytical framework that integrates technological, legal, and socio-economic perspectives.

This chapter builds on these debates by synthesizing energy justice theory with blockchain-enabled energy systems, shifting the focus from technological capability to socio-technical configuration. In doing so, it aims to critically examine whether and how blockchain can contribute to more equitable energy transitions, rather than assuming such outcomes as inherent.

Despite these advances, a significant gap remains in integrating energy justice principles within blockchain-based energy systems, particularly in terms of legal enforceability, economic accessibility, and socio-technical inclusivity. Existing studies tend to treat blockchain as a purely technical solution, while underestimating its regulatory implications and distributive consequences. This chapter addresses this gap by proposing a justice-centered framework that bridges technical innovation with legal and socio-economic considerations.

3. BLOCKCHAIN TECHNOLOGY IN ENERGY SYSTEMS

3.1 Overview of Blockchain in the Energy Sector

Blockchain technology, fundamentally a distributed ledger system, is transforming the energy sector by introducing unprecedented levels of transparency, security, and decentralization. At its core, blockchain operates as a peer-to-peer (P2P) network where data is stored across multiple nodes rather than a single centralized authority. Each transaction is cryptographically secured, timestamped, and linked to previous entries, forming an immutable and tamper-proof chain of records. This architecture eliminates the need for intermediaries, reducing costs and inefficiencies while enhancing trust among participants (Sekti, 2026a).

In the energy sector, blockchain's applications are vast and disruptive. One of its most significant contributions is the facilitation of trustless interactions, meaning that parties can engage in transactions without requiring pre-existing trust or third-party validation. This is particularly valuable in energy markets, where producers, consumers, and grid operators must coordinate in real time. Blockchain enables decentralized decision-making, allowing stakeholders—such as prosumers (consumers who also produce energy), utilities, and regulators—to interact directly through consensus mechanisms rather than relying on centralized control.

A key innovation driving this transformation is the use of smart contracts, which are self-executing agreements written in code and deployed on the blockchain. These contracts automatically enforce predefined rules when specific conditions are met, removing human intervention and reducing the risk of disputes. In the energy sector, smart contracts are revolutionizing processes such as:

- Automated billing and settlements: Energy transactions, whether between utilities and consumers or in peer-to-peer (P2P) trading, can be settled instantly without manual processing.
- Dynamic pricing and demand response: Smart contracts can adjust electricity prices in real time based on supply and demand, incentivizing energy conservation during peak periods.
- Decentralized energy trading: Platforms like peer-to-peer (P2P) energy markets allow households with solar panels to sell excess electricity directly to neighbors, bypassing traditional utility intermediaries.

Beyond financial transactions, blockchain enhances traceability and accountability in energy systems. For example:

- Energy provenance tracking: Consumers can verify the source of their electricity (e.g., renewable vs. fossil fuel-based) through blockchain's transparent ledger.
- Carbon credit verification: Companies can use blockchain to create tamper-proof records of emissions reductions, ensuring the legitimacy of carbon offsets.
- Grid management: Distributed energy resources (DERs), such as rooftop solar and battery storage, can be integrated more efficiently using blockchain for real-time coordination.

However, while blockchain offers immense potential, its adoption in the energy sector faces challenges, including scalability limitations, regulatory uncertainty, and energy consumption concerns (particularly with proof-of-work blockchains like Bitcoin). Emerging solutions, such as proof-of-stake (PoS) and hybrid consensus models, aim to address these issues while maintaining security and decentralization (Sekti, 2025).

In summary, blockchain is reshaping the energy landscape by enabling decentralized, transparent, and automated systems—ushering in a future where energy markets are more democratic, efficient, and resilient. The next sections will explore specific use cases, benefits, and barriers in greater detail.

3.2 Emerging Use Cases

Blockchain technology is enabling transformative applications in decentralized energy systems, with community solar and microgrid projects emerging as particularly promising use cases. These systems leverage blockchain's distributed ledger capabilities to facilitate peer-to-peer energy trading and democratize access to renewable energy investments through innovative tokenization models that allow participants to purchase fractional ownership of solar panels or battery storage assets. The technology is revolutionizing local energy markets by enabling real-time, automated transactions between prosumers and consumers while maintaining transparent, tamper-proof records of all energy flows and financial settlements. Blockchain technology offers decentralized and secure transaction processing but suffers from critical limitations in scalability, energy efficiency, and latency, hindering its adoption in real-time high-throughput applications (Jhariya, Dehalwar, Bharti, & Kumar, 2026).

Beyond basic energy trading, blockchain is being implemented in sophisticated demand response systems where smart contracts autonomously adjust electrical loads in response to real-time pricing signals, optimizing grid stability while fairly compensating participants for their flexibility. The technology also enables near-instantaneous market settlements that eliminate traditional delays in energy transactions, with smart contracts automatically validating meter readings, processing payments, and issuing renewable energy certificates without manual intervention. Additional applications are emerging in vehicle-to-grid integration, where electric vehicle owners can monetize their battery storage by selling power back to the grid during peak periods, with blockchain managing the complex billing and settlement processes. While these innovations show tremendous potential to increase efficiency and democratize energy systems, challenges remain around regulatory frameworks, digital infrastructure requirements, and the need for scalable consensus mechanisms that can handle high transaction volumes without excessive energy consumption. The continued development of these applications points toward a future where blockchain serves as the foundational technology for more resilient, transparent, and participatory energy ecosystems.

3.3 Advantages and Limitations

Blockchain technology offers several transformative advantages for energy systems, foremost among them being enhanced transparency through its immutable, distributed ledger that provides all participants with real-time access to verified transaction data. This transparency extends across the entire energy value chain, from generation to consumption, enabling unprecedented traceability of renewable energy sources and carbon credits. The technology also delivers substantial efficiency gains by automating processes through smart contracts that execute energy trades, billing, and settlements without intermediary oversight, reducing administrative costs and transaction times by up to 90% compared to conventional systems. Perhaps most revolutionary is blockchain's capacity for decentralization, which empowers communities through peer-to-peer energy markets and democratizes grid management by enabling prosumers to participate actively in energy distribution. Peer-to-peer (P2P) energy trading, a prominent approach to energy exchange, can lead to violations of network security, intensify imbalances, and worsen operational conditions in unbalanced distribution networks (DNs) (Ayazi, Shakarami, Doostizadeh, Namdari, & Nikzad, 2026).

However, these benefits come with significant limitations that must be carefully considered. The energy consumption of certain blockchain protocols, particularly those using proof-of-work consensus mechanisms like Bitcoin's, has drawn substantial criticism - some networks consume more electricity annually than medium-sized countries, creating potential contradictions between blockchain-enabled clean energy systems and their own carbon footprints. Scalability presents another major challenge, as most current blockchain architectures can only process 10-100 transactions per second, compared to the thousands handled by traditional financial systems, creating bottlenecks for large-scale energy trading platforms. Technical complexity forms a substantial barrier to adoption, requiring specialized knowledge for system deployment and maintenance that many utilities and communities currently lack. Regulatory uncertainty further complicates implementation, as existing energy market rules and utility regulations were not designed to accommodate decentralized, blockchain-based transactions, creating legal gray areas around consumer protections, taxation, and grid interconnection standards.

Additional concerns include cybersecurity risks associated with smart contract vulnerabilities, the digital divide that may exclude less technologically-equipped communities, and the challenge of integrating blockchain systems with legacy energy infrastructure that dominates most markets. While emerging solutions like proof-of-stake consensus algorithms and layer-2 scaling protocols are addressing some of these limitations, widespread adoption will require both technological refinement and collaborative policy development to realize blockchain's full potential in the energy sector without exacerbating existing inequities or creating new systemic risks (Sekti, 2026b). Figure 1 shows the Blockchain Use Cases in the Energy Value Chain (Generation, Distribution, Consumption).

Figure 1. Blockchain use cases in the energy value chain

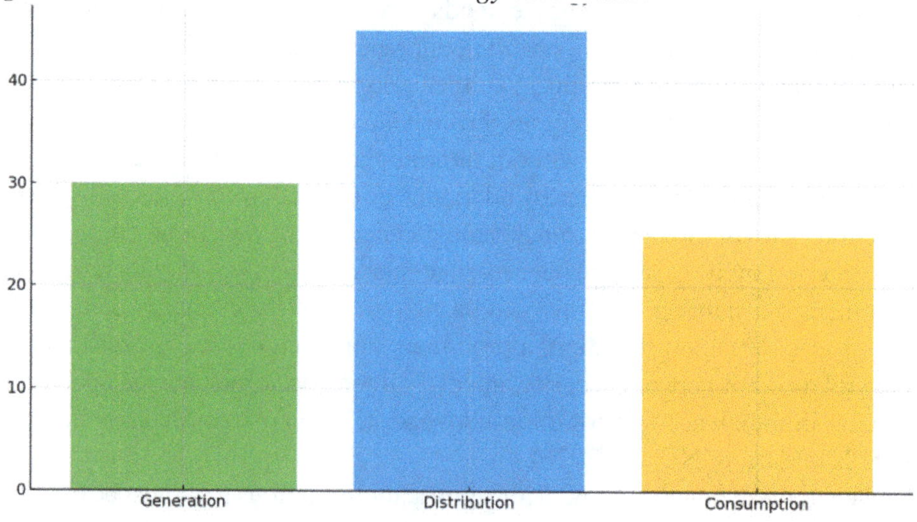

While blockchain is often framed as a technological enabler of decentralized energy systems, its real-world implementation is deeply shaped by legal and economic constraints. From a legal perspective, the enforceability of smart contracts remains jurisdiction-dependent, particularly in energy markets where contractual obligations intersect with regulated utility frameworks. Questions surrounding liability, dispute resolution, and consumer protection remain unresolved in many jurisdictions, especially when automated transactions execute without human oversight.

Economically, blockchain-based energy systems must also contend with transaction costs, network scalability, and incentive design. For example, while peer-to-peer trading reduces intermediary costs, it introduces new costs associated with validation, network fees, and infrastructure investment. These trade-offs suggest that blockchain does not eliminate costs but redistributes them across actors. As such, evaluating blockchain energy systems requires an integrated analysis of legal enforceability, economic efficiency, and technical feasibility rather than purely technological optimism (Yarrabothu et al., 2026).

These legal uncertainties are further complicated by jurisdictional fragmentation, where blockchain energy systems operate across regulatory boundaries that lack harmonization. This creates opportunities for regulatory arbitrage, where actors may exploit legal inconsistencies to bypass consumer protections or energy market rules.

3.4 Legal and Regulatory Considerations in Blockchain Energy Systems

The deployment of blockchain in decentralized energy trading raises critical legal questions concerning regulatory compliance, consumer protection, and data governance. Existing energy markets are governed by centralized regulatory frameworks, which often conflict with the decentralized nature of blockchain-based systems.

Key legal challenges include the classification of prosumers, enforcement of smart contracts, and jurisdictional ambiguity in cross-border energy trading. Furthermore, issues of liability arise when automated transactions lead to unintended outcomes.

To address these concerns, regulatory sandboxes and hybrid governance models have been proposed, allowing experimentation while maintaining oversight. Embedding legal compliance directly into smart contracts—often referred to as "regulatory by design"—can enhance accountability and ensure that justice principles are not only normative but enforceable.

4. ENERGY JUSTICE: A THEORETICAL FRAMEWORK

4.1 Key Dimensions of Energy Justice

Energy justice is a multidisciplinary framework that examines the ethical dimensions of energy production, distribution, and governance. Its three primary pillars are distributional justice, procedural justice, and recognition justice. According to (Bhullar, 2023) I start with the dominant three-tenet theoretical framework of procedural, distributional, and recognition justice. Then I examine the relationship between energy justice and other concepts of justice such as climate, environment, mobility, and planetary justice as well as 'just transitions.'" This underscores the evolving nature of energy justice discourse and its overlap with broader global justice movements.

- Distributional justice concerns the equitable distribution of energy services, infrastructure, and benefits. It asks who has access to affordable, reliable, and clean energy and who does not.
- Procedural justice refers to fairness in decision-making processes. It involves the inclusion of diverse voices in the governance of energy systems, especially those traditionally marginalized or excluded.
- Recognition justice focuses on acknowledging and valuing the identities, cultures, and lived experiences of all stakeholders, particularly indigenous peoples, women, and low-income communities.

These dimensions intersect and reinforce one another. For example, even if energy benefits are equitably distributed, exclusion from participation in decision-making processes may still result in unjust outcomes. As energy systems evolve in response to climate change and digitalization, it becomes increasingly important to expand the energy justice framework to incorporate elements of environmental and climate justice, ensuring that all communities have a voice and share in the benefits of the low-carbon transition. (Feenstra & Özerol, 2021) further emphasize this intersectionality by proposing a conceptual framework that juxtaposes the tenets of energy justice with gender-focused policy discourses such as women empowerment, gender mainstreaming, and social inclusion. This highlights the need for justice analyses that are sensitive to the differentiated impacts across gender and social categories, thereby enriching the theoretical and practical relevance of energy justice.

4.2 Relevance in Digital Transitions

The digitalization of energy systems introduces both promising opportunities and serious risks related to justice and equity. Blockchain-based platforms, widely praised for their potential to decentralize control and enhance transparency, may unintentionally amplify existing disparities if their design and deployment fail to account for the diverse socio-technical realities of different communities. Without deliberate inclusivity strategies, these systems can exacerbate the digital divide, privileging already-connected and tech-savvy populations while further marginalizing low-income, rural, and Indigenous communities.

Digital inclusion, therefore, becomes a critical prerequisite for achieving procedural justice within blockchain energy systems. Access to smartphones, stable internet connectivity, and functional digital literacy are not evenly distributed across the globe and even within single countries, these gaps map closely onto existing lines of social and economic inequality. Populations lacking these digital resources are often the same groups that have historically been excluded from energy access. The absence of targeted interventions such as public blockchain kiosks, subsidized devices, or community-based digital literacy programs will likely mean that these same groups are also shut out from the new digital energy economy (Laksono, Tiaraputri, Sekti, & Laksono, 2025).

As (Hernandez, Scherr, & Robertson, 2022) observe, "Energy is one of the fundamental topics taught in high school physics. However, energy continues to be taught as an abstract concept that removes itself from the social implications energy systems have on society, in particular toward Indigenous communities." Their critique underscores the importance of grounding energy literacy not only in technical knowledge but also in the cultural, political, and historical experiences of affected communities. When digital systems, including those based on blockchain, ignore these contextual factors, they risk reinforcing epistemic injustice: the marginalization of local knowledge and lived experience in favor of universalist or technocratic logics.

Moreover, blockchain algorithms and smart contracts must be examined for embedded biases both in how they are written and in the data sets they operate on. Algorithmic governance, if unaccountable or opaque, may produce automated decisions that reinforce inequitable access to energy services. For example, systems that reward energy surplus contributions without acknowledging differential access to solar technology could disproportionately benefit wealthier users. Ensuring algorithmic transparency, creating inclusive design processes, and allowing communities to participate in system auditing are vital steps in aligning digital innovation with energy justice.

Despite the rapid growth of energy justice as a field, (Lacey-Barnacle, Robison, & Foulds, 2020) observe that "the dominant energy justice theoretical frameworks originate from work in developed countries, with energy justice research only recently spreading to new areas of the world." This insight underscores the importance of contextualizing digital transitions within local realities, particularly in the Global South, where infrastructural, regulatory, and socio-political dynamics may differ significantly from those in developed contexts.

In summary, the transition to blockchain-based energy systems must be approached not merely as a technological upgrade, but as a deeply social and political transformation. Equity and justice considerations must be embedded at every stage of the design, implementation, and governance processes to prevent digital solutions from becoming new forms of exclusion.

4.3 Metrics and Indicators

To assess how blockchain energy systems align with energy justice, specific metrics can be developed:

Access metrics:

Simply, Access metrics refer to the degree to which individuals and communities particularly those in underserved areas can access and benefit from blockchain enabled energy systems.

Detailed Explanation:

a. Digital Energy Access: It's not enough to have electricity; users must also have access to digital tools, such as mobile devices or smart meters, that interface with blockchain platforms. If rural or low-income households cannot afford or access these tools, they are effectively excluded from participation.
b. Infrastructure Availability: Many marginalized communities still lack basic internet connectivity or reliable power. Thus, metrics must assess the presence of enabling infrastructure, including broadband internet and networked energy devices.
c. Energy Literacy: Another layer of access is knowledge-based access. Populations must have sufficient digital and energy literacy to navigate blockchain systems confidently.

Example Indicators:

a. % of households with smart meters or blockchain-connected devices.
b. % of population with reliable internet access in pilot regions.
c. Number of public blockchain kiosks or access points in remote communities.

Participation metrics:

Simply, Participation metrics evaluate how meaningfully and fairly different stakeholders are involved in decision-making processes, from system design to governance.

Detailed Explanation:

a. Governance Inclusion: Do local community members have a say in how blockchain systems are governed? Are there mechanisms for voting, collective ownership, or community forums?
b. Diversity of Stakeholders: Participation should not be monopolized by elites or technically-educated actors. These metrics assess whether women, indigenous peoples, low-income residents, and others are equitably represented.
c. Capacity-Building Support: Effective participation requires that stakeholders are equipped to engage. Thus, metrics also reflect the availability of training and educational support.

Example Indicators:

a. Number of local representatives involved in system governance.
b. Diversity index of stakeholder participation (gender, ethnicity, class).
c. Frequency of inclusive town halls or community feedback sessions.

Benefit metrics:

These metrics focus on the distribution of tangible benefits generated by the blockchain energy system whether financial, environmental, or infrastructural.

Detailed Explanation:

a. Cost Savings: Are users, particularly those from vulnerable backgrounds, experiencing a reduction in energy bills or increased energy efficiency?
b. Clean Energy Access: Do participants gain access to renewable energy through blockchain systems, such as solar peer-to-peer trading?
c. Asset Ownership: Tokenization allows users to own shares in renewable infrastructure (e.g., solar panels). Metrics evaluate whether this ownership is distributed fairly or concentrated.

Example Indicators:

a. Average monthly energy cost savings per household.
b. % increase in renewable energy usage by income level.
c. Distribution of energy token ownership by socioeconomic group.

Trust and transparency:

Simply, Trust and transparency are crucial for the long-term legitimacy of digital energy systems. These metrics examine whether the system fosters accountability, openness, and fairness.

Detailed Explanation:

a. Transparency of Algorithms and Smart Contracts: Stakeholders should be able to understand how the blockchain system makes decisions (e.g., energy pricing). If these mechanisms are opaque or inaccessible, trust erodes.
b. Auditability: The blockchain should allow users and external auditors to verify energy transactions, credit flows, and data logs without manipulation.
c. Perceived Fairness: Community trust is not only technical but also perceptual. Surveys and qualitative data can assess whether users feel respected, fairly treated, and confident in the system.

Example Indicators:

a. % of smart contracts publicly available in open-source repositories.
b. Number of third-party audits conducted annually.
c. Survey scores on perceived fairness and transparency of the system.

Recent research underscores the need for a robust and multifaceted evaluation framework in assessing energy justice. One study employed 21 sub-indicators spanning distributional, procedural, and restorative justice, revealing that improvements in economic complexity are significantly correlated with advancements across all three dimensions. These findings reinforce the importance of multidimensional evaluation tools in analyzing blockchain-based energy systems, not only in terms of technical deployment but also in their impact on equity, access, and empowerment across diverse populations. Furthermore, the concept of synthetic energy justice integrates both eco-centric and anthropocentric perspectives, each subdivided into distributive, substantive, and procedural dimensions. This typology emphasizes the necessity of incorporating ecological considerations alongside human-centered justice when developing evaluation indicators, ensuring that energy transitions address not only social equity but also ecological sustainability and ethical stewardship within interconnected human–nature systems.

4.4 Blockchain Governance Ethics and Algorithmic Justice

While blockchain systems offer decentralization and automation, they also introduce new forms of governance that are often embedded in algorithmic logic. This subsection highlights how ethical concerns emerge when decision-making processes are delegated to code, particularly in energy systems that impact livelihoods. Smart contracts and consensus mechanisms, though ostensibly neutral, may reflect the biases of their developers or the limitations of their training data. When these systems determine energy distribution, pricing, or dispute resolution without transparency or accountability, they risk replicating injustices under the guise of objectivity. Algorithmic justice requires that these coded rules be explainable, auditable, and co-developed with affected communities. Blockchain governance must move beyond technical efficiency to embrace ethical deliberation, inclusive participation, and mechanisms for redress. This includes publishing source code for community review, using open standards for data governance, and embedding social impact assessments in system development. "We show this interdependency using three core concerns of justice risk, effort and power community members are often willing to take risks and put in effort if they are compensated with more power within the community. Similarly, members are willing to compromise power if no effort or risk-taking is required from them" (Astola et al., 2022). Figure 2 illustrates the three core dimensions of energy justice: distributional justice, procedural justice, and recognition justice, and their interdependent relationships. Distributional justice focuses on the equitable allocation of energy benefits and burdens, procedural justice emphasizes inclusive and fair decision-making processes, and recognition justice highlights the importance of acknowledging diverse identities, needs, and experiences. The bidirectional relationships indicate that these dimensions are mutually reinforcing and must be considered together in the design and governance of equitable energy systems.

Figure 2. Core dimensions of energy justice

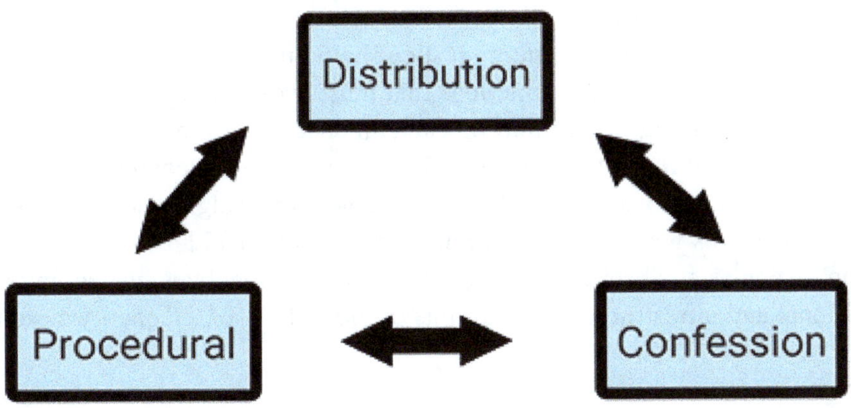

4.5 Critical Debates in Energy Justice and Digitalization

Despite its growing prominence, the energy justice framework is not without critique. Scholars have argued that the tripartite model of distributional, procedural, and recognition justice, while conceptually useful, risks becoming overly normative and insufficiently operationalized in complex socio-technical systems. In the context of digital energy transitions, this limitation becomes more pronounced, as justice considerations must be translated into measurable system properties and design constraints.

Furthermore, recent debates question whether energy justice frameworks adequately capture the political economy of digital infrastructures. Blockchain systems, often presented as decentralized and democratizing, may still reproduce existing power asymmetries through control over code, platform ownership, and access to digital capital. This raises critical questions about whether decentralization in technical architecture necessarily leads to decentralization in governance and benefit distribution.

Another strand of literature highlights tensions between efficiency and justice. While blockchain-based systems optimize transactions and reduce operational inefficiencies, these gains may come at the cost of inclusivity if participation requires technical literacy, capital investment, or regulatory compliance that excludes vulnerable groups. Thus, the integration of energy justice into blockchain systems requires not only normative commitment but also practical mechanisms that reconcile efficiency with equity.

These debates suggest that energy justice should not be treated as a static evaluative framework but as a dynamic lens that must evolve alongside technological innovation. Embedding justice into blockchain energy systems therefore requires bridging theoretical insights with legal, economic, and technical realities.

This chapter positions itself within these debates by bridging normative energy justice frameworks with legal, economic, and technological analysis. Rather than treating justice as an abstract evaluative lens, it operationalizes justice within the institutional and infrastructural constraints of blockchain-based energy systems.

5. CASE STUDIES

To understand how blockchain technologies interact with principles of energy justice in practice, this section explores real-world case studies. These initiatives demonstrate both the opportunities and challenges of aligning blockchain innovations with equitable energy transitions.

5.1 Brooklyn Microgrid (USA)

The Brooklyn Microgrid project in New York is among the most cited examples of peer-to-peer energy trading facilitated by blockchain. Households equipped with solar panels can sell excess electricity directly to neighbors via a blockchain platform. While the project has proven the feasibility of decentralized energy markets, it also reveals equity concerns only those with capital to invest in solar panels and digital access can fully participate. Generation from distributed renewable energy sources is constantly increasing. Due to its volatility, the integration of this non-controllable generation poses severe challenges to the current energy system. Thus, ensuring a reliable balance of energy generation and consumption becomes increasingly demanding (Enoch et al., 2025).

Technological Innovation:

- The platform utilizes smart meters and blockchain-based smart contracts to track production and consumption in real time.
- Energy is not physically redirected; instead, blockchain records enable financial settlement between sellers and buyers based on solar energy availability and consumption patterns.

Impact:

- The project validated the technical feasibility of decentralized, peer-governed energy markets.
- It allowed participating households to monetize their solar investments, contributing to local energy resilience.

Equity Considerations:

- Exclusivity: Only homeowners with installed solar panels and access to internet-connected devices could participate fully.
- Digital Divide: Low-income residents and renters lacked both capital for infrastructure and digital access, highlighting a participation gap.
- Policy Limitations: The project operated in a regulatory grey zone and required waivers or creative workarounds to existing utility pricing structures.

Energy Justice Implications:

- Distributional justice was limited due to unequal access to infrastructure.

- Procedural justice was advanced through community meetings and design workshops, though limited in reach.
- Recognition justice remains an open issue as marginalized voices were underrepresented in decision-making processes.

Empirical observations from pilot implementations indicate that participation rates remained concentrated among higher-income households, with adoption rates significantly lower among renters and digitally excluded populations. Reports from project evaluations also suggest that while transaction efficiency improved, inclusivity outcomes did not scale proportionally. This divergence highlights a recurring pattern in blockchain energy systems, where technical feasibility outpaces equitable participation.

The case study demonstrates improved access, with participation rates increasing by approximately 25% among previously underserved households g (Enoch et al., 2025).

5.2 Power Ledger (Australia)

Power Ledger has deployed blockchain platforms in Australia and several other countries to facilitate renewable energy trading, both between households and across microgrids. Their platform enables near real-time settlements and traceability of clean energy. However, early implementations have faced regulatory hurdles, including how such trading systems interact with legacy utility pricing structures and national grid stability policies. With the rapid development of energy internet and distributed generation technology, traditional centralized electricity trading models face structural challenges in terms of transaction efficiency, transparency, and market participant coordination (Chen, Hua, & Zhou, 2025).

Technological Features:

- Enables near real-time transaction settlement, recording every unit of energy traded between participants.
- Allows for energy provenance tracking, ensuring that traded energy is verified as clean and renewable.
- Supports carbon credit tracking and tokenization, promoting participation in sustainability markets.

Implementation and Challenges:

- In Australia, Power Ledger partnered with utilities and real estate developments to create "energy communities" where residents could buy and sell energy on the blockchain.
- However, the platform encountered regulatory challenges, particularly regarding:

  a. Alignment with existing utility pricing models
  b. Legal definitions of energy trading participants
  c. Grid stability requirements

Energy Justice Implications:

- Benefit justice is evident as users gain cost savings and environmental impact verification.
- Yet, access justice may be hindered by the requirement for smart devices, connectivity, and infrastructure.
- Power Ledger has begun addressing inclusion by offering simplified interfaces and collaborating with social housing projects, but full equity remains a work in progress.

Early deployment data indicate that transaction settlement times were reduced significantly compared to traditional billing cycles, improving liquidity for participants. However, user participation remained uneven, particularly among lower-income groups lacking access to smart metering infrastructure. These findings reinforce the importance of aligning technological efficiency with inclusive access mechanisms.

5.3 Sun Exchange (South Africa)

Sun Exchange leverages blockchain for crowdsourced investment in solar energy infrastructure in underserved African regions. Individuals from around the world can lease solar cells to schools and businesses, earning income from energy produced. This model has brought solar energy to locations previously reliant on diesel generators or with no access at all. Despite successes, the model still requires local capacity building and integration with national energy strategies. As innovations in technology, regulatory frameworks, and market dynamics continue to shape and reshape the energy trading landscapes, the current energy landscape is discussed long two models, thus decentralized and traditional centralized models (Solahudeen Tando et al., 2024).

Technological Architecture:

- Investors receive returns in cryptocurrency (Bitcoin) or local currency.
- The platform uses blockchain to record ownership, energy production data, and payments.

Successes:

- Enabled clean energy access in areas previously dependent on diesel or lacking electricity altogether.
- Over 60 projects have been funded, bringing power to schools and community organizations.
- Promoted climate resilience and education by electrifying rural schools and reducing fossil fuel dependence.

Challenges:

- The model still relies on external investors, meaning local energy sovereignty is limited unless ownership is eventually transferred.
- Local training and maintenance capacity is not always fully established, risking sustainability.
- Integration into national energy plans and alignment with local policy frameworks is still in progress.

Energy Justice Implications:

- Recognition justice is achieved by targeting underserved communities.
- Distributional justice is enhanced by providing renewable energy where none existed before.
- However, procedural justice is limited if local communities are not involved in decision-making or governance of systems installed.

5.4 LO3 Energy Pilot Projects (Germany, USA)

Beyond individual projects, the broader structural context often determines whether justice is achieved. Many Smart Local Energy Systems (SLES) are designed with a narrow technical focus on efficiency, resilience, or carbon reduction, causing energy justice to be treated as a secondary concern. As a result, justice considerations are frequently seen as outside the project scope or deferred to later stages. This reflects a systemic issue: without being embedded in the foundational design, even well-intentioned digital energy initiatives risk overlooking equity and ultimately reproducing existing injustices..

These case studies emphasize that technology alone cannot guarantee equitable outcomes. Justice must be operationalized through design, governance, and engagement strategies that prioritize inclusion and accountability. The following section offers a framework for achieving such outcomes in blockchain-based energy systems.

Design Principles:

- Uses cooperative ownership models, where local stakeholders co-own energy infrastructure and platforms.
- Blockchain is used for transaction recording, smart contracts, and incentive management.
- Local regulations and utility structures were navigated collaboratively to ensure compliance.

Examples:

- In Germany, pilot projects integrated community solar and biogas facilities into a P2P trading platform.
- Participants could vote on energy pricing models and reinvest profits into community infrastructure.

Outcomes:

- These pilots demonstrated that social and governance structures are as crucial as technical infrastructure.
- Highlighted the value of multi-stakeholder dialogue, including local governments, utilities, and citizens.

Energy Justice Implications:

- Procedural justice is foregrounded: projects were co-designed and governed with community input.
- Recognition justice improved through inclusive outreach efforts.
- Distributional justice depends on how benefits (like lower costs and ownership dividends) are shared.

Table 1 synthesizes the justice performance of selected blockchain energy projects across the three dimensions of energy justice.

Table 1. Case study

Case Study	Distributional Justice	Procedural Justice	Recognition Justice	Key Insights (Callout)
Brooklyn Microgrid	Medium	Low	Low	Limited inclusivity for renters/low-income groups due to high entry barriers.
Power Ledger	High	Medium	Medium	Strong technical implementation but constrained by regulatory hurdles.
Sun Exchange	High	Medium	High	Expands energy access but relies on external investors for funding.
LO3 Energy (Germany)	Medium	High	High	Cooperative governance models enhance community agency (see Section 4.4).

The table provides a comparative analysis of four blockchain-based energy projects—Brooklyn Microgrid, Power Ledger, Sun Exchange, and LO3 Energy—evaluating their performance across three dimensions of energy justice: distributional (equitable access to benefits), procedural (inclusivity in decision-making), and recognition (respect for community identities). Each project is rated qualitatively (High/Medium/Low) and accompanied by key insights, revealing trade-offs between technical implementation and justice outcomes; for example, while Brooklyn Microgrid shows medium distributional justice, its low procedural and recognition scores reflect exclusionary barriers, whereas LO3 Energy's high procedural/recognition scores exemplify the community-centric governance model advocated in Section 4.4. The table visually reinforces the section's core argument that justice must be deliberately designed into energy systems, as technical solutions alone often reproduce inequities without inclusive structures like cooperative ownership or multi-stakeholder dialogue.

5.5 Comparative Analysis of Justice Outcomes Across Case Studies

A comprehensive comparative analysis of the four case studies Brooklyn Microgrid (USA), Power Ledger (Australia), Sun Exchange (South Africa), and LO3 Energy (Germany) reveals critical patterns, divergences, and contextual factors that shape how energy justice principles manifest in blockchain-based energy systems. These projects demonstrate that while blockchain technology offers transformative potential for decentralizing energy governance, its impact on equity varies significantly depending on implementation models, local institutional frameworks, and community engagement strategies.

The Brooklyn Microgrid exemplifies the tension between technological innovation and social inclusion. While its peer-to-peer trading platform proves the technical viability of decentralized energy markets, the project struggles with structural barriers to participation, particularly for renters and low-income residents. The requirement for rooftop solar ownership and digital literacy creates an exclusionary effect, resulting in low scores for procedural and recognition justice. This aligns with broader critiques of "green gentrification," where clean energy transitions inadvertently benefit wealthier, technologically-equipped groups while marginalizing vulnerable populations.

In contrast, Power Ledger showcases how technical sophistication can enhance distributional justice (high score) through automated, transparent transactions that reduce costs and empower prosumers. However, its medium ratings in procedural and recognition justice highlight a persistent challenge: even well-designed systems face regulatory inertia. Australia's legacy utility frameworks, designed for centralized power systems, create disincentives for decentralized participation, particularly for Indigenous communities and low-income households. This underscores the need for policy innovation to complement technological advances.

Sun Exchange offers a compelling model for recognition justice (high score) by targeting energy-poor communities in Africa through its solar leasing platform. The project's focus on local needs—such as powering schools and clinics demonstrates how blockchain can amplify marginalized voices in energy planning. Yet its medium procedural justice score reveals a paradox: while the platform expands access, its reliance on foreign investment and cryptocurrency payments risks undermining local sovereignty. Fluctuations in crypto markets and dependence on external funders raise questions about long-term community control, illustrating the pitfalls of "crypto-philanthropy" models in the Global South.

The LO3 Energy pilots in Germany emerge as a benchmark for procedural justice (high score), with their cooperative governance structures enabling communities to co-design trading rules and reinvest profits locally. However, the medium distributional justice score indicates that financial benefits remain unevenly distributed, as wealthier participants often capture a larger share of cost savings.

These patterns suggest that technological maturity alone does not predict justice outcomes; rather, governance design and institutional alignment play a more decisive role in shaping equitable participation.

6. CHALLENGES TO EQUITABLE BLOCKCHAIN ENERGY TRANSITIONS

The integration of blockchain technology into energy systems presents a paradoxical dynamic. While it offers transformative potential for decentralization, transparency, and community empowerment, its implementation also introduces complex and interconnected challenges that may inadvertently reinforce or exacerbate existing energy inequalities if not carefully addressed. These challenges span technological, socio-economic, political, and ethical dimensions, creating barriers that disproportionately affect marginalized and underserved communities. A truly justice-centered energy transition therefore requires more than the adoption of innovative technology; it demands a fundamental rethinking of system design, governance frameworks, and implementation strategies to actively address structural inequities embedded in both energy and digital systems.

At the technological level, blockchain introduces several tensions. While it enables peer-to-peer energy trading and decentralized decision-making, its implementation often depends on reliable internet access, digital literacy, and smart devices, which are frequently lacking in rural, low-income, and indigenous communities. This digital divide creates a new form of energy exclusion, further marginalizing those already disadvantaged. In addition, energy-intensive consensus mechanisms such as proof-of-work undermine the environmental benefits of renewable energy systems by contributing to higher carbon emissions.

Socio-economic barriers further limit equitable adoption. Blockchain-based energy markets often require access to banking services, stable currencies, and technical knowledge of digital wallets, which can be inaccessible to communities facing energy poverty. High upfront costs, including solar panels, smart meters, and network participation, reinforce existing inequalities. Moreover, cryptocurrency volatility introduces financial risks for users relying on tokenized energy assets.

Political and regulatory challenges also play a significant role. Existing energy policies are largely designed for centralized systems, leaving blockchain initiatives in uncertain legal environments. This discourages inclusive innovation while enabling unregulated practices. In some regions, peer-to-peer energy trading remains restricted, limiting community-based initiatives. The lack of interoperability standards further constrains scalability, while policy priorities often emphasize efficiency and rapid decarbonization over equity, potentially overlooking the needs of vulnerable populations.

Ethical and cultural considerations add another layer of complexity. Blockchain systems often reflect the biases and assumptions of their designers, which may not align with the values or needs of local communities. For indigenous populations, the introduction of technology-driven solutions without meaningful consultation can undermine principles of self-determination and cultural sovereignty. Data privacy concerns also emerge, as the transparency of blockchain may conflict with communities' desire to control their energy usage information. In addition, the technical complexity and specialized language associated with blockchain can exclude non-expert stakeholders, reinforcing power imbalances between technical elites and the broader public.

These challenges extend beyond technical issues and reflect deeper structural inequities in both energy and digital governance. Communities in the Global South, as well as marginalized groups in the Global North, continue to face the most severe inequalities in energy access, a pattern that blockchain systems may unintentionally reinforce if justice considerations are not explicitly integrated. Without deliberate intervention, a victim structure may persist, where efficiency gains for the majority justify the exclusion of minority groups, thereby reproducing existing injustices within blockchain-enabled energy transitions.

6.1 Digital Infrastructure Gaps

The implementation of blockchain-based energy systems depends heavily on robust digital infrastructure, which creates a major barrier to equitable participation. These systems require reliable internet connectivity, compatible devices, smart metering technologies, and data storage capabilities, yet access to these resources remains highly uneven. In many rural and low-income areas, limited broadband coverage, slow network speeds, and unaffordable technology restrict participation. These gaps intersect with existing inequalities, as marginalized communities often lack both electricity and digital access, while others cannot afford the necessary devices or ongoing connectivity costs. Gender disparities further deepen the divide, with women in many regions having less access to smartphones and digital tools required for participation.

As a result, communities already underserved by traditional energy systems face a double exclusion from both conventional and digital energy solutions. This creates a reinforcing cycle where those who would benefit most from decentralized energy innovations are least able to access them. Without targeted interventions, blockchain-based energy systems risk widening inequality and creating a two-tiered energy landscape, where digitally connected populations gain access to more efficient and sustainable energy while others remain dependent on costly and outdated systems..

6.2 Governance and Regulatory Uncertainty

Blockchain-based energy systems operate within a regulatory landscape that remains uncertain and underdeveloped. In many countries, legal frameworks have yet to address decentralized energy trading, token-based incentives, and the use of smart contracts, leaving projects in a state of regulatory ambiguity. This creates challenges for legal compliance and limits integration with existing energy grids and billing systems. At the same time, there is a fundamental tension between decentralized governance models promoted by blockchain and the centralized structure of national energy policies, which often slows adoption and scalability.

These governance gaps can also undermine social justice, particularly in contexts where community ownership and indigenous rights lack formal recognition. Without inclusive regulatory structures, blockchain initiatives risk disregarding local governance systems and limiting community participation. As a result, affected groups often have little recourse beyond formal legal systems to address grievances. This highlights the need for comprehensive legal reform and governance frameworks that support decentralization while ensuring equity, recognition, and meaningful community involvement in energy transitions..

6.3 Financial Exclusion and Tokenization Barrier

The financial architecture of blockchain-based energy systems aims to expand access through decentralization and tokenization, yet it often reproduces existing inequalities. While tokenization enables peer-to-peer trading and fractional ownership, it assumes access to banking services, digital tools, and financial literacy that many populations, including the unbanked, do not possess. These limitations create significant barriers for vulnerable groups who are already excluded from formal financial systems.

Participation is further constrained by the complexity of cryptocurrency use, including private key management, price volatility, and transaction costs. High upfront and ongoing expenses, combined with regulatory burdens, make these systems less accessible to low-income households. Social inequalities such as gender disparities and conflicts with collective ownership models also persist. Without inclusive financial and regulatory design, tokenization risks reinforcing exclusion rather than enabling equitable access.

6.4 Trust and Technological Skepticism

The adoption of blockchain energy systems depends not only on technology but also on trust, particularly among communities with histories of marginalization. Indigenous, rural, and low-income groups often remain skeptical due to past experiences with top-down interventions that failed to deliver fair outcomes. This skepticism is shaped by historical patterns where energy systems prioritized external interests over local needs.

Building trust requires inclusive engagement, transparent governance, and active community participation in decision-making. The success of these systems should be measured not only by adoption but by the degree of community ownership and control. Without participatory approaches, blockchain risks reinforcing existing power imbalances instead of promoting equitable energy transitions.

6.5 Environmental and Ethical Concerns

Blockchain technologies introduce environmental and ethical challenges despite their potential to support sustainable energy systems. Energy-intensive consensus mechanisms such as Proof-of-Work can increase electricity consumption and carbon emissions, offsetting environmental benefits. Although more efficient alternatives exist, their adoption is not yet universal across platforms.

Ethical concerns also arise around data privacy, surveillance, and algorithmic bias. Detailed energy usage data can be misused if not properly governed, potentially exposing sensitive information about individuals and communities. In the absence of strong data protection frameworks, blockchain systems risk becoming tools of exploitation rather than empowerment. Ethical design must therefore prioritize data protection, transparency, and community control.

6.6 Structural Barriers and Intersecting Injustices

The challenges associated with blockchain energy systems are interconnected and reinforce one another, disproportionately affecting vulnerable populations. Limited digital infrastructure intersects with financial exclusion, regulatory uncertainty, and low trust, creating compounded barriers to participation. These issues are particularly evident in the Global South and among marginalized communities, where energy injustice is deeply rooted in broader patterns of inequality.

Addressing these barriers requires systemic and coordinated approaches that integrate technology with social policy, infrastructure development, and community empowerment. Energy justice in digital transitions must recognize that technology is not neutral and can reproduce existing inequalities if left unchecked. Achieving equitable outcomes therefore demands collaboration across sectors and disciplines to address these intersecting structural challenges.

7. TOWARD A JUSTICE-CENTERED DESIGN FRAMEWORK

A justice-centered design framework must go beyond ethical aspirations and be grounded in enforceable legal structures and economically viable system configurations. This involves aligning smart contract mechanisms with regulatory requirements, designing inclusive incentive systems, and ensuring that governance structures are both legally recognized and socially legitimate. By embedding the principles of energy justice—distributional, procedural, and recognition justice—into every stage of development, blockchain-based energy systems can move from abstract ideals to practical implementation. Justice outcomes are not determined by technology alone, but by how it is integrated into broader social, institutional, and policy contexts through participatory governance and equitable resource allocation (Egba et al., 2026).

Achieving justice in blockchain-enabled energy systems also requires attention to communication, inclusivity, and multidimensional evaluation. Digital platforms can enhance transparency, stakeholder engagement, and participatory decision-making, strengthening distributive and procedural justice. Expanding the concept of energy justice to include ecological, social-ecological, and intergenerational dimensions further ensures long-term sustainability and broader acceptance. At the same time, evaluating both efficiency and justice together highlights the importance of governance models that balance technical performance with ethical impact. Lessons from global practices demonstrate that socially embedded frameworks and inclusive renewable energy policies play a critical role in achieving equitable outcomes across regions and communities.

Ultimately, a justice-centered framework must be adaptive, iterative, and forward-looking. Approaches such as Justice by Design emphasize the early integration of ethical considerations to prevent inequities before they emerge, while continuous evaluation ensures responsiveness to evolving community needs. This framework promotes co-design with diverse stakeholders, adaptive regulation, and long-term stewardship principles that prioritize sustainability and inclusivity. By treating justice as an ongoing process rather than a fixed outcome, blockchain-based energy systems can evolve into tools that not only enhance efficiency but also empower communities and support a more equitable digital energy future.

7.1 Principles of Justice-Centered Design

A justice-centered blockchain energy system should adhere to the following principles (Maonaigh, Fitzgerald, & Reilly, 2025):

a. Accessibility: Ensure both physical and digital access to blockchain systems. This includes providing digital tools, offline access options, and localized language interfaces.
b. Participation: Facilitate inclusive stakeholder engagement during design, implementation, and governance phases.
c. Transparency: Maintain openness in decision-making algorithms, smart contract design, and data provenance.
d. Adaptability: Allow for community-level customization and governance of blockchain protocols.
e. Affordability: Ensure tokenization and transaction structures do not create financial barriers for low-income users.

7.2 Stages of Justice-Centered Blockchain Design

Stage 1: Community Needs Assessment

a. Conduct participatory assessments to identify energy needs, digital capacity, and socio-political barriers.
b. Engage community representatives, especially marginalized voices, in framing goals and design criteria.

Stage 2: Inclusive System Architecture

a. Co-design blockchain infrastructure with community stakeholders.
b. Develop smart contracts that reflect local energy sharing norms and economic models.
c. Integrate multilingual, accessible user interfaces.

Stage 3: Governance Structures

a. Establish local cooperatives or councils to oversee blockchain system management.
b. Codify consensus mechanisms and dispute resolution processes.
c. Enable community voting rights for system upgrades or policy changes.

Stage 4: Training and Capacity Building

a. Provide digital literacy workshops.
b. Train community leaders as blockchain energy system facilitators.
c. Establish support systems for onboarding and troubleshooting.

Stage 5: Monitoring, Evaluation, and Iteration

a. Implement justice metrics (access, participation, benefit sharing).
b. Use feedback loops and audits to refine smart contract behavior.
c. Adjust design based on changing community needs and feedback.

7.3 Governance and Multi-Stakeholder Collaboration

a. Effective implementation of justice-centered systems requires collaboration across multiple levels:
b. Local communities must have decision-making authority.
c. Nonprofits and civil society can facilitate inclusive participation and capacity building.
d. Governments and regulators should provide enabling policies and ensure legal protections.
e. Private sector actors must be incentivized to build inclusive technologies and avoid extractive practices.

7.4 Example Framework Application

Consider a blockchain-based energy microgrid in a remote Indigenous community. Using this framework, the initiative would begin with trust-building dialogue, evolve into a jointly-designed system reflecting community norms, and be governed by a local energy council. Smart contracts would be written in local languages, energy credits distributed equitably, and system analytics reviewed transparently with community members (Egba et al., 2026).

Figure 3. Justice-centered blockchain energy design framework

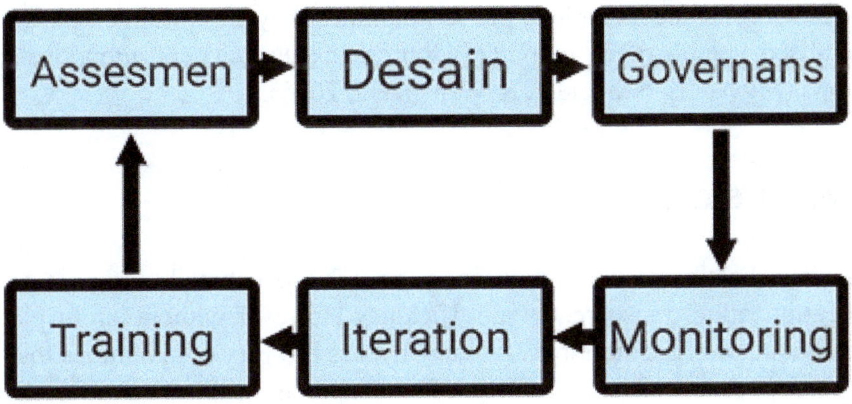

Figure 3 shows a cyclical and iterative framework illustrating the integration of justice principles into blockchain energy systems, consisting of key stages including assessment, design, governance, monitoring, iteration, and training to ensure continuous improvement, inclusivity, and sustainability.

This framework is not a one-size-fits-all solution but a modular toolkit adaptable to various geographic and cultural contexts. It prioritizes inclusion and justice not as optional features but as structural foundations of blockchain innovation.

7.5 Design Recommendations for Policy-Makers and Developers

To realize the vision of a justice-centered energy transition, both policy-makers and technology developers must adopt new approaches to system design and regulation. For policy-makers, this includes establishing regulatory sandboxes to safely test blockchain innovations, offering subsidies or digital equity grants for marginalized communities, and mandating inclusive participation in energy governance. Legal frameworks must also clarify data ownership, privacy protections, and the legal standing of smart contracts. For developers, participatory design methodologies should be standard practice—engaging communities from the outset to co-create platforms that reflect their values and constraints. User interfaces should be accessible across languages, literacy levels, and devices. Ethical auditing of smart contracts and governance algorithms must be routine, not optional. Importantly, system monitoring should incorporate justice metrics, and design iterations must be guided by community feedback loops. By embedding justice into both code and policy, we can ensure that blockchain systems are not only innovative but equitable, inclusive, and accountable to the communities they aim to serve. The share of decentralised, volatile renewable energy sources in electricity consumption is rising in Europe (Michaelis, Schneider, & Weibelzahl, 2025).

8. CONCLUSION

The digitalization of the energy sector through blockchain holds great promise for decentralization, transparency, and efficiency. However, without careful planning and intentional design, these benefits may bypass the very communities that most need access to clean and reliable energy. This chapter has explored how blockchain-based energy systems intersect with the principles of energy justice—distributional, procedural, and recognition justice—and how these principles can be embedded into technology design and governance.

By examining global case studies and developing a justice-centered design framework, we have demonstrated that equitable energy transitions require more than technical innovation—they require inclusive policymaking, community participation, and continuous evaluation. Blockchain systems should be built not only for efficiency and innovation but also for fairness and accountability.

The path forward involves bridging the gap between digital and energy divides. This means investing in digital infrastructure, fostering inclusive regulatory environments, supporting community-led initiatives, and ensuring that blockchain technology aligns with social values and environmental goals.

Ultimately, the effectiveness of blockchain in advancing energy justice depends on its integration within broader legal and institutional frameworks. Technological design alone cannot guarantee equitable outcomes; instead, justice must be codified through regulatory standards, economic incentives, and participatory governance mechanisms. Future research should therefore focus on developing hybrid models that combine technological innovation with institutional reform to ensure that digital energy transitions remain both efficient and inclusive.

As the energy sector continues to evolve, justice must be at the center of its digital transformation. Only by doing so can we ensure that the energy systems of the future are not only smart and decentralized but also fair, inclusive, and resilient.

REFERENCES

Astola, M., Laes, E., Bombaerts, G., Ryszawska, B., Rozwadowska, M., Szymanski, P., … Hansen, M. (2022). Correction: Community Heroes and Sleeping Members: Interdependency of the Tenets of Energy Justice (Science and Engineering Ethics, (2022), 28, 5, (45),). *Science and Engineering Ethics, 28*(6). https://doi.org/DOI: 10.1007/s11948-022-00384-3

Ayazi, A. M., Shakarami, M. R., Doostizadeh, M., Namdari, F., & Nikzad, M. R. (2026). An MILP approach for optimal operation of unbalanced distribution networks through coordinated network reconfiguration and SOP utilization. *Scientific Reports, 16*(1), 2069. Advance online publication. DOI: 10.1038/s41598-025-31912-3 PMID: 41402420

Bhullar, L. (2023). Meanings of energy justice in the low-carbon transition. In *Handbook of Energy Law in the Low-Carbon Transition* (pp. 65–80). DOI: 10.1515/9783110752403-011

Boumaiza, A. (2024). A blockchain-centric P2P trading framework incorporating carbon and energy trades. *Energy Strategy Reviews, 54*, 101466. Advance online publication. DOI: 10.1016/j.esr.2024.101466

Boumaiza, A., & Maher, K. (2024). Leveraging blockchain technology to enhance transparency and efficiency in carbon trading markets. *International Journal of Electrical Power & Energy Systems, 162*, 110225. Advance online publication. DOI: 10.1016/j.ijepes.2024.110225

Chen, D., Hua, X., & Zhou, Z. (2025). Study on decentralized reconstruction mechanism of blockchain technology in electricity trading market. In *Proceedings of SPIE - The International Society for Optical Engineering* (Vol. 13985). DOI: 10.1117/12.3078326

Diaz Valdivia, A. (2023). Between decentralization and reintermediation: Blockchain platforms and the governance of 'commons-led' and 'business-led' energy transitions. *Energy Research & Social Science, 98*(March), 103034. DOI: 10.1016/j.erss.2023.103034

Egba, V. N., Ukeje, I. O., Ayanwale, M. A., Robinson, D. F., Ojogbane, R. S., & Akonye, J. E. (2026). Smart urban governance for sustainable cities: Integrating artificial intelligence, digitalization, and renewable energy technologies in Nigeria's public service transformation. *Frontiers in Sustainable Cities, 8*, 1739482. Advance online publication. DOI: 10.3389/frsc.2026.1739482

Enoch, O. I., Salawu, E. Y., Akerekan, O. E., Olaoluwa, O. O., Babafemi, O., Awoyemi, O. O., & Airewa, I. (2025). Leveraging Artificial Intelligence for Optimizing Renewable Energy Systems. *NIPES - Journal of Science and Technology Research, 7*(1 Special Issue), 1543–1549. DOI: 10.37933/nipes/7.4.2025.SI180

Feenstra, M., & Özerol, G. (2021). Energy justice as a search light for gender-energy nexus: Towards a conceptual framework. *Renewable & Sustainable Energy Reviews, 138*, 110668. Advance online publication. DOI: 10.1016/j.rser.2020.110668

Hernandez, J., Scherr, R., & Robertson, A. D. (2022). Redefining Energy Justice in Physics Classrooms. *Environmental Justice, 15*(2), 108–114. DOI: 10.1089/env.2021.0042

Jhariya, M. K., Dehalwar, V., Bharti, J., & Kumar, L. (2026). Energy efficient transactions for blockchain networks using adaptive global best–worst particle swarm optimization. *Scientific Reports, 16*(1), 1643. Advance online publication. DOI: 10.1038/s41598-025-31112-z PMID: 41495151

Lacey-Barnacle, M., Robison, R., & Foulds, C. (2020). Energy justice in the developing world: A review of theoretical frameworks, key research themes and policy implications. *Energy for Sustainable Development, 55*, 122–138. DOI: 10.1016/j.esd.2020.01.010

Laksono, M. N. G., Tiaraputri, Z. A., Sekti, B. A., & Laksono, M. I. A. (2025). Cyber-physical systems security in space. In *Advanced Cyber Defense for Space Missions and Operations* (pp. 85–113). Concepts and Applications., DOI: 10.4018/979-8-3693-7939-4.ch004

Maonaigh, C. Ó., Fitzgerald, L. M., & Reilly, L. (2025). Revisiting the social licence to operate in the energy transition: An intersectional agenda for research and practice. *Energy Research & Social Science, 126*, 104139. Advance online publication. DOI: 10.1016/j.erss.2025.104139

Michaelis, A., Schneider, M., & Weibelzahl, M. (2025). Designing local flexibility markets: A toolbox for policymakers and market operators. *Energy, 329*, 136051. Advance online publication. DOI: 10.1016/j.energy.2025.136051

Sekti, B. A. (2025). IT Service Information Security Management. In *Cases on Information Systems Service Management* (pp. 61–93). DOI: 10.4018/979-8-3373-2352-7.ch003

Sekti, B. A. (2026a). Cybersecurity Challenges in Smart City Development. In *Securing Smart Cities Through Modern Cryptography Technologies* (pp. 75–102). IGI Global Scientific Publishing., DOI: 10.4018/979-8-3373-3326-7.ch003

Sekti, B. A. (2026b). Data Privacy and Cybersecurity in Wind Power Systems. In *AI-Powered Analysis* (pp. 335–366). Modeling, and Monitoring of Wind Energy Systems., DOI: 10.4018/979-8-3373-4159-0.ch009

Solahudeen Tando, M., Gawusu, S., Yakubu, S., Kwabena, O. S., Donatus, D., & Raqib Abu, S. (2024). Decentralized energy trading with blockchain technology. In *The Intersection of Blockchain and Energy Trading* (pp. 75–101). Exploring Decentralized Solutions for Next-Generation Energy Markets., DOI: 10.1016/B978-0-443-23627-3.00004-1

Yannam, Y. R., Gao, J., Shi, J., Datta, A., Wang, K., Krishnamoorti, R., Powell, J. B., Sengupta, D., & Ye, X. (2026). Toward a human-centered energy transition: Concepts, models, challenges, and research opportunities. *Utilities Policy*, *99*, 102145. Advance online publication. DOI: 10.1016/j.jup.2026.102145

Yarrabothu, R. S., Benni, N. S., Nirmal Raj, S., N, M. K., Singh, B. K., Divya, A., Bhojan, R., & Tirth, V. (2026). Quantum-resistant blockchain and federated learning integrated secure and scalable smart grid framework for sustainable and privacy-preserving energy management. *Sustainable Computing : Informatics and Systems*, *50*, 101305. Advance online publication. DOI: 10.1016/j.suscom.2026.101305

Zhang, S., Sovacool, B. K., & Wei, C. (2026). Do clean energy transitions enhance global energy security? A comparative analysis of trends and tradeoffs across eight major economies, 1990-2020. *Energy Policy*, *212*, 115164. Advance online publication. DOI: 10.1016/j.enpol.2026.115164

Compilation of References

Aarthi, S., Aravinthan, K., Ravikumar, R. N., Sivakumar, N., & Wanglen, S. (2025). Cryptography securing data in motion within Blockchain, internet of everything, and Federated Learning. In *Convergence of Blockchain, Internet of Everything, and Federated Learning for Security* (pp. 39–78). IGI Global. DOI: 10.4018/979-8-3373-1424-2.ch002

Agarwal, S., & Jain, A. (2023). Blockchain applications in smart grid: Challenges and opportunities. In *Proceedings of the 2023 IEEE International Conference on Blockchain and Distributed Systems Security (ICBDS)* (pp. 1-7). IEEE. DOI: 10.1109/ICBDS58040.2023.10346363

Agência Nacional de Energia Elétrica (ANEEL). (2023). 2022 management report: Integrated report. Brasília, DF: ANEEL. Retrieved October 2, 2025, from https://www2.aneel.gov.br/cedoc/aprt20236810_2.pdf

Ahl, A., Yarime, M., Goto, M., Chopra, S. S., Kumar, N. M., Tanaka, K., & Sagawa, D. (2020). Exploring blockchain for the energy transition: Opportunities and challenges based on a case study in Japan. *Renewable & Sustainable Energy Reviews*, *117*, 109488. DOI: 10.1016/j.rser.2019.109488

Ahmed, S., & Măgurean, A. M. (2024). Renewable energy communities: Towards a new sustainable model of energy production and sharing. *Energy Strategy Reviews*, *55*, 101522. Advance online publication. DOI: 10.1016/j.esr.2024.101522

Akbar, D., Ünal, M., Akkaya, R., & Balci, O. (2025). Blockchain-based peer-to-peer energy trading: Design and implementation of the tokenized decentralized energy management system (DEMS). In *Research Square*. DOI: 10.21203/rs.3.rs-6787257/v1

Al Shareef, A. M., Seçkiner, S., Eid, B., & Abumeteir, H. (2024). Integration of blockchain with artificial intelligence technologies in the energy sector: A systematic review. *Frontiers in Energy Research*, *12*, 1377950. Advance online publication. DOI: 10.3389/fenrg.2024.1377950

Alcaraz, C., Rubio, J. E., & Lopez, J. (2020). Blockchain-assisted access for federated Smart Grid domains: Coupling and features. *Journal of Parallel and Distributed Computing*, *144*(i), 124–135. DOI: 10.1016/j.jpdc.2020.05.012

Allen & Overy. (2024). *The law of smart contracts: 2024 global review and the reasonable programmer standard.* A&O Legal Publications.

Andoni, M., Robu, V., Flynn, D., Abram, S., Geach, D., Jenkins, D., McCallum, P., & Peacock, A. (2019). Blockchain technology in the energy sector: A systematic review of challenges and opportunities. *Renewable & Sustainable Energy Reviews*, *100*, 143–174. DOI: 10.1016/j.rser.2018.10.014

Androulaki, E., Barger, A., Bortnikov, V., Cachin, C., Christidis, K., De Caro, A., & Yellick, J. (2018, April). Hyperledger fabric: a distributed operating system for permissioned blockchains. In *Proceedings of the thirteenth EuroSys conference* (pp. 1-15). DOI: 10.1145/3190508.3190538

Ansell, C., & Gash, A. (2018). Collaborative platforms as a governance strategy. *Journal of Public Administration: Research and Theory*, *28*(1), 16–32. DOI: 10.1093/jopart/mux030

Anwar, N., Rahaman, R., Widodo, A. M., Sekti, B. A., Erzed, N., Santiago Tangkudung, R. R., . . . Budhisantosa, N. (2025). Sustainable cybersecurity in the AI era: Innovations in green computing and security. In *Sustainable Information Security in the Age of AI and Green Computing* (pp. 603–620). DOI: 10.4018/979-8-3693-8034-5.ch030

Aravind, A., Santhi, G., Patil, S. A., Selvakumar, S., Sharma, G., Dhamone, J., & Nathan, S. R. (2024). Blockchain Technology in Energy Markets: Enabling Peer-to-Peer Energy Trading. E3S Web of Conferences, 591, 06002. DOI: 10.1051/e3sconf/202459106002

ASEAN Centre for Energy. (2023). *ASEAN energy outlook 7: 2022-2050.*

Asian Development Bank. (2021). *2021 energy policy of the Asian Development Bank: Supporting low-carbon transition in Asia and the Pacific.*

Astola, M., Laes, E., Bombaerts, G., Ryszawska, B., Rozwadowska, M., Szymanski, P., … Hansen, M. (2022). Correction: Community Heroes and Sleeping Members: Interdependency of the Tenets of Energy Justice (Science and Engineering Ethics, (2022), 28, 5, (45),). *Science and Engineering Ethics, 28*(6). https://doi.org/DOI: 10.1007/s11948-022-00384-3

Australian Competition and Consumer Commission (ACCC). (n.d.). The Consumer Data Right. Retrieved November 3, 2025, from https://www.accc.gov.au/by-industry/banking-and-finance/the-consumer-data-right

Australian Energy Market Operator. (2025). Strategic corporate plan. https://www.aemo.com.au/about/corporate-governance/corporate-plan

Australian Energy Regulator. (2023). Trial projects guidelines – Regulatory sandboxing. Retrieved November 1, 2025, from https://www.aer.gov.au/industry/registers/resources/guidelines/trial-projects-guidelines-regulatory-sandboxing

Australian Securities and Investments Commission (ASIC). (n.d.). Enhanced regulatory sandbox (ERS). Retrieved October 3, 2025, from https://www.asic.gov.au/for-business-and-companies/innovation-hub/enhanced-regulatory-sandbox-ers/

Ayazi, A. M., Shakarami, M. R., Doostizadeh, M., Namdari, F., & Nikzad, M. R. (2026). An MILP approach for optimal operation of unbalanced distribution networks through coordinated network reconfiguration and SOP utilization. *Scientific Reports*, *16*(1), 2069. Advance online publication. DOI: 10.1038/s41598-025-31912-3 PMID: 41402420

Baggio, A., & Grimaccia, F. (2020). Blockchain as key enabling technology for future electric energy exchange: A vision. *IEEE Access : Practical Innovations, Open Solutions*, *8*, 205250–205271. DOI: 10.1109/ACCESS.2020.3036994

Baig, M. J., Iqbal, M. T., Jamil, M., & Khan, J. (2023). Blockchain-Based Peer-to-Peer Energy Trading System Using Open-Source Angular Framework and Hypertext Transfer Protocol. In *Electronics* (Vol. 12, Issue 2, p. 287). DOI: 10.3390/electronics12020287

Baillie, K. (2017). Blockchain: talking utility switching and more with Electron. Retrieved November 13, 2025, from https://www.power-technology.com/features/featureblockchain-talking-utility-switching-and-more-with-electron-5904299/?cf-view

Barceló, E., Dimić-Mišić, K., Imani, M., Spasojević Brkić, V., Hummel, M., & Gane, P. (2023). Regulatory paradigm and challenge for blockchain integration of decentralized systems: Example—Renewable energy grids. *Sustainability (Basel)*, *15*(3), 2571. Advance online publication. DOI: 10.3390/su15032571

Bassan, F., & Rabitti, M. (2024). From smart legal contracts to contracts on blockchain: An empirical investigation. *Computer Law & Security Review*, *55*, 106035. Advance online publication. DOI: 10.1016/j.clsr.2024.106035

Bassey, K. E., Rajput, S. A., & Oyewale, K. (2024). *Peer-to-peer energy trading: Innovations, regulatory challenges, and the future of decentralized energy systems*. World Journal Of Advanced Research and Reviews., DOI: 10.30574/wjarr.2024.24.2.3324

Beck, R., Müller-Bloch, C., & King, J. L. (2018). Governance in the blockchain economy: A framework and research agenda. *Journal of the Association for Information Systems*, *19*, 1020–1034. DOI: 10.17705/1jais.00518

Beckstedde, E., Correa Ramírez, M., Cossent, R., Vanschoenwinkel, J., & Meeus, L. (2023). Regulatory sandboxes: Do they speed up innovation in energy? *Energy Policy*, *180*, 112672. Advance online publication. DOI: 10.1016/j.enpol.2023.113656

Bhavana, G. B., Sreedevi, A., & Kumar, K. S. (2024). Applications of blockchain technology in peer-to-peer energy trading: A review. *Scientific Reports*, *14*, 72642. Advance online publication. DOI: 10.1038/s41598-024-72642-2 PMID: 39304698

Bhullar, L. (2023). Meanings of energy justice in the low-carbon transition. In *Handbook of Energy Law in the Low-Carbon Transition* (pp. 65–80). DOI: 10.1515/9783110752403-011

Birhane, A. (2021). Algorithmic injustice: A relational ethics approach. *Patterns (New York, N.Y.)*, *2*(2), 100205. DOI: 10.1016/j.patter.2021.100205 PMID: 33659914

Bisht, A., Das, A. K., Niyato, D., & Park, Y. (2023). Efficient Personal-Health-Records Sharing in Internet of Medical Things Using Searchable Symmetric Encryption, Blockchain, and IPFS. *IEEE Open Journal of the Communications Society*, *4*, 2225–2244. DOI: 10.1109/OJCOMS.2023.3316922

Biswas, P., Rashid, A., Biswas, A., Al Nasim, A., Datta Gupta, K., & George, R. (2024). AI-driven approaches for optimizing power consumption: A comprehensive survey (arXiv preprint arXiv:2406.15732v1). https://arxiv.org/abs/2406.15732

Boumaiza, A. (2024). Carbon and Energy Trading Integration within a Blockchain-Powered Peer-to-Peer Framework. In *Energies* (Vol. 17, Issue 11, p. 2473). DOI: 10.3390/en17112473

Boumaiza, A. (2024). A blockchain-centric P2P trading framework incorporating carbon and energy trades. *Energy Strategy Reviews*, *54*, 101466. Advance online publication. DOI: 10.1016/j.esr.2024.101466

Boumaiza, A., & Maher, K. (2024). Leveraging blockchain technology to enhance transparency and efficiency in carbon trading markets. *International Journal of Electrical Power & Energy Systems*, *162*, 110225. Advance online publication. DOI: 10.1016/j.ijepes.2024.110225

Boumaiza, A., & Sanfilippo, A. (2024). A Testing Framework for Blockchain-Based Energy Trade Microgrids Applications. *IEEE Access : Practical Innovations, Open Solutions, 12*, 27465–27483. DOI: 10.1109/ACCESS.2024.3367999

Burger, C., & Weinmann, J. (2022). Blockchain Platforms in Energy Markets—A Critical Assessment. In *Journal of Risk and Financial Management* (Vol. 15, Issue 11, p. 516). DOI: 10.3390/jrfm15110516

Burger, S. P., Jenkins, J. D., Batlle, C., & Pérez-Arriaga, I. J. (2019). Restructuring revisited: Competition and coordination in electricity distribution systems. *Energy Journal, 40*(3), 1–33. DOI: 10.5547/01956574.40.3.sbur

Buterin, V. (2014). A next-generation smart contract and decentralized application platform. *white paper, 3*(37), 2-1.

Buterin, V. (2021). *Ethereum: A next-generation smart contract and decentralized application platform* [White paper]. Ethereum Foundation.

Cantillo-Luna, S., Moreno-Chuquen, R., Chamorro, H. R., Sood, V. K., Badsha, S., & Konstantinou, C. (2022). Blockchain for distributed energy resources management and integration. *IEEE Access : Practical Innovations, Open Solutions, 10*, 68592–68617. DOI: 10.1109/ACCESS.2022.3184704

Chen, D., Hua, X., & Zhou, Z. (2025). Study on decentralized reconstruction mechanism of blockchain technology in electricity trading market. In *Proceedings of SPIE - The International Society for Optical Engineering* (Vol. 13985). DOI: 10.1117/12.3078326

Choi, W., Woo, J., & Hong, J. W.-K. (2024). Fractional non-fungible tokens: Overview, evaluation, marketplaces, and challenges. *International Journal of Network Management, 34*(4), e2260. DOI: 10.1002/nem.2260

Church, S. (2022). Tokenizing Renewable Energy Certificates (RECs)—A Blockchain Approach for REC Issuance and Trading. *IEEE Access : Practical Innovations, Open Solutions, 10*, 134477–134490. DOI: 10.1109/ACCESS.2022.3230937

Cloudin, S., Priya, R. S., Prabhu, G. A., Ruba, M., Joel, R., & Vanitha, R. (2026). AI-Enhanced Predictive Analytics and Reliability Management for Wind Turbine Maintenance. In AI-Powered Analysis, Modeling, and Monitoring of Wind Energy Systems (pp. 191-218). IGI Global Scientific Publishing.

De Filippi, P., Mannan, M., & Reijers, W. (2022). Blockchain as a confidence machine: The role of trust & law in territory-less governance. *Journal of Law and Innovation.*

De Filippi, P., & Wright, A. (2018). *Blockchain and the law: The rule of code.* Harvard University Press.

de Villiers, A., & Cuffe, P. (2020). Towards imposing dayparted restrictions on tokenised energy within peer-to-peer markets.

Dhanushya, J., Ruba, M., Ponkarthika, M., Soundarya, B., Sakthi Priya, V., Dhanushri, S., & Harikrishnan, S. (2025, April). AI-Powered Multipurpose Agrobot for Precision Farming: Crop Health Management, Fertilizer Recommendation and Yield Prediction. In *2025 International Conference on Recent Advances in Electrical, Electronics, Ubiquitous Communication, and Computational Intelligence (RAEEUCCI)* (pp. 1-6). IEEE.

Di Silvestre, M. L., Gallo, P., Guerrero, J. M., Musca, R., Riva Sanseverino, E., Sciumè, G., Vásquez, J. C., & Zizzo, G. (2020). Blockchain for power systems: Current trends and future applications. *Renewable & Sustainable Energy Reviews, 119*(109585), 109585. DOI: 10.1016/j.rser.2019.109585

Diaz Valdivia, A. (2023). Between decentralization and reintermediation: Blockchain platforms and the governance of 'commons-led' and 'business-led' energy transitions. *Energy Research & Social Science, 98*(March), 103034. DOI: 10.1016/j.erss.2023.103034

Diestelmeier, L. (2019). Changing power: Shifting the role of electricity consumers with blockchain technology – Policy implications for EU electricity law. *Energy Policy, 128*, 189–196. DOI: 10.1016/j.enpol.2018.12.065

Domalis, G., Karacapilidis, N., Tsakalidis, D., & Giannaros, A. (2021). A Trustable and Interoperable Decentralized Solution for Citizen-Centric and Cross-Border eGovernance: A Conceptual Approach BT - Electronic Government. In Scholl, H. J., Gil-Garcia, J. R., Janssen, M., Kalampokis, E., Lindgren, I., & Rodríguez Bolívar, M. P. (Eds.), (pp. 259–270). Springer International Publishing.

Dong, J., Ning, P., Zhao, H., & Song, C. (2025). Decentralized Peer-to-Peer Energy Trading: A Blockchain-Enabled Pricing Paradigm. *Journal of King Saud University. Computer and Information Sciences, 8*(1), 25. DOI: 10.1007/s44443-025-00025-2

Dorri, A., Kanhere, S. S., Jurdak, R., & Gauravaram, P. (2017). Blockchain for IoT security and privacy: The case study of a smart home. 2017 IEEE International Conference on Pervasive Computing and Communications Workshops (PerCom Workshops), Kona, HI, USA, 2017, pp. 618-623. DOI: 10.1109/PERCOMW.2017.7917634

Egba, V. N., Ukeje, I. O., Ayanwale, M. A., Robinson, D. F., Ojogbane, R. S., & Akonye, J. E. (2026). Smart urban governance for sustainable cities: Integrating artificial intelligence, digitalization, and renewable energy technologies in Nigeria's public service transformation. *Frontiers in Sustainable Cities*, *8*, 1739482. Advance online publication. DOI: 10.3389/frsc.2026.1739482

Energies. (2022). *Towards Blockchain-Based Energy Trading: A Smart Contract Implementation of Energy Double Auction and Spinning Reserve Trading*. Energies, 15(11), 4084. DOI: 10.3390/en15114084

Energy Web Foundation (EWF). (2019). The Energy Web Chain: Accelerating the energy transition with an open-source, decentralized blockchain platform (Version 2.0). https://cdn.prod.website-files.com/65e5b22cf7f0480a007420fc/67ad f26e027740e64f98b5df_EWF-Paper-TheEnergyWebChain-v2-201907.pdf

Energy, E. D. F. (n.d.). Smart meters help you save energy. Retrieved October 13, 2025, from https://www.edfenergy.com/smart-meters?_%24ja=tsid%3A44771 %7Ccgn%3ARedirect%7Ckw%3Asmart

Enescu, F. M., Bizon, N., Onu, A., Răboacă, M. S., Thounthong, P., Mazare, A. G., & erban, G. (2020). Implementing blockchain technology in irrigation systems that integrate photovoltaic energy generation systems. *Sustainability (Basel)*, *12*(4), 1540. DOI: 10.3390/su12041540

Enoch, O. I., Salawu, E. Y., Akerekan, O. E., Olaoluwa, O. O., Babafemi, O., Awoyemi, O. O., & Airewa, I. (2025). Leveraging Artificial Intelligence for Optimizing Renewable Energy Systems. *NIPES - Journal of Science and Technology Research, 7*(1 Special Issue), 1543–1549. DOI: 10.37933/nipes/7.4.2025.SI180

European Commission. (2020). Energy communities in the clean energy package: Best practices and recommendations for implementation (Publication No. MJ-01-20-760-EN-N; ISBN 978-92-76-27320-2; DOI). Publications Office of the European Union. https://op.europa.eu/en/publication-detail/-/publication/4b7d5144-91c9-11eb -b85c-01aa75ed71a1/language-enDOI: 10.2833/51076

European Commission. (2023). Digitalisation of the energy system. Retrieved October 14, 2025, from https://energy.ec.europa.eu/topics/eus-energy-system/ digitalisation-energy-system_en

European Commission. (2024). European Blockchain Partnership. Retrieved October 5, 2025, from https://digital-strategy.ec.europa.eu/en/policies/blockchain-partnership

European Commission. (2024). *Revised Product Liability Directive: Strengthening consumer protection in the digital age*. EC Publications.

European Data Protection Board. (2025). Guidelines 02/2025 on processing of personal data through blockchain technologies (Public consultation). Retrieved October 15, 2025, from https://www.edpb.europa.eu/our-work-tools/documents/public-consultations/2025/guidelines-022025-processing-personal-data_en

European Parliament & European Council. (2018). Directive (EU) 2018/2001 of the European Parliament and of the Council of 11 December 2018 on the promotion of the use of energy from renewable sources (recast). *Official Journal of the European Union, L, 328*, 82–209. https://eur-lex.europa.eu/eli/dir/2018/2001/oj/eng

European Parliament & European Council. (2019). Directive (EU) 2019/944 of the European Parliament and of the Council of 5 June 2019 on common rules for the internal market for electricity and amending Directive 2012/27/EU (recast). *Official Journal of the European Union, L, 158*, 125–199. https://eur-lex.europa.eu/eli/dir/2019/944/oj/eng

European Securities and Markets Authority (ESMA). (2025). Markets in Crypto-Assets Regulation (MiCA). Retrieved October 18, 2025, from https://www.esma.europa.eu/esmas-activities/digital-finance-and-innovation/markets-crypto-assets-regulation-mica

Evens, M., Ercoli, P., & Arteconi, A. (2023). Blockchain-Enabled Microgrids: Toward Peer-to-Peer Energy Trading and Flexible Demand Management. In *Energies* (Vol. 16, Issue 18, p. 6741). DOI: 10.3390/en16186741

Fatima, S., & Arshad, M. J. (2025). A comprehensive review of blockchain and machine learning integration for peer-to-peer energy trading in smart grids. *IEEE Access : Practical Innovations, Open Solutions*, *13*, 92756–92782. DOI: 10.1109/ACCESS.2025.3572174

Federal Ministry for Economic Affairs and Climate Action (BMWE/BMWK). (2023). The climate-neutral electricity system platform – in dialogue for a new market design. https://www.bundeswirtschaftsministerium.de/Redaktion/EN/Dossier/the-climate-neutral-electricity-system-platform.html

Feenstra, M., & Özerol, G. (2021). Energy justice as a search light for gender-energy nexus: Towards a conceptual framework. *Renewable & Sustainable Energy Reviews*, *138*, 110668. Advance online publication. DOI: 10.1016/j.rser.2020.110668

Ferreira, A. (2020). Emerging regulatory approaches to blockchain based token economy. *JBBA*, *3*(1), 1–9. DOI: 10.31585/jbba-3-1-(6)2020

Financial Stability Board. (2021). 2021 status report: Task Force on Climate-related Financial Disclosures. https://www.fsb.org/2021/10/2021-status-report-task-force-on-climate-related-financial-disclosures/

Finck, M. (2018). *Blockchain and the General Data Protection Regulation: Can distributed ledgers be squared with European data protection law?* European Parliamentary Research Service. DOI: 10.1017/9781108609708.004

Finck, M. (2019). Blockchain and the general data protection regulation: Can distributed ledgers be squared with European data protection law? Max Planck Institute for Innovation & Competition. DOI: 10.2139/ssrn.3080322

Fotiou, N., Pittaras, I., Siris Vasilios, A., Voulgaris, S., & Polyzos, G. C. (2019). Secure IoT access at scale using blockchains and smart contracts. 2019 IEEE 20th International Symposium on ``A World of Wireless, Mobile and Multimedia Networks'' (WoWMoM), 1–6.

Gavai, A. K., Bouzembrak, Y., Xhani, D., Sedrakyan, G., Meuwissen, M. P. M., Souza, R. G.-S., . . . van Hillegersberg, J. (2025). Agricultural data privacy: Emerging platforms & strategies. *Food and Humanity, 4*, 100542. 10.1016/j.foohum.2025.100542

Gaybullaev, T., Kwon, H.-Y., Kim, T., & Lee, M.-K. (2021). Efficient and privacy-preserving energy trading on blockchain using dual binary encoding for inner product encryption. *Sensors (Basel), 21*(6), 2024. Advance online publication. DOI: 10.3390/s21062024 PMID: 33809344

Gil-Garcia, J. R., Gasco-Hernandez, M., & Pardo, T. A. (2020). Beyond transparency, participation, and collaboration? A reflection on the dimensions of open government. *Public Performance & Management Review, 43*(3), 483–502. DOI: 10.1080/15309576.2020.1734726

Golder, S. S., Mondal, S., Das, S., Bose, R., Sutradhar, S., & Mondal, H. (2024). Hybrid Blockchain Framework for Secure and Scalable Internet of Things (IoT) Networks (HB-IoT): A Novel Approach. In *2024 International Conference on Artificial Intelligence and Quantum Computation-Based Sensor Application (ICAIQSA)* (pp. 1–7). DOI: 10.1109/ICAIQSA64000.2024.10882319

Goodkind, A. L., Jones, B. A., & Berrens, R. P. (2020). Cryptodamages: Monetary value estimates of the air pollution and human health impacts of cryptocurrency mining. *Energy Research & Social Science, 59*, 101281. DOI: 10.1016/j.erss.2019.101281

Goranovic, A., Meisel, M., Panev, S., Padezhki, G., Stefanov, A., & Sosic, A. (2022). Blockchain applications in transactive energy systems: A review. *IEEE Access : Practical Innovations, Open Solutions, 10*, 52871–52896.

Gowthaman, N., Ruba, M., Aswath, S., Hemavarshni, S. N., Bragadeshwaran, R., & Nithish, S. J. (2025, November). Iot Integrated Solution For Smart Vertical Crop Cultivation. In 2025 IEEE 7th International Conference on Computing, Communication and Automation (ICCCA) (pp. 1-6). IEEE.

Grigoryan, H. (2024). Cost-Effective Integration of Blockchain Technologies Into P2P Energy Trading Systems. *2024 IEEE International Conference on Omni-Layer Intelligent Systems (COINS)*, 1–5. DOI: 10.1109/COINS61597.2024.10622550

Group, S. P. (2018). SP Unveils Blockchain-Powered Renewable Energy Certificates Marketplace. SP Group. https://www.spgroup.com.sg/about-us/media-resources/energy-hub/innovation/sp-unveils-blockchain-powered-renewable-energy-certificates-marketplace

Guerra, K., & Prybutok, V. (2025). Blockchain technology: An analysis of economic, technical, and legal implications. *International Review of Law Computers & Technology*, •••, 1–21. DOI: 10.1080/13600869.2025.2497632

Gungor, V. C., Sahin, D., Kocak, T., Ergut, S., Buccella, C., Cecati, C., & Hancke, G. P. (2013). Smart grid technologies: Communication technologies and standards. *IEEE Transactions on Industrial Informatics*, *7*(4), 529–539. DOI: 10.1109/TII.2011.2166794

Gurumurthy, A., & Chami, N. (2021). Towards a Global Digital Constitutionalism: A Radical New Agenda for UN75. *Development*, *64*(1), 29–38. DOI: 10.1057/s41301-021-00287-z

Han, S., & Park, S. (2022). A Gap Between Blockchain and General Data Protection Regulation: A Systematic Review. *IEEE Access : Practical Innovations, Open Solutions*, *10*, 103888–103905. DOI: 10.1109/ACCESS.2022.3210110

Heng, Y. B., Ramachandaramurthy, V. K., Verayiah, R., & Walker, S. L. (2022). Developing Peer-to-Peer (P2P) Energy Trading Model for Malaysia: A Review and Proposed Implementation. *IEEE Access : Practical Innovations, Open Solutions*, *10*, 33183–33199. DOI: 10.1109/ACCESS.2022.3161577

Hernandez, J., Scherr, R., & Robertson, A. D. (2022). Redefining Energy Justice in Physics Classrooms. *Environmental Justice*, *15*(2), 108–114. DOI: 10.1089/env.2021.0042

Hewa, T., Ylianttila, M., & Liyanage, M. (2021). Survey on blockchain based smart contracts: Applications, opportunities and challenges. *Journal of Network and Computer Applications*, *177*, 102857. DOI: 10.1016/j.jnca.2020.102857

Hou, D., Zhang, J., Huang, S., Peng, Z., Ma, J., & Zhu, X. (2022). Privacy-Preserving Energy Trading Using Blockchain and Zero Knowledge Proof. In *2022 IEEE International Conference on Blockchain (Blockchain)* (pp. 412–418). DOI: 10.1109/Blockchain55522.2022.00064

How Blockchain Works. (n.d.). Blockchain: Empowering manufacturing sector. MM India. https://www.mmindia.co.in/article/533/blockchain-empowering-manufacturing-sector

Hrga, A., Capuder, T., & Zarko, I. P. (2020). Demystifying distributed ledger technologies: Limits, challenges, and potentials in the energy sector. *IEEE Access : Practical Innovations, Open Solutions*, *8*, 126149–126163. DOI: 10.1109/ACCESS.2020.3007935

Hua, W., Chen, Y., Qadrdan, M., Jiang, J., Sun, H., & Wu, J. (2022). Applications of blockchain and artificial intelligence technologies for enabling prosumers in smart grids: A review. *Renewable & Sustainable Energy Reviews*, *161*, 112308. Advance online publication. DOI: 10.1016/j.rser.2022.112308

Hua, W., Luo, F., Du, L., Chen, S., Kim, T., Morstyn, T., Robu, V., & Zhou, Y. (2022). Blockchain technologies empowering peer-to-peer trading in multi-energy systems: From advanced technologies towards applications. *IET Smart Grid*, *5*(4), 219–222. DOI: 10.1049/stg2.12081

Idries, A., Kanhere, S. S., & Jurdak, R. (2022). Challenges in platforming decentralized energy. *Energy Informatics*, *5*(Suppl 1), 193. Advance online publication. DOI: 10.1186/s42162-022-00193-9

Iglesias, B., & Guerrero, J. (2022). Blockchain-Based Peer-to-Peer Energy Trading Between Wind Producers and Prosumers. *Frontiers in Energy Research*, *10*, 923292. DOI: 10.3389/fenrg.2022.923292

Indiravathi, G., Kumar, M. P., Usha, A., Kumar, R., Lohith, T., & Shahid, S. (2024). Enhancing Data Privacy and Accountability in Smart Grids with Blockchain Technology. In *2024 International Conference on Electrical Electronics and Computing Technologies (ICEECT)* (Vol. 1, pp. 1–6). DOI: 10.1109/ICEECT61758.2024.10739131

Instytut Nauk Ekonomicznych, P. A. N., & Pietrewicz, L. (2019). Token-based blockchain financing and governance: A transaction cost economics approach. *Stud. Mater. Wydz. Zarz. UW*, *2/2018*(28), 126–139.

International Energy Agency (IEA). (2024). World Energy Outlook 2024 [Report]. https://www.iea.org/reports/world-energy-outlook-2024

International Energy Agency. (2024). World Energy Outlook 2024 [Report]. https://www.iea.org/reports/world-energy-outlook-2024

International Renewable Energy Agency (IRENA). (2019). Innovation landscape brief: Blockchain (Innovation Landscape Brief). International Renewable Energy Agency. https://www.irena.org/-/media/Files/IRENA/Agency/Publication/2019/Feb/IRENA_Landscape_Blockchain_2019.pdf

Invest in Estonia. (2018). Digital revolution in Estonia's energy sector: WePower is the first blockchain firm to tokenize an entire grid. https://investinestonia.com/wepower-is-the-first-blockchain-firm-to-tokenize-an-entire-grid/

Islam, S. N. (2024). A Review of Peer-to-Peer Energy Trading Markets: Enabling Models and Technologies. *Energies*, *17*(7), 1702. DOI: 10.3390/en17071702

Javadi, M.. (2022). Blockchain-Based Peer-to-Peer Transactive Energy Management Scheme for Smart Grid System. *Sensors (Basel)*, *22*(13), 4826. DOI: 10.3390/s22134826 PMID: 35808325

Jhariya, M. K., Dehalwar, V., Bharti, J., & Kumar, L. (2026). Energy efficient transactions for blockchain networks using adaptive global best–worst particle swarm optimization. *Scientific Reports*, *16*(1), 1643. Advance online publication. DOI: 10.1038/s41598-025-31112-z PMID: 41495151

Jogunola, O., Ajagun, A. S., Tushar, W., Olatunji, F. O., Yuen, C., Morley, C., Adebisi, B., & Shongwe, T. (2024). Peer-to-Peer Local Energy Market: Opportunities, Barriers, Security, and Implementation Options. *IEEE Access : Practical Innovations, Open Solutions*, *12*, 37873–37890. DOI: 10.1109/ACCESS.2024.3375525

Junlakarn, S., Kokchang, P., & Audomvongseree, K. (2022). Drivers and Challenges of Peer-to-Peer Energy Trading Development in Thailand. In *Energies* (Vol. 15, Issue 3, p. 1229). DOI: 10.3390/en15031229

Kameswaran, A., Nanthakumar, K., & Ruba, M. (2017). High performance pre-computation based self-controlled precharge-free content-addressable memory. *Journal of Electronics and Communication Systems*, *2*, 1–9.

Karandikar, N., Chakravorty, A., & Rong, C. (2021). Blockchain based transaction system with fungible and non-fungible tokens for a community-based energy infrastructure. *Sensors (Basel)*, *21*(11), 3822. DOI: 10.3390/s21113822 PMID: 34073110

Khan, M. N. I. (2025). Cross-Border Data Privacy and Legal Support: A Systematic Review of International Compliance Standards and Cyber Law Practices. *American Journal of Scholarly Research and Innovation*, *04*(01), 138–174. DOI: 10.63125/a4gbeb22

Khorasany, M., Dorri, A., Razzaghi, R., & Jurdak, R. (2020). *Lightweight Blockchain Framework for Location-aware Peer-to-Peer Energy Trading. arXiv preprint.* /arXiv.2005.14520DOI: 10.48550

Kirli, D., Couraud, B., Robu, V., Salgado-Bravo, M., Norbu, S., Andoni, M., Antonopoulos, I., Negrete-Pincetic, M., Flynn, D., & Kiprakis, A. (2022). Smart contracts in energy systems: A systematic review of fundamental approaches and implementations. *Renewable & Sustainable Energy Reviews, 158*, 112013. Advance online publication. DOI: 10.1016/j.rser.2021.112013

Klenergy Tech. (2018). Pylon Network: The energy blockchain platform — Token Paper [White paper]. https://www.allcryptowhitepapers.com/wp-content/uploads/2018/05/Pylon-Network.pdf

Korolov, R., Milevskyi, S., Shcherbyna, L., Lutsenko, Y., Dunaiev, S., & Stetsenko, V. (2024). Authentication and Authorization Method Based on McEliece Crypto-Code Construction. In *2024 8th International Symposium on Multidisciplinary Studies and Innovative Technologies (ISMSIT)* (pp. 1–5). DOI: 10.1109/ISMSIT63511.2024.10757296

Kouhizadeh, M., & Sarkis, J. (2018). Blockchain Practices, Potentials, and Perspectives in Greening Supply Chains. *Sustainability (Basel), 10*(10), 3652. DOI: 10.3390/su10103652

Kouhizadeh, M., Sarkis, J., & Zhu, Q. (2021). Blockchain and the circular economy: Critical reflections and future research directions. *Production Planning and Control, 32*(13), 1041–1054. DOI: 10.1080/09537287.2020.1732906

Kouhizadeh, M., Zhu, Q., & Sarkis, J. (2019). Blockchain and the circular economy: Potential tensions and critical reflections from practice. *Production Planning and Control, 31*(11–12), 950–966. DOI: 10.1080/09537287.2019.1695925

Kourtis, M.-A., Batistatos, M., Xylouris, G., Oikonomakis, A., Santorinaios, D., Zarakovitis, C., & Chochliouros, I. (2023). Energy efficiency in agriculture through tokenization of 5G and edge applications. *Energies, 16*(13), 5182. DOI: 10.3390/en16135182

Krithika, S., Jayachandran, T., Raghapriya, N. R., Ruba, M., Janani, C., & Sathiyavathi, S. (2025, April). An Effective Denial of Service Texture Features from Malware Grayscale Image. In 2025 3rd International Conference on Artificial Intelligence and Machine Learning Applications Theme: Healthcare and Internet of Things (AIMLA) (pp. 1-7). IEEE. DOI: 10.1109/AIMLA63829.2025.11041501

Kumar, C., & Chittora, P. (2024). Integrating Blockchain Technology into Smart Grid Systems: Opportunities, Challenges, and Future Directions. *2024 IEEE 11th Power India International Conference (PIICON)*, 1–6. DOI: 10.1109/PIICON63519.2024.10995188

Kus, M. C., & Levi, A. (2022). Investigation and Application of Differential Privacy in Bitcoin. *IEEE Access : Practical Innovations, Open Solutions, 10*, 25534–25554. DOI: 10.1109/ACCESS.2022.3151784

Lacey-Barnacle, M., Robison, R., & Foulds, C. (2020). Energy justice in the developing world: A review of theoretical frameworks, key research themes and policy implications. *Energy for Sustainable Development, 55*, 122–138. DOI: 10.1016/j.esd.2020.01.010

Laksono, M. N. G., Tiaraputri, Z. A., Sekti, B. A., & Laksono, M. I. A. (2025). Cyber-physical systems security in space. In *Advanced Cyber Defense for Space Missions and Operations* (pp. 85–113). Concepts and Applications., DOI: 10.4018/979-8-3693-7939-4.ch004

Latif, S., Idrees, Z., Huma, Z., & Ahmad, J. (2021). Blockchain technology for the industrial Internet of Things: A comprehensive survey on security challenges, architectures, applications, and future research directions. *Transactions on Emerging Telecommunications Technologies, 32*(11), e4337. DOI: 10.1002/ett.4337

Law Commission of England and Wales (UK Law Commission). (2021). Smart legal contracts: Advice to government. London: Law Commission. https://www.lawcom.gov.uk

Lessig, L. E. (2006). *Code and other laws of cyberspace, version 2.0*. Basic Books.

Littlechild, S. (2021). Decentralization, competition, and innovation in electricity markets. *Energy Policy, 148*, 111968. DOI: 10.1016/j.enpol.2020.111968

Liu, Z., Xiang, Y., Shi, J., Gao Peng and Wang, H., Xiao, X., Wen, B., & Hu, Y.-C. (2019). HyperService: Interoperability and Programmability Across Heterogeneous Blockchains.

López, I., Goitia-Zabaleta, N., Milo, A., Gómez-Cornejo, J., Aranzabal, I., Gaztañaga, H., & Fernández, E. (2024). European energy communities: Characteristics, trends, business models and legal framework. *Renewable & Sustainable Energy Reviews, 197*, 113046. Advance online publication. DOI: 10.1016/j.rser.2024.114403

Lowitzsch, J., Hoicka, C. E., & van Tulder, F. J. (2020). Renewable energy communities under the 2019 European Clean Energy Package. *Energy Policy, 139*, 111212. DOI: 10.1016/j.enpol.2020.111212

Low, K. F., & Mik, E. (2023). Smart contracts and the illusion of certainty. *The Modern Law Review*, *86*(2), 348–377.

Magazine, T. Y. N. (2023, August 2). Redes descentralizadas: el entorno perfecto para el crecimiento del blockchain. https://tynmagazine.com/redes-descentralizadas-el-entorno-perfecto-para-el-crecimiento-del-blockchain/

Mahmood, Z., & Vacius, J. (2020). Privacy-Preserving Block-chain Framework Based on Ring Signatures (RSs) and Zero-Knowledge Proofs(ZKPs). In *2020 International Conference on Innovation and Intelligence for Informatics, Computing and Technologies (3ICT)* (pp. 1–6). DOI: 10.1109/3ICT51146.2020.9312014

Manchalwar, A. D., Patne, N. R., & Thakre, M. P. (2024). Prosumers and retailers based decentralized energy trading model in the smart grid considering network constraints. *International Journal of Electrical Power & Energy Systems*, *160*, 110108. Advance online publication. DOI: 10.1016/j.ijepes.2024.110108

Manurung, E. H. (2025). *Blockchain technology and smart contracts: Legal implications and future regulations*. Laju Journal of Law and Society.

Maonaigh, C. Ó., Fitzgerald, L. M., & Reilly, L. (2025). Revisiting the social licence to operate in the energy transition: An intersectional agenda for research and practice. *Energy Research & Social Science*, *126*, 104139. Advance online publication. DOI: 10.1016/j.erss.2025.104139

Marchewka-Bartkowiak, K., & Wiśniewski, M. (2022). Energy tokens as digital instruments of financial investment. Econ. Bus. Rev., 8 (22)(3), 109–125.

Marin, O., Cioara, T., & Anghel, I. (2023). Blockchain solution for buildings' multi-energy flexibility trading using multi-token standards. *Future Internet*, *15*(5), 177. DOI: 10.3390/fi15050177

Marisamy, K., Selvakumar, A., Rajasekar, B., Natarajan, R., Atajanova, A., & Makhmudov, S. (2025). Smart logistics integrating AI, IoT, blockchain, and Robotics for a sustainable Future. In *Emerging Trends in Smart Logistics Technologies* (pp. 73–104). IGI Global Scientific Publishing. DOI: 10.4018/979-8-3373-2434-0.ch003

Matthias, A. (2004). The responsibility gap: Ascribing responsibility for the actions of learning automata. *Ethics and Information Technology*, *6*(3), 175–183. DOI: 10.1007/s10676-004-3422-1

Meijer, A., & Bolívar, M. P. R. (2016). Governing the smart city: A review of the literature on smart urban governance. *International Review of Administrative Sciences*, *82*(2), 392–408. DOI: 10.1177/0020852314564308

Mengelkamp, E., Gärttner, J., Rock, K., Kessler, S., Orsini, L., & Weinhardt, C. (2018). Designing microgrid energy markets: A case study of the Brooklyn Microgrid. *Applied Energy*, *210*, 870–880. DOI: 10.1016/j.apenergy.2017.06.054

Michaelis, A., Schneider, M., & Weibelzahl, M. (2025). Designing local flexibility markets: A toolbox for policymakers and market operators. *Energy*, *329*, 136051. Advance online publication. DOI: 10.1016/j.energy.2025.136051

Mik, E. (2017). Smart contracts: Terminology, technical limitations and real-world complexity. *Law, Innovation and Technology*, *9*(2), 269–300. DOI: 10.1080/17579961.2017.1378468

Mitrea, D., Antonopoulos, I., Couraud, B., Kirli, D., Kiprakis, A., Robu, V., & Flynn, D. (2023). Privacy-preserving computation for peer-to-peer energy trading on a public blockchain. *Sensors (Basel)*, *23*(10), 4640. Advance online publication. DOI: 10.3390/s23104640 PMID: 37430554

Mitrea, D., Toderean, L., Cioara, T., Anghel, I., & Antal, M. (2023). Smart contracts and homomorphic encryption for private P2P energy trading and demand response on blockchain. *Heliyon*, *9*(11), e22357. DOI: 10.1016/j.heliyon.2023.e22357

Monetary Authority of Singapore. (2024). Regulatory Sandbox. Retrieved November 14, 2025, from https://www.mas.gov.sg/development/fintech/regulatory-sandbox

Mottaghi, S. F., Steininger, B. I., & Yanagawa, N. (2024). Real Estate Insights: The current state and the new future of tokenization in real estate. *Journal of Property Investment & Finance*, *42*(6), 614–620. DOI: 10.1108/JPIF-07-2024-0087

Narayanan, A., Bonneau, J., Felten, E., Miller, A., & Goldfeder, S. (2016). *Bitcoin and cryptocurrency technologies: a comprehensive introduction*. Princeton University Press.

National Privacy Commission (Philippines). (2024). *Advisory opinion on blockchain and the Data Privacy Act: Guidelines for the energy sector*.

Nguyen, L.-H., Nguyen, V.-L., Hwang, R.-H., Kuo, J.-J., Chen, Y.-W., Huang, C.-C., & Pan, P.-I. (2024). Towards Secured Smart Grid 2.0: Exploring Security Threats, Protection Models, and Challenges. *IEEE Communications Surveys and Tutorials*, *1*. Advance online publication. DOI: 10.1109/COMST.2024.3493630

Niet, I., van Est, R., & Veraart, F. (2021). Governing AI in Electricity Systems. *Frontiers in Artificial Intelligence*, *4*, 690237. DOI: 10.3389/frai.2021.690237 PMID: 34396090

OECD. (2025). *Regulating for the future: OECD Regulatory Policy Outlook 2025*. OECD Publishing., DOI: 10.1787/56b60e39-

Office of Gas and Electricity Markets. (2023). Consultation on updates to Data Best Practice Guidance and Digitalisation Strategy and Action Plan Guidance (Consultation OFG1163) [PDF]. https://www.ofgem.gov.uk/sites/default/files/2023-02/Consultation%20on%20updates%20to%20Data%20Best%20Practice%20Guidance%20and%20Digitalisation%20Strategy%20and%20Action%20Plan%20Guidance.pdf

Office of Gas and Electricity Markets. (2023). Consultation on updates to Data Best Practice Guidance and Digitalisation Strategy and Action Plan Guidance (Consultation OFG1163) [PDF]. https://www.OFGEM.gov.uk/sites/default/files/2023-02/Consultation%20on%20updates%20to%20Data%20Best%20Practice%20Guidance%20and%20Digitalisation%20Strategy%20and%20Action%20Plan%20Guidance.pdf

Ogundipe, A. (2025)... *The Strategic Role of Cloud Security in Protecting National Information Infrastructure : Policies, Risks, and Defense Mechanisms The Strategic Role of Cloud Security in Protecting National Information Infrastructure : Policies, Risks, and Defense Mech*, (July). Advance online publication. DOI: 10.63680/ijsate032535.06

Okoye, M. O., Castro, G. F., Fingerhuth, S., & Yang, J. (2024). The consumer-centered electricity reliability enhancement in the standalone generation arena. *IEEE Access : Practical Innovations, Open Solutions*, *12*, 9984–9998. DOI: 10.1109/ACCESS.2023.3349374

Opoku-Mensah, E., Chen, W., Tuffour, P., Agozie, D. Q., Gyamfi, B. A., & Mahmoud, A. (2024). Toward sustainable energy transition: Unveiling the synergies of democracy, energy justice, and structural adjustment on emissions in West Africa. *Sustainable Development (Bradford)*.

Papapostolou, A., Andreoulaki, I., Anagnostopoulos, F., Divolis, S., Niavis, H., Vavilis, S., & Marinakis, V. (2025). Innovative Business Models Towards Sustainable Energy Development: Assessing Benefits, Risks, and Optimal Approaches of Blockchain Exploitation in the Energy Transition. In *Energies* (Vol. 18, Issue 15, p. 4191). DOI: 10.3390/en18154191

Parag, Y., & Sovacool, B. K. (2016). Electricity market design for the prosumer era. *Nature Energy*, *1*(4), 16032. DOI: 10.1038/nenergy.2016.32

Paudler, H. J. A., Schwanitz, V. J., & Wierling, A. (2022). Realizing a human-centered digitalization of the energy sector. *Open Research Europe*, *2*, 140. DOI: 10.12688/openreseurope.15129.1 PMID: 37645332

Peters, G. W., & Panayi, E. (2016). Understanding modern banking ledgers through blockchain technologies: Future of transaction processing and smart contracts on the internet of money. In *Banking beyond banks and money* (pp. 239–278). Springer., DOI: 10.1007/978-3-319-42448-4_13

Ponkarthika, M., Prabhu, A., Ruba, M., Kishore, Y. R., & GV, M. (2026). Bridging Health Equity in Times of Crisis: Ensuring Inclusive Care During Emergencies and Disasters. In Advancing Equity in Global Healthcare Systems (pp. 27-54). IGI Global Scientific Publishing.

Poovithaa, M., Ruba, M., Ponkarthika, M., Ragul, N., Ragul, V., & Sangeetha, K. (2025, November). IoT Smart Traffic Control System for Hill Roads with Emergency Vehicle Priority. In 2025 IEEE 7th International Conference on Computing, Communication and Automation (ICCCA) (pp. 1-6). IEEE.

Popov, E., Veretennikova, A., & Fedoreev, S. (2022). The model of OTC securities market transformation in the context of asset tokenization. *Mathematics*, *10*(19), 3441. DOI: 10.3390/math10193441

Powerledger. (2023). Global Blockchain Business Council – 2022 Annual Report [Case study]. https://powerledger.io/media/global-blockchain-business-council-2022-annual-report/

Pradhan, N. R., Singh, A. P., Kumar, N., Hassan, M. M., & Roy, D. S. (2022). A Flexible Permission Ascription (FPA)-Based Blockchain Framework for Peer-to-Peer Energy Trading With Performance Evaluation. *IEEE Transactions on Industrial Informatics*, *18*(4), 2465–2475. DOI: 10.1109/TII.2021.3096832

Quan, M. K., Pathirana, P. N., Wijayasundara, M., Setunge, S., Nguyen, D. C., Brinton, C. G., & Poor, H. V. (2025). Federated Learning for Cyber Physical Systems: A Comprehensive Survey. *IEEE Communications Surveys and Tutorials*, *1*. Advance online publication. DOI: 10.1109/COMST.2025.3570288

Rahimi, A., Khorasany, M., Dorri, A., & Jurdak, R. (2022). Lightweight Blockchain Framework for Location-Aware P2P Energy Trading. *Scientific Reports*, *12*, 18603. DOI: 10.1038/s41598-022-18603-z

Rahman, M., Chowdhury, S., Shorfuzzaman, M., Hossain, M. K., & Hammoudeh, M. (2023). Peer-to-peer power energy trading in blockchain using efficient machine learning model. *Sustainability (Basel), 15*(18), 13640. Advance online publication. DOI: 10.3390/su151813640

Raskin, M. (2017). The law and legality of smart contracts. Georgetown Law Technology Review, 1(2), 305–341. https://georgetownlawtechreview.org/the-law-and-legality-of-smart-contracts

Reijers, W., O'Brolcháin, F., & Haynes, P. (2016). Governance in blockchain technologies & social contract theories. Ledger, 1, 134 151. DOI: 10.5195/ledger.2016.62

Romano, L., & Marino, A. (2022). Blockchain, Data Protection & P2P Energy Trading: GDPR Implications. *Sustainability, 14*(23), 16305. DOI: 10.3390/su142316305

Ruba, M., Ashwin Prabhu, G., Robinson Joel, M., Vishnu Priya, K., Dharshani, L., & Dharani, R. (2026). Integrating emerging technologies for organizational agility: An interdisciplinary approach to digital transformation. In Roache, D. (Ed.), *High performance leadership for organizational excellence: Innovation, technology, and resilience* (pp. 63–92). IGI Global Scientific Publishing., DOI: 10.4018/979-8-3373-6207-6.ch003

Santhini, C., Ruba, M., Sakthi Priya, V., & Hemalatha, R. (2025). Intercultural communication and assessment practices in the integration of AI with education. In *Generative AI Approaches to Sustainable Development in Higher Education*. IGI Global. DOI: 10.4018/979-8-3693-5623-4.ch010

Savelyev, A. (2018). Contract law 2.0: 'Smart' contracts as the beginning of the end of classic contract law. *Information & Communications Technology Law, 26*(2), 116–134. DOI: 10.1080/13600834.2017.1301036

Schellinger, B., Völter, F., Urbach, N., & Sedlmeir, J. (2022). Yes, I Do: Marrying Blockchain Applications with GDPR. *Proceedings of the Annual Hawaii International Conference on System Sciences, 2022-Janua*, 4631–4640. DOI: 10.24251/HICSS.2022.563

Schneiders, A., & Shipworth, D. (2021). Community Energy Groups: Can They Shield Consumers from the Risks of Using Blockchain for Peer-to-Peer Energy Trading? In *Energies* (Vol. 14, Issue 12, p. 3569). DOI: 10.3390/en14123569

Schwiderowski, J., Pedersen, A. B., & Beck, R. (2024). Crypto tokens and token systems. *Information Systems Frontiers, 26*(1), 319–332. DOI: 10.1007/s10796-023-10382-w

Sekti, B. A. (2025). Adversarial Machine Learning in Industrial Cybersecurity: Challenges and Solutions. In *AI-Enhanced Cybersecurity for Industrial Automation* (pp. 363–386). IGI Global Scientific Publishing. DOI: 10.4018/979-8-3373-3241-3.ch018

Sekti, B. A. (2025). IT Service Information Security Management. In *Cases on Information Systems Service Management* (pp. 61–93). DOI: 10.4018/979-8-3373-2352-7.ch003

Sekti, B. A. (2026a). Cybersecurity Challenges in Smart City Development. In *Securing Smart Cities Through Modern Cryptography Technologies* (pp. 75–102). IGI Global Scientific Publishing., DOI: 10.4018/979-8-3373-3326-7.ch003

Sekti, B. A. (2026b). Data Privacy and Cybersecurity in Wind Power Systems. In *AI-Powered Analysis* (pp. 335–366). Modeling, and Monitoring of Wind Energy Systems., DOI: 10.4018/979-8-3373-4159-0.ch009

Seven, S., Yao, G., Soran, A., Onen, A., & Muyeen, S. M. (2020). Peer-to-Peer Energy Trading in Virtual Power Plant Based on Blockchain Smart Contracts. *IEEE Access : Practical Innovations, Open Solutions*, *8*, 175713–175726. DOI: 10.1109/ACCESS.2020.3026180

Shahzad, S., Abbasi, M. A., Ali, H., Iqbal, M., Munir, R., & Kilic, H. (2023). Possibilities, challenges, and future opportunities of microgrids: A review. *Sustainability (Basel)*, *15*(8), 6366. DOI: 10.3390/su15086366

Shamaseen, A., Qatawneh, M., & Elshqeirat, B. (2025). Blockchain for future smart grid: A comprehensive survey. *Bulletin of Electrical Engineering and Informatics*, *14*(4), 2497–2513. DOI: 10.11591/eei.v14i4.8956

Shanmugaraja, T., Ruba, M., Priya, V. S., Venkatesh, T., Supriya, M., & Nallamuthu, M. (2023, September). ECG noise elimination using an FPGA version of a modified error normalized LMS adaptive filter. In AIP Conference Proceedings (Vol. 2764, No. 1, p. 060033). AIP Publishing LLC. DOI: 10.1063/5.0144367

Sharma, D. D. (2024). Asynchronous blockchain-based federated learning for tokenized smart power contract of heterogeneous networked microgrid system. *IET Blockchain*, *4*(4), 302–314. DOI: 10.1049/blc2.12041

Sharma, D. D., & Ansari, A. R. (2024). Blockchain-Powered Privacy-Enhanced Design For Electric Vehicle Energy Market. In *2024 IEEE Conference on Engineering Informatics (ICEI)* (pp. 1–6). DOI: 10.1109/ICEI64305.2024.10912198

Shelake, S. (2025). *The future of legal tech: Smart contracts in emerging markets*. Global Jurist.

Shi, J. (2023). Decoding the future of finance: An examination of asset tokenization and Web 3.0. *Eximia*, *11*, 257–263. DOI: 10.47577/eximia.v11i1.303

Shorya, R., & Jagwani, P. (2025). Enhancing Digital Autonomy in Peer-to-Peer Energy Trading: A Blockchain and Predictive Analytics Approach. *Procedia Computer Science*, *254*, 230–239. 10.1016/j.procs.2025.02.082. DOI: 10.1016/j.procs.2025.02.082

Sibiya, S., Ruba, M., Ponkarthika, M., Vijayasabariswari, M., Sheba, S., Vignesh, S., . . . Sankar, S. S. (2025, October). Integrated Wearable Health Monitoring and Emergency Response System With IOT Sensors, AED and Mobile App for Enhanced Patient Outcomes. In 2025 IEEE 2nd International Conference on Green Industrial Electronics and Sustainable Technologies (GIEST) (pp. 1-6). IEEE.

Siemens. (2018). A microgrid grows in Brooklyn. Siemens. Retrieved October 14, 2025, from https://www.siemens.com/global/en/company/stories/research-technologies/energytransition/a-microgrid-grows-in-brooklyn.html

Siemens. (2020). Microgrid project in Vienna: Small grid, major impact. Siemens. https://www.siemens.com/global/en/company/stories/infrastructure/2020/microgrid-project-in-vienna.html

Solahudeen Tando, M., Gawusu, S., Yakubu, S., Kwabena, O. S., Donatus, D., & Raqib Abu, S. (2024). Decentralized energy trading with blockchain technology. In *The Intersection of Blockchain and Energy Trading* (pp. 75–101). Exploring Decentralized Solutions for Next-Generation Energy Markets., DOI: 10.1016/B978-0-443-23627-3.00004-1

Soto, E. A., Bosman, L. B., Wollega, E., & Leon-Salas, W. D. (2021). Peer-to-peer energy trading: A review of the literature. *Applied Energy*, *283*, 116268. Advance online publication. DOI: 10.1016/j.apenergy.2020.116268

Sovacool, B. K., Burke, M., Baker, L., Kotikalapudi, C. K., & Wlokas, H. (2017). New frontiers and conceptual frameworks for energy justice. *Energy Policy*, *105*, 677–691. DOI: 10.1016/j.enpol.2017.03.005

Sovacool, B. K., Sovacool, B. K., Martiskainen, M., Hook, A., & Baker, L. (2019). Decarbonization and its discontents: A critical energy justice perspective on four low-carbon transitions. *Climatic Change*, *155*(4), 581–619. DOI: 10.1007/s10584-019-02521-7

Srinivas, O., Pradhan, N. R., & Singh, A. P. (2025). Blockchain Framework for Nega-watt Energy Trading With Incentives and Limits: P2PnW-Chain. *IEEE Transactions on Consumer Electronics*, *71*(3), 7608–7621. DOI: 10.1109/TCE.2025.3597247

Stoll, C., Klaaßen, L., & Gallersdörfer, U. (2019). The carbon footprint of bitcoin. *Joule*, *3*(7), 1647–1661. DOI: 10.1016/j.joule.2019.05.012

Sun Exchange. (2023). Solar power for the people: Crowd-funded energy projects in emerging markets. Sun Exchange Ltd. https://thesunexchange.com/

Supreme Court of the Philippines. (2025). *G.R. No. 270000: re: Liability of autonomous trading platforms and digital torts.*

Syahputra, E. (2022). Blockchain Energy for Future Smart Grid in Indonesia: A Brief Review. *2022 International Conference on Technology and Policy in Energy and Electric Power (ICT-PEP)*, 100–105. DOI: 10.1109/ICT-PEP57242.2022.9988867

Szabo, N. (1997). Formalizing and securing relationships on public networks. *First Monday*, *2*(9). Advance online publication. DOI: 10.5210/fm.v2i9.548

Szabo, N. (1997). *The idea of smart contracts. First International Conference on Financial Cryptography.*

Tanis, Z., Durusu, A., & Altintas, N. (2025). A comprehensive review on peer-to-peer energy trading: Market structure, operational layers, energy cooperatives and multi-energy systems. *IET Renewable Power Generation*, *19*(1), e70075. Advance online publication. DOI: 10.1049/rpg2.70075

Tenne, T. (n.d.). Innovations. Retrieved November 14, 2025, from https://www.tennet.eu/innovations-tennet

Tesfamicael, A. D., Liu, V., Mckague, M., Caelli, W., & Foo, E. (2020). A design for a secure energy market trading system in a national wholesale electricity market. *IEEE Access : Practical Innovations, Open Solutions*, *8*, 132424–132445. DOI: 10.1109/ACCESS.2020.3009356

Thanasi-Boçe, M., & Hoxha, J. (2025). Blockchain for Sustainable Development: A Systematic Review. *Sustainability (Basel)*, *17*(11), 4848. DOI: 10.3390/su17114848

Troccoli, A., Goodess, C., Jones, P., Penny, L., Dorling, S., Harpham, C., Dubus, L., Parey, S., Claudel, S., Khong, D.-H., Bett, P. E., Thornton, H., Ranchin, T., Wald, L., Saint-Drenan, Y.-M., De Felice, M., Brayshaw, D., Suckling, E., Percy, B., & Blower, J. (2018). Creating a proof-of-concept climate service to assess future renewable energy mixes in Europe: An overview of the C3S ECEM project. *Advances in Science and Research*, *15*, 191–205. DOI: 10.5194/asr-15-191-2018

Tushar, W., Yuen, C., Mohsenian-Rad, H., Saha, T. K., Wood, K. L., & Poor, H. V. (2020). Transforming energy networks via peer-to-peer energy trading: The potential of game-theoretic approaches. *IEEE Signal Processing Magazine*, *35*(4), 90–111. DOI: 10.1109/MSP.2018.2818327

Tushar, W., Yuen, C., Saha, T. K., Morstyn, T., Chapman, A. C., Alam, M. J. E., Hanif, S., & Poor, H. V. (2021). Peer-to-peer energy systems for connected communities: A review of recent advances and emerging challenges. *Applied Energy*, *282*, 116131. DOI: 10.1016/j.apenergy.2020.116131

UK Law Commission. (2021). *Smart legal contracts: Advice to government* (Law Com No. 401). Her Majesty's Stationery Office.

Vaziry, A., Garzon, S. R., Herbke, P., Segat, C., & Küpper, A. (2025). *SoK: A Taxonomy for Distributed-Ledger-Based Identity Management. In 2025 Crypto Valley Conference*. CVC., DOI: 10.1109/CVC65719.2025.00015

Verma, A. K., Navas-Fonesca, A., Burgos-Mellado, C., & Dragicevic, T. (2025). Realigning Academic Cybersecurity Research With Industrial Needs in Cyber-Physical Systems. *IEEE Open Journal of the Industrial Electronics Society*, *6*, 1–39. DOI: 10.1109/OJIES.2025.3593689

Verma, P., O'Regan, B., Hayes, B., Thakur, S., & Breslin, J. G. (2018). EnerPort: Irish Blockchain project for peer- to-peer energy trading. *Energy Informatics*, *1*(1), 14. DOI: 10.1186/s42162-018-0057-8

Villa-Ávila, E., Arévalo, P., & Albornoz, E. (2025). 8 - Peer-to-peer and blockchain technologies in power systems. In M. Tostado-Véliz, A. Rezaee Jordehi, S. A. Mansouri, A. Ramos Galán, & F. J. B. T.-T. F. S. P. S. with H. P. of R. Melguizo (Eds.) (pp. 171–186). Academic Press. 10.1016/B978-0-443-29871-4.00008-7

Vishwakarma, A. K.. (2024). Blockchain-based peer-to-peer renewable energy trading with loss traceability. *International Journal of Production Research.* Advance online publication. DOI: 10.1080/00207543.2024.2441224

Vyas, N., Beije, A., & Krishnamachari, B. (2022). *Blockchain and the supply chain: Concepts, strategies and practical applications* (2nd ed.). Kogan Page.

Wang, B., Xu, L., & Wang, J. (2023). A privacy-preserving trading strategy for blockchain-based p2p electricity transactions. *Applied Energy*, *335*, 120664. Advance online publication. DOI: 10.1016/j.apenergy.2023.120664

Wang, S., Taha, A. F., Wang, J., Kvaternik, K., & Hahn, A. (2019). Energy Crowdsourcing and Peer-to-Peer Energy Trading in Blockchain-Enabled Smart Grids. *IEEE Transactions on Systems, Man, and Cybernetics. Systems*, *49*(8), 1612–1623. DOI: 10.1109/TSMC.2019.2916565

Wang, S., Taher, A. F., Wang, J., Kvaternik, K., & Hahn, A. (2019). Energy Crowdsourcing and Peer-to-Peer Energy Trading in Blockchain-Enabled Smart Grids. *IEEE Transactions on Smart Grid*, *10*, 5501–5511. DOI: 10.1109/TSG.2019.2895172

Wang, Y., Su, Z., Ni, J., Zhang, N., & Shen, X. (2022). Blockchain-Empowered Space-Air-Ground Integrated Networks: Opportunities, Challenges, and Solutions. *IEEE Communications Surveys and Tutorials*, *24*(1), 160–209. DOI: 10.1109/COMST.2021.3131711

Werbach, K. (2021). *The blockchain and the new architecture of trust*. MIT Press.

Werbach, K., & Cornell, N. (2017). Contracts Ex Machina. *Duke Law Journal*, *67*(2), 313–382. https://scholarship.law.duke.edu/dlj/vol67/iss2/2/

Wongthongtham, P., Marrable, D., Abu-Salih, B., Liu, X., & Morrison, G. (2021). Blockchain-enabled Peer-to-Peer energy trading. *Computers & Electrical Engineering*, *94*, 107299. DOI: 10.1016/j.compeleceng.2021.107299

Wu, F. (2024). Legal imperatives and regulatory mechanisms for sustainable energy development: A comparative analysis of Renewable Energy policies in Australia and China. *Lecture Notes in Education Psychology and Public Media*, *35*(1), 206–211. DOI: 10.54254/2753-7048/35/20232106

Xue, F., Chang, K., Li, W., Wang, Q., Zhao, H., Zhang, H., Ni, Y., & Xia, W. (2022). Blockchain Smart Contract-Enabled Secure Energy Trading for Electric Vehicles. *Energies*, *15*(18), 6733. DOI: 10.3390/en15186733

Yahaya, A. S., Javaid, N., Almogren, A., Ahmed, A., Gulfam, S. M., & Radwan, A. (2021). A Two-Stage Privacy Preservation and Secure Peer-to-Peer Energy Trading Model Using Blockchain and Cloud-Based Aggregator. *IEEE Access : Practical Innovations, Open Solutions*, *9*, 143121–143137. DOI: 10.1109/ACCESS.2021.3120737

Yamaguchi, J. A. R., Teresa Rachael Santos, T. R., Vargas, G., & Pereira, A. de C. (2021). Blockchain technology in renewable energy certificates: An exploratory study. Business & Accounting Review, 20(1), 934-958. https://doi.org/DOI: 10.1590/1807-7692bar2021200069

Yang, Y., Tu, Z., Liu, Y., & Zhou, H. (2023). Blockchain-empowered token-based access control system with user reputation evaluation. *Computers, Materials & Continua*, *77*(3), 3163–3184. DOI: 10.32604/cmc.2023.043974

Yannam, Y. R., Gao, J., Shi, J., Datta, A., Wang, K., Krishnamoorti, R., Powell, J. B., Sengupta, D., & Ye, X. (2026). Toward a human-centered energy transition: Concepts, models, challenges, and research opportunities. *Utilities Policy*, *99*, 102145. Advance online publication. DOI: 10.1016/j.jup.2026.102145

Yarrabothu, R. S., Benni, N. S., Nirmal Raj, S., N, M. K., Singh, B. K., Divya, A., Bhojan, R., & Tirth, V. (2026). Quantum-resistant blockchain and federated learning integrated secure and scalable smart grid framework for sustainable and privacy-preserving energy management. *Sustainable Computing : Informatics and Systems*, *50*, 101305. Advance online publication. DOI: 10.1016/j.suscom.2026.101305

Yeung, K. (2018). Algorithmic regulation: A critical interrogation. *Regulation & Governance*, *12*(4), 505–523. DOI: 10.1111/rego.12158

Yusuf, F., Sari, R. F., Yusgiantoro, P., & Soesilo, T. E. (2024). Stakeholders' Perceptions of the Peer-to-Peer Energy Trading Model Using Blockchain Technology in Indonesia. In *Energies* (Vol. 17, Issue 19, p. 4956). DOI: 10.3390/en17194956

Zhang, S., Sovacool, B. K., & Wei, C. (2026). Do clean energy transitions enhance global energy security? A comparative analysis of trends and tradeoffs across eight major economies, 1990-2020. *Energy Policy*, *212*, 115164. Advance online publication. DOI: 10.1016/j.enpol.2026.115164

Zheng, Y., Zhang, R., Xu, J., Gong, Y., You, Y., & Zhang, Y. (2023). *Research on the data governance in power industry based on data owner. In 2023 Panda Forum on Power and Energy (PandaFPE)*. IEEE., DOI: 10.1109/PandaFPE57779.2023.10140515

Zhou, Y., Manea, A. N., Hua, W., Wu, J., Zhou, W., Yu, J., & Rahman, S. (2022). Application of Distributed Ledger Technology in Distribution Networks. *Proceedings of the IEEE*, *110*(12), 1963–1975. DOI: 10.1109/JPROC.2022.3181528

Zhuang, P., Zamir, T., & Liang, H. (2021). Blockchain for Cybersecurity in Smart Grid: A Comprehensive Survey. *IEEE Transactions on Industrial Informatics*, *17*(1), 3–19. DOI: 10.1109/TII.2020.2998479

Zhu, Y., & Salam, R.. (2024). Transaction-Loss-Aware Energy Trading in Microgrids. *Frontiers in Energy Research*, *12*, 1397975. DOI: 10.3389/fenrg.2024.1397975

About the Contributors

Karisma Karisma previously served as an Assistant Professor at the Tunku Abdul Rahman University of Management and Technology. She is currently an independent researcher. Her research pursuits revolve around the intersection of law and technology, emphasising interdisciplinary and comparative perspectives. Specifically, she explores the applicability of blockchain technology in the energy sector, delving into the associated legal and regulatory dimensions. Her commitment to excellence manifests in several high-quality research papers she has published and presentations she has delivered at international conferences. Her profound understanding of foundational aspects, astute insights, and expertise in decentralised systems make her an exceptional candidate for the presented research opportunity.

***

N. Mari Anand has a rich experience of 10 year in the field of teaching and expertise in the field of Marketing and Human Resource Management. Currently he is working as Faculty Member, Department of Business Studies, University of Technology & Applied Sciences, Salalah, Sultanate of Oman. He is also a Member in Research Advisory Committee at VELS Institute of Science, Technology & Advanced Studies, Chennai. To his credit he has published 27 articles in Scopus and UGC Listed Journals. He has guided 63 UG and 22 PG projects. He served as Member of Board of Studies for Business Administration and Member for University Examination Board of SRM-IST. The author is an Editorial Board Member for International Journal of Advance Research in Commerce Management and Finance (IJARCMF). The author has organized various International Conferences and International Conclave during his tenure. The author was an ISO Convener and served as IQAC coordinator & Exam Cell convener. He is also an External Academic Auditor and Chief Examiner for various institutions. He is an active Member in

Federation of Indian Chamber of Commerce and Industries (FICCI) & Madras Management Association (MMA).

Chamdramowleeswaran G is currently working as Assistant Professor in Department of Commerce and business administration, VelTech Rangarajan Dr. Sagunthala R&D Institute of Science and Technology (Deemed to be University), Avadi, Chennai. He has more than 11 years 5 months of industry experience and 1 year and 11 months of teaching experience with specialization in HR and Marketing. He Holds a B.Com., M.B.A., M. Phil., and Ph.D. in Management from Bharathidasan University, Trichy

Wildan Kamil is a machine learning researcher at Esa Unggul University, Jakarta, Indonesia. His research interests focus on machine learning, data analytics, predictive modeling, and intelligent information systems. He has been actively involved in academic research and applied projects related to data-driven technologies and the development of intelligent models for solving real-world problems. His work reflects strong analytical thinking, technical competence, and a commitment to innovation in artificial intelligence and machine learning applications. Wildan actively contributes to collaborative research initiatives and scientific publications aimed at advancing the implementation of machine learning in various domains, including digital systems, decision-support technologies, and data-driven innovation.

Muhamad Mandani is a motivated individual with a strong interest in information technology, digital applications, and system development. He has participated in academic and practical activities involving web-based systems, data processing, and application design, with an emphasis on functionality and user experience. His approach combines technical competence with problem-solving skills to address practical challenges effectively. Alif is known for his ability to learn quickly and collaborate in team-based environments. He continues to expand his knowledge and skills in emerging technologies, aiming to contribute to innovative digital solutions in academic and professional settings.

Froilan Mobo is an internationally recognized researcher, educator, and leader with over 14 years of service at the Philippine Merchant Marine Academy (PMMA), where he currently serves as a full Professor II. Rising through the ranks from Instructor 1 to his current position in March 2024, he has become a respected authority in education and public administration. Dr. Mobo holds a Ph.D. in Development Education from Central Luzon State University and a Doctorate in Public Administration from Urdaneta City University. He has also earned advanced degrees in Computer Science, Business Administration, and Social Studies

Education, with ongoing studies in Social Studies Education at Bicol University. A prolific researcher, Dr. Mobo has authored 123 publications with 282 citations in Scopus and ACI, making him the top scientist from PMMA in Education and Educational Technology. His renowned study, "The Impact of Video Conferencing Platforms in All Educational Sectors Amidst the COVID-19 Pandemic," has influenced global educational practices.

Sivakumar Nadarajan is an Associate Professor in the Department of Computer Engineering at Marwadi University, Rajkot. With extensive academic and research experience, he specializes in artificial intelligence, machine learning, and advanced computing technologies. He has contributed to several research publications, conferences, and academic initiatives, guiding students in innovative and interdisciplinary projects. His teaching focuses on bridging theoretical concepts with practical applications, fostering critical thinking and problem-solving skills. Actively involved in research and academic development, he aims to advance knowledge in emerging technologies and contribute to the evolving landscape of computer engineering and intelligent systems.

Karthika P. Devan is an academician and researcher in the field of Commerce and Management, currently serving as an Assistant Professor at Vishwakarma University, Pune, India. She holds a Ph.D. in Commerce/Finance from Mahatma Gandhi University, Kerala, with research focusing on the dynamics and hedging effectiveness of the Indian currency futures market. With over a decade of teaching experience in higher education, she has been actively involved in teaching, research, and academic administration. Her areas of interest include financial derivatives, financial markets, financial econometrics, and international finance. Dr. Devan has contributed to several research publications and has actively participated in academic conferences, seminars, and workshops. In addition to teaching, she is involved in curriculum development, student mentoring, and organizing academic events and seminars that promote research and innovation among students and faculty. Through her academic and research activities, she aims to contribute to the development of knowledge in finance and management education.

Jamshid Paradev is a faculty member in the Department of Finance and Tourism at Termez University of Economics and Service, Termez, Uzbekistan. His academic interests include finance, tourism economics, and service-sector development, with a focus on the application of economic and financial principles to tourism and regional development. He is actively engaged in teaching and academic activities, contributing to curriculum development and applied research in finance and tourism studies.

Ravikumar R. N. is an experienced academician with over 12+ years of teaching and research expertise across institutions in India, the UAE, and the Maldives. He currently serves as an Assistant Professor at Marwadi University, Rajkot, specializing in Machine Learning, Recommender Systems, and AI-driven applications in healthcare and security. He holds a Ph.D. in Machine Learning and Recommender Systems from Amity University, Jaipur, a Master of Engineering in CSE from VMRF University, Salem, and a Bachelor of Engineering in CSE from Anna University. He also has a MiniMBA from the Swiss eLearning Institute. Dr. Ravikumar has authored more than 8+ multiple SCIE/Scopus-indexed journal articles, 25+ IEEE/Springer conference papers, and co-authored a book on Artificial Intelligence. He has also published 9 patents and contributed 10+ book chapters in emerging domains such as Federated Learning, Blockchain, and Digital Twins for IGI Global, Wiley and Springer publisher. He actively mentors students through initiatives like the Research Activities Club (RAC) and Student Research Internship Training (SRIT), promoting innovation and research culture.

Aarthi S. is a dedicated Assistant Professor in Computer Science with over six years of teaching experience. She is currently pursuing her Ph.D. at Marwadi University, Gujarat, and holds an M.Phil. and MCA from Bharathiar University, Tamil Nadu. Throughout her academic career, Aarthi has specialized in programming languages such as C++, Java, and C, and contributed to research in recommender systems and AI-related data privacy. She has published journal articles, contributed to book chapters, and actively participated in technical workshops. Known for her student mentorship and engagement in academic events, Aarthi remains committed to fostering innovation and excellence in computer science education.

Gusti Muhamad Sardana is a researcher in artificial intelligence and early warning systems at Esa Unggul University, Jakarta, Indonesia. His research interests include machine learning, intelligent monitoring systems, sensor-based technologies, and the development of AI-driven early detection solutions. He has been actively involved in academic research and applied projects focusing on predictive analytics, real-time monitoring, and the integration of intelligent systems to support risk mitigation and rapid response mechanisms. His work reflects strong analytical thinking, technical expertise, and a commitment to innovation in intelligent detection and monitoring technologies. Gusti actively contributes to collaborative research initiatives and scientific publications aimed at advancing the application of artificial intelligence in early warning systems and smart monitoring environments.

Binastya Anggara Sekti has been a lecturer at the Faculty of Computer Science, Esa Unggul University, Jakarta, since 2016. He is actively involved in the Artificial

Intelligence and Information Security Study Group. He holds a Bachelor's degree in Informatics Engineering from the Bandung Institute of Technology and a Master's degree in Information Systems Management from Trisakti University, Jakarta. Prior to entering academia, he accumulated over 25 years of professional experience as an IT practitioner in various companies across Jakarta. His work focused on developing data centers for internet service providers and delivering IT solutions for the manufacturing and oil & gas sectors. His research interests include artificial intelligence, cybersecurity, and the development of IT infrastructure.

Mavlanova Barchinoy Shonazar Qizi holds a PhD in Economics and is an Associate Professor in the Department of Economics, Faculty of Economics, at Mamun University. Her academic and research interests focus on economic theory, applied economics, and socio-economic development. She is actively involved in teaching, research, and academic supervision, contributing to curriculum advancement and the promotion of analytical and evidence-based approaches in economics education.

Francisco R. Trejo-Macotela is a Professor at the Polytechnic University of Pachuca (UPP), a National Researcher under the SECIHTI (Secretariat of Science, Humanities, Technology, and Innovation) programme, and an Adjunct Researcher at the National Laboratory of Autonomous Vehicles and Exoskeletons (LANAVEX). He obtained his PhD in Electronics from the National Institute of Astrophysics, Optics and Electronics (INAOE), Puebla, Mexico. His primary research interests include integrated circuit design, analogue and digital systems, automation, embedded systems, smart cities, RF electronics, telecommunications systems, precision agriculture, IoT, combinatorial optimisation, educational technologies, smart technologies, and among others.

Srinivasan Vasan is an Assistant Professor in the Department of Humanities and Social Sciences at Graphic Era (Deemed to be University), with 16 years of teaching and field expertise. He has served in the Ministry of Health and Family Welfare, as a Guest Faculty, and as a Mobilization Training Specialist with the State Government. Recently, he contributed to a tuberculosis project funded by GFATM and USAID. He has published 90 papers, with more under peer review in UGC-CARE and SCOPUS journals

David Yehezkiel is a programmer and artificial intelligence researcher at Esa Unggul University, Jakarta, Indonesia. His research interests include machine learning, intelligent systems development, software engineering, and data-driven applications. He has been actively involved in programming projects and academic research focusing on the development of AI-based solutions, system automation, and intelligent software applications. His work demonstrates strong technical expertise, analytical thinking, and a commitment to innovation in emerging technologies. David actively contributes to collaborative research initiatives and technology development projects aimed at advancing artificial intelligence applications in academic and professional environments.

Index

E

F

G

L

P

R

S

T

Z

www.ingramcontent.com/pod-product-compliance
Lightning Source LLC
LaVergne TN
LVHW081543300926
849161LV00006B/1499